Mosquito Gene Drives and the Malaria Eradication Agenda

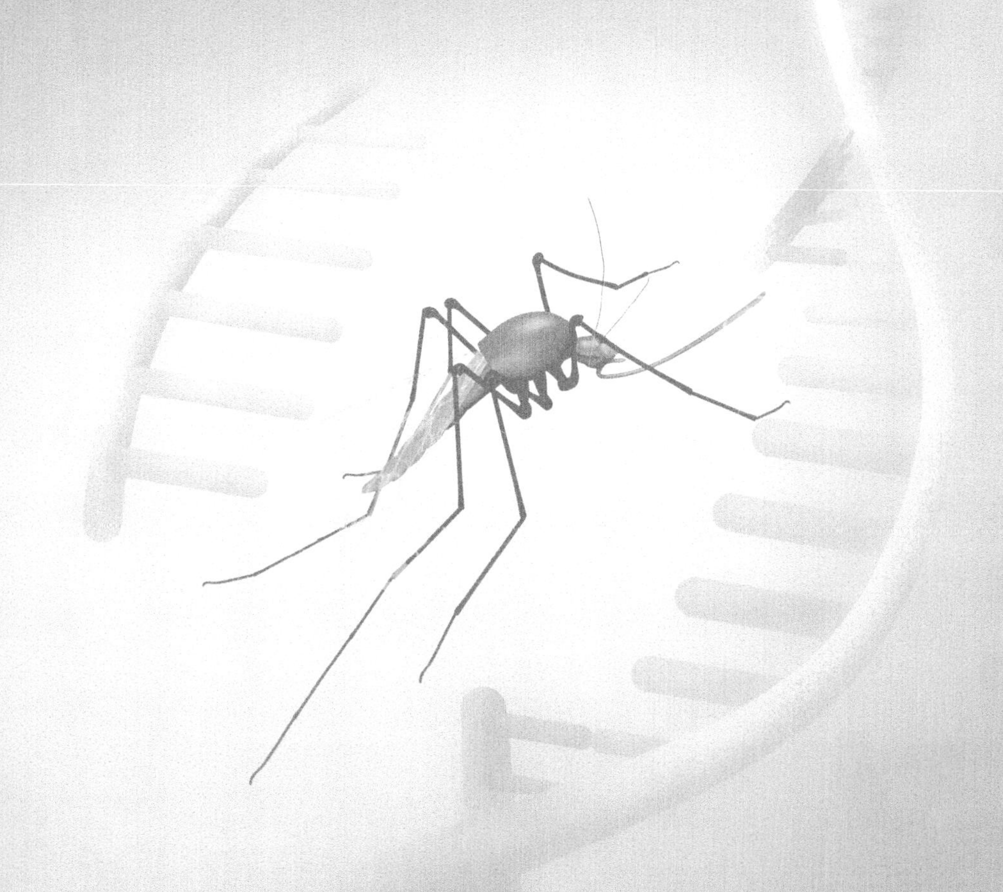

Mosquito Gene Drives and the Malaria Eradication Agenda

edited by

Rebeca Carballar-Lejarazú

Published by

Jenny Stanford Publishing Pte. Ltd.
101 Thomson Road
#06-01, United Square
Singapore 307591

Email: editorial@jennystanford.com
Web: www.jennystanford.com

British Library Cataloguing-in-Publication Data
A catalogue record for this book is available from the British Library.

Mosquito Gene Drives and the Malaria Eradication Agenda

Cover image: Courtesy of Bryn Hobson.

ISBN 978-981-4968-33-1 (Hardcover)
ISBN 978-1-003-30877-5 (eBook)

Contents

Preface xi

SECTION I THE MALARIA CHALLENGE

1 Current Scenario of Malaria and the Transformative Power of Gene Drive-Based Technologies **3**
George Dimopoulos
1.1 Implementation of Malaria Control: Priorities and Constraints 3
1.2 Malaria Prevention by Novel Control Methods 6
1.3 Population Suppression vs Population Modification 7

SECTION II MOSQUITO GENETIC MANIPULATION FOR MALARIA

2 Transgenesis and Paratransgenesis for the Control of Malaria **21**
Sibao Wang and Marcelo Jacobs-Lorena
2.1 Transgenesis 21
2.1.1 Tissue-Specific Promoters 23
2.1.2 Effector Genes 25
2.1.2.1 Peptides/proteins 25
2.1.2.2 Antibodies 26
2.1.2.3 Mosquito immune genes 27
2.2 Paratransgenesis 27
2.3 Prospects 29

3 Gene Drives for *Anopheles* Mosquitoes **39**
Jackson Champer
3.1 Introduction to Gene Drives and Their Characteristics 39
3.1.1 The Concept of Gene Drive and Its Applications 39

3.1.2 Outcomes of a Gene Drive Strategy 40
3.1.3 Confinement of a Gene Drive to Target Populations 44
3.2 Types of Gene Drive 46
3.2.1 Chromosomal Rearrangements 46
3.2.2 Transposons 47
3.2.3 Homing Drives 47
3.2.4 X-Shredders 51
3.2.5 RNAi-Based Toxin-Antidote Drives 53
3.2.6 CRISPR-Based Toxin-Antidote Drives 54
3.2.7 Wolbachia 55

4 Gene Drive Applications for Malaria Control 65
Vanessa Macias and Anthony James
4.1 Introduction 65
4.2 Gene-Drive Applications 78
4.2.1 Population Suppression 83
4.2.2 Population Modification (Replacement/Alteration) 89
4.2.3 Considerations 96
4.2.4 Pathways to Deployment 99

SECTION III GENE DRIVE MOSQUITO TRIALS

5 Large Cage Trials of Gene Drive Mosquitoes: Does Size Matter? 115
Mark Q. Benedict
5.1 Introduction 115
5.2 Large-Cage Trials Are Widely Advised 118
5.3 Case Studies 120
5.3.1 Case Study 1: Success after Only Small Cage Studies: *Culex pipiens* in Burma 121
5.3.2 Case Study 2: Success after Preliminary Cage Studies: *Culex quinquefasciatus* in Florida 121
5.3.3 Case Study 3: Failure without Cage Trials: *Aedes aegypti* in Florida 122
5.3.4 Case Study 4: Failure without Large Cage Trials: *Anopheles gambiae* in Burkina Faso 123

5.3.5 Case Study 5: Indoor Cage Trials Were Encouraging but Outdoor Cage Studies in Mexico Ended Development 124
5.3.6 Case Study 6: *Culex tarsalis* in California, USA 124
5.3.7 Case Study 7: *Anopheles gambiae* Ag(DSM)2 126
5.3.8 Case Study 8: *Anopheles albimanus* in El Salvador 126
5.4 Colonization Considerations 127
5.5 Defining 'Large' 129
5.6 Conclusion 133

6 Field Trial Site Selection for Mosquitoes with Gene Drive: Geographic, Ecological, and Population Genetic Considerations **141**
Gregory C. Lanzaro, Melina Campos, Marc Crepeau, Anthony Cornel, Abram Estrada, Hans Gripkey, Ziad Haddad, Ana Kormos, and Steven Palomares
6.1 Introduction 141
6.2 Defining the Goal 143
6.3 Framework for Field Site Selection 146
6.3.1 Physical Features 148
6.3.2 Biological Features 153
6.4 Evaluation of Potential Island Field Sites 172
6.4.1 Evaluation of Tier 1 Criteria 173
6.4.2 Other Considerations – Tier 2 Criteria 176
6.4.3 Other Considerations – Tier 3 Criteria 179
6.5 Conclusion 181

7 Modeling Priorities as Gene Drive Mosquito Projects Transition from Lab to Field **197**
John M. Marshall and Ace R. North
7.1 Introduction 197
7.2 Model Building 200
7.2.1 Population Genetics Models 200
7.2.2 Mosquito Vector Models 204
7.2.2.1 Mosquito life cycle 204
7.2.2.2 Spatial population structure 205
7.2.2.3 Density dependence 208

7.2.2.4 Movement ecology 210
7.2.2.5 Dry season ecology 211
7.2.3 Malaria Transmission Models 212
7.3 Model Application 215
7.3.1 Target Product Profiles 216
7.3.2 Monitoring and Surveillance 218
7.3.3 Risk and Regulatory Considerations 220
7.3.4 Cage Trials 221
7.3.5 Field Trial Design 222
7.3.6 Intervention Design 223
7.4 Conclusion 224

Section IV Risk Assessment and Community Engagement

8 Probabilistic Ecological Risk Assessment: An Overview of the Process 241
Keith R. Hayes, Geoffrey R. Hosack, and Adrien Ickowicz
8.1 Introduction 241
8.2 Stakeholders, Planning, and Problem Formulation 243
8.3 The GMO and Receiving Environment 245
8.4 Estimating the Probability and Consequences of Adverse Outcomes 247
8.4.1 Models and Risk Assessment 247
8.4.2 Probabilistic Risk Assessment Methods 250
8.5 Risk Calculations 253
8.6 Monitoring, Management, and Acceptability 255
8.6.1 Risk Acceptance 255
8.6.2 Monitoring and Management 256
8.7 Concluding Remarks 256

9 Community Engagement and Mosquito Gene Drives 269
Ana Kormos
9.1 Introduction 269
9.2 Defining Engagement 271
9.3 Importance of Engagement 272
9.4 General Engagement Considerations 275
9.4.1 Funding 275
9.4.2 Timeline 275

9.4.3 Risks 276
9.5 Implementing Engagement 277
9.5.1 Identifying Stakeholders 277
9.5.2 Identifying a Model/Strategy for Engagement 278
9.5.3 Understanding Responsibilities 278
9.6 RBM for Engagement 278
9.7 RBM, Context, and Concepts 279
9.8 Applying the RBM 283
9.8.1 Commitment to the Model 283
9.8.2 Interdisciplinary Approach 284
9.8.3 Build on Existing Strengths and Resources 286
9.8.4 Environment for Building/Strengthening Relationships 288
9.8.5 Relationship-Based Engagement Planning, Development, and Implementation 289
9.8.6 Continuous Evaluation and Improvement 290
9.8.7 Capacity Building 291
9.9 Conclusion 292

SECTION V POLICY, REGULATORY, AND ETHICAL CONSIDERATIONS

10 Review of International Regulatory Instruments and Processes 299
Felicity Keiper and Ana Atanassova
10.1 Introduction 299
10.2 International Regulatory Framework and Current Developments 301
10.2.1 CBD 301
10.2.1.1 Biotechnology provisions 302
10.2.1.2 Developments under the CBD related to gene drives 303
10.2.2 Cartagena Protocol 305
10.2.2.1 Definitions & scope 305
10.2.2.2 National implementation 306
10.2.2.3 Risk assessment 308
10.3 Regulatory and Policy Developments 311
10.3.1 International Developments 311
10.3.1.1 Scientific community 311

10.3.1.2 LM (and non-LM) mosquitoes 313
10.3.1.3 World Health Organization 315
10.3.1.4 Organization for Economic Co-operation and Development 317
10.3.2 Regional Developments 318
10.3.2.1 African Union 318
10.4 Conclusion 321

11 Gene Drive Mosquitoes: Ethical and Political Considerations 329
Daniel Edward Callies and Athmeya Jayaram
11.1 Introduction 329
11.2 Slippery Slope of Research 330
11.3 Precautionary Principle 332
11.4 Conventional Alternatives 334
11.5 Environmental Ethics 336
11.6 Hubris 339
11.7 Public Participation 341
11.8 Distributive Justice 344
11.9 The Dual Use Dilemma 347
11.10 Conclusion 349

Index 353

Preface

Renewed efforts by national governments and international organizations to reduce the spread and impact of malaria were initiated early in the 21st century. A global reduction in incidence and mortality was achieved in the first 15 years using anti-malarial drugs, residual insecticides, and long-lasting insecticide bed nets. However, progress has slowed and more than 3 billion people continue to be at risk for malaria infection and more than 241 million cases and over 267,00 deaths were reported in 2020. Despite the success of the initial efforts, no significant progress in reducing global malaria incidence has been achieved since 2015 due to increases in insecticide and drug resistance. Emerging technologies such as synthetic gene drive system may offer powerful tools to rapidly spread beneficial genetic traits to modify or suppress vector mosquito populations for malaria control.

In editing and organizing this book, I have made every attempt to cover all the major areas of importance to mosquito gene-drive system, to facilitate the identification of research and development needs to enable 'gene drive technologies' and their possible application to the control and eradication of malaria and other vector-borne diseases. This includes the use of mathematical models and the application of social science disciplines to assess and strengthen the capacity of biosafety and regulatory professionals in risk assessment of genetic biocontrol technologies. While this book will serve as a foundation and guide to gene drive emerging technologies, new systems appearing close to the time of publication may have been omitted. In terms of strategies for the use of gene drive organisms, there are numerous possibilities for both basic and applied purposes and this book provides only a sampling of strategies for vector mosquitoes and malaria control.

The contributors to this book are held in high regard in their fields of expertise with decades of extensive experience in developing genetic technologies for use in insects and the application of those technologies to explore the physiology and genetics to control mosquito populations and their ability to transmit human pathogens such as malaria-causing parasites. Also included are experts in developing the adequate framework in ethics, regulatory affairs, community engagement and risk assessment to deploy these emerging technologies. The target audience includes scientists, health professionals, ecologists, those involved in basic and translational research, educators, and graduate students that have the interest in applying the CRISPR/Cas9 gene drive technology, those organizing or participating in community engagement, field trials, regulatory affairs, and ethics, and stakeholders and decision-makers in malaria endemic countries.

I am grateful to Bryn Hobson, a talented graphic designer, who contributed to the book's cover design. Bryan has experience working as a designer for tech companies and enjoys challenging himself with unique art projects in his free time. I also thank those who provided enormous assistance throughout the development and production of this book, including Arvind Kanswal, all the Jenny Stanford staff, and Thai Binh Pham and Taylor Tushar from the University of California Irvine.

Section I

The Malaria Challenge

Chapter 1

Current Scenario of Malaria and the Transformative Power of Gene Drive-Based Technologies

George Dimopoulos
Department of Molecular Microbiology and Immunology, Malaria Research Institute, Bloomberg School of Public Health, Johns Hopkins University, Baltimore, Maryland, USA
gdimopo1@jhu.edu

1.1 Implementation of Malaria Control: Priorities and Constraints

While malaria is frequently highlighted as one of today's most serious infectious diseases, it does not represent a recent public health crisis, but rather has afflicted humankind for thousands of years. The first descriptions of malaria are found in ancient Chinese medical records from 2700 BC, and Hippocrates linked malaria with the inhalation of evaporating water from the swamps in the 3rd century BC. This explanation was maintained until the late 19th century, when Charles Louis Alphonse Laveran discovered the causative agent and Ronald Ross and Giovanni Battista Grassi

Mosquito Gene Drives and the Malaria Eradication Agenda
Edited by Rebeca Carballar-Lejarazú

ISBN 978-981-4968-33-1 (Hardcover), 978-1-003-30877-5 (eBook)
www.jennystanford.com

showed that the *Plasmodium* parasite was transmitted between humans via *Anopheles* mosquitoes [1, 2]. Malaria control with drugs was begun already in the 2nd century BC by the Chinese, who used *Artemisia annua* for treatment [3]. Much later, in the 16th century, Pierre Joseph Pelletie and Joseph Bienaimé Caventou isolated the antimalarial quinine from *Cinchona succirubra* bark, which the Spanish had previously adopted from native people of Peru for malaria treatment, and in 1934 Hans Andersag synthesized chloroquin. The active substance artemisinin, from the plant *Artemisia annua*, was isolated by Youyou Tu in 1970 and became the basis of a variety of antimalarial artemisinin-related drugs used today. The high resistance rates to other drugs like chloroquine, amodiaquine, and sulfadoxine-pyrimethamine have resulted in a WHO recommendation for artemisinin-based combination therapies (ACT) to reduce the emergence of resistance and achieve a high cure rate of *P. falciparum* malaria [4–8]. Nevertheless, the parasite appears to be highly competent in developing resistance to the newer drugs and this has led to a continuous quest of developing additional artemisia-based drugs and other molecules with different parasite targets.

An enormous effort has been invested in the development of an affordable preventive malaria vaccine over the past decades, and a variety of epitopes have been explored up to the stage of clinical trials. Nevertheless, an antimalaria vaccine with efficacy above 50% has not been achieved to date. The difficulties with antimalaria vaccine development largely stem from the complexity of the organism in terms of genetics and life cycle, enabling it to evade the host immune system. The morphologically different stages and ability of antigenic variation renders it difficult to identify an effective epitope [9–12].

The discoveries by Ronald Ross and Giovanni Battista Grassi, regarding mosquito-meditated transmission of the parasite, led to the implementation of mosquito avoidance as a malaria preventive measure in the late 19th century. Veiled hats and gloves became commonly used personal mosquito protectants, and mosquito nets were used for mosquito proofing of houses, outdoor activities, and cradles to reduce exposure to mosquitoes. The burning of brimstone as a means to fumigate houses and kill mosquitoes was used in the same period in Brazil as a countermeasure for yellow fever [13–15]. Pyrethrum powder was also used, by either spraying or burning, as an insecticide [16, 17]. Reducing mosquito populations by either applying petrol to stagnant water or drainage became widely used in the same period [18]. Dichlorodiphenyltrichloroethane, commonly known as DDT, was synthesized in 1874 but was first used for the control of mosquitoes and other insects in 1939; during and after the Second World War, together with other chlorinated hydrocarbons, it became the backbone of mosquito-targeted malaria control and had a huge demographic and socioeconomic impact, significantly decreasing the global prevalence of a disease to which two-thirds of the world's population had become exposed since the Middle Ages [7, 8, 19–23].

Approximately 220 million people are still infected today, and 435,000 people died from this mosquito-borne disease in 2017, despite the use of various malaria preventive and control methods over hundreds of years and more than a century of coordinated global control efforts using modern tools, together with research into and development of new strategies for prevention, diagnosis, and disease treatment. In Africa and some regions of Asia, malaria represents the most commonly occurring disease and reaches a mortality rate of 11–30% for some severe forms of the disease [24]. However, the global malaria prevalence has decreased significantly in this century, sustaining the hope of achieving global eradication of the disease [8].

It is important to note that the number of deaths has been more than halved despite the rapid growth of the population in most endemic regions. A major contributor to this achievement is vector control through insecticide-treated bed nets and indoor residual spraying, as well as combinatorial drug treatment [25, 26]. Economic development, resulting in better housing and widespread urbanization in malaria-endemic regions, has also contributed to the decline. The decrease has, nevertheless, been more prominent outside sub-Saharan Africa, where at least 90% of malaria deaths occur today, and the prevalence has increased in some parts of the broader African region since 2015 [26–28]. The remaining malaria mortality mostly reflects a lack of access to diagnosis and treatment, insecticide and drug resistance, and a general failure to sustain intervention operations [24, 25, 29]. The recently emerged malaria vaccine offers some promise, but its low and complex efficacy profile suggests limited epidemiological impact [30–34].

The stalled progress in reducing malaria-related deaths since 2015 [24, 35] taken together with epidemiological models, predict a reversal of the trend toward elimination achieved with currently used control strategies [36, 37]. Malaria control efforts aimed at the reduction of disease incidence and transmission, eventually leading to elimination, require reducing transmission to levels that are less than self-sustaining. Given the absence of high-efficacy vaccines, this public health goal can realistically only be achieved by reducing human exposure to infectious vectors [38].

Despite the efficacy and broad implementation of long-lasting insecticide-treated nets and indoor residual spraying across sub-Saharan Africa, these control strategies, together with effective drug treatment, are predicted to be insufficient to reach the goal of malaria elimination from hyperendemic areas, even if applied at very high coverage [37–39]. It is almost inevitable that insecticide resistance will arise [29], as will resistance to drugs in both mass drug administration [MDA] and seasonal malaria chemoprevention [SMC] campaigns. While MDA can result in a large short-term reduction in the human parasite reservoir and thereby decrease malaria prevalence [40], these beneficial effects are transient because of malaria re-importation from untreated areas unless drugs are administered indefinitely at high coverage levels [41]. Similarly, SMC can also effectively reduce disease incidence, but only in areas with highly seasonal transmission [42]. Malaria control with an eye toward elimination, therefore, becomes a Sisyphean endeavor with the available control methods, requiring an endless and intense use of partially effective strategies. The current situation calls for novel and innovative vector-targeted intervention technologies that will lessen the need for a continuous, intense, costly, and thus far only partially effective effort.

1.2 Malaria Prevention by Novel Control Methods

A variety of novel vector-targeted control technologies are being developed, such as new chemical and biological mosquitocides, along with novel delivery systems that rely on the insect's feeding behavior and transgenic bacteria that produce *Plasmodium*-blocking proteins in the mosquito midgut [43–45]. The successful implementation of genetically modified [paratransgenesis] parasite-blocking mosquito symbionts in mosquito populations could represent a control strategy that is at least partially self-sustainable and may reduce disease transmission without also affecting the actual mosquito population. The antiparasitic and mosquitocidal drug ivermectin, along with other mosquitocidal and transmission-blocking drug candidates, are also under development and could contribute significantly to malaria control if implemented alongside a mass treatment intervention campaign [46]. However, although promising and potentially effective, these novel malaria control approaches will always be limited by the need for continuous application as well as extensive compliance and participation by the endemic population. An ideal malaria intervention method would require only a single or, at most, a few applications within a defined time period; be cost-effective, self-propagating, and self-sustainable; and therefore have a long-lasting epidemiological impact, with no harm to the environment or non-target species. Such transformative malaria vector control technologies are currently being developed on the basis of genetic modifications of mosquitoes in ways that result in either reduced vector competence or mosquito density. A key component of these population modification and suppression strategies that renders them self-propagating is gene-drive technology, which enables the non-Mendelian spread and fixation of transgenes in mosquito populations even when a certain fitness cost is imposed by the transgene [47–49].

1.3 Population Suppression vs Population Modification

While several novel and innovative approaches have been proposed and are being developed for malaria vector control, most do not fulfill the requirement of a single or a few applications, cost-effectiveness, and having a long-lasting epidemiological impact without negatively affecting the environment and non-target species [36]. However, the self-propelling and self-sustainable features of a malaria control strategy that would eliminate the need for continuous application, as in the case of insecticide spraying, have been enabled by the application of advances in genetics and synthetic biology to mosquito vectors. The spread of recombinant genes and mutations in a mosquito population has become possible through the advent of gene-drive technology, that enables the super-Mendelian inheritance of genes, thereby resulting in a spread and fixation of recombinant genes and mutations in mosquito populations despite the lack of a selective advantage. The concept has undergone significant development over the past decade and eventually matured in the most recent generation which is based on clustered regularly interspaced short palindromic repeat [CRISPR]-mediated gene editing [50].

The CRISPR-based gene drive technology has mainly been employed for the development of two types of malaria control strategies: population modification, which modifies a malaria-transmitting mosquito population into one that is unable to support infection with the malaria parasite; and mosquito population suppression, which essentially eliminates the vector from the application area [51, 52]. Proof-of-principle studies have demonstrated that both population modification and suppression approaches are feasible [53–55]. The population modification approach relies on both the genetic engineering of mosquitoes to render them refractory to the malaria parasite and the spread of the engineered genes or mutations through a gene drive system. There are many varied and versatile ways in which a mosquito can be engineered for refractoriness to the malaria parasite. One can either overexpress anti-*Plasmodium* factors [antagonists] or mutate or disrupt *Plasmodium* host factors [agonists] at a stage and tissue at which the disruption will inhibit infection. There are both endogenous [mosquito-encoded] and exogenous [derived from other organisms or synthetic] antagonists to choose from that have been studied in the context of engineered refractoriness to the malaria parasite. Most endogenous antagonists are components of the mosquito's innate immune system that are naturally involved in suppressing malaria parasite infection [56].

Overexpression of antagonists by using appropriate tissue- and stage-specific promoters with activation patterns that coincide with the spatiotemporal location of the parasite results in mosquitoes that are super-immune to infection with the malaria parasite. Similarly, overexpression of exogenous antagonists also results in refractoriness to the malaria parasite that operates through blocking mechanisms that do not normally function in the mosquito. Examples of exogenous antagonists are single-chain antibodies that target the parasite and antimicrobial peptides from other organisms [56–58]. The best-characterized and most commonly used promoters for driving anti-*Plasmodium* transgenes are the blood meal-inducible midgut-specific carboxypeptidase and fat body-specific vitellogenin promoters. A mosquito-encoded *Plasmodium* agonist that has been explored as a possible producer of transgenesis-mediated refractoriness to *Plasmodium* is the fibrinogen-related protein 1 [FREP1], which is required for mosquito midgut infection by the ookinete-stage parasite. CRISPR/CAS9-based FREP1 disruption results in strong refractoriness to the malaria parasite [59]. Another candidate *Plasmodium* host factor is c-type lectin 4, which helps protect the ookinete against melanotic encapsulation-mediated killing [60].

Two advantageous features of a population modification strategy are that [**a**] this approach does not involve the removal of a mosquito species from its ecological niche and food chain and [**b**] it can be rendered long-lasting through the use of gene drive technology, avoiding the need to repeat the release of the engineered mosquitoes [48]. Should vector mosquitoes of the same species be introduced into the treatment area, they would, through mating with the sympatric transgenic mosquitoes, produce offspring resistant to the malaria parasite. The main challenges involved in the development and implementation of mosquito-population modification strategies for malaria control are to engineer resistance so that [**a**] mosquitoes will be refractory to diverse *Plasmodium* species and strains; [**b**] the malaria parasite will not easily develop resistance to the refractory mechanisms; and [**c**] the genetic modification will not render the mosquitoes significantly less competitive than their wild-type counterparts. Another major challenge is achieving public acceptance since the genetically modified organisms will remain in the local environment.

Genetic-based population suppression approaches, as opposed to insecticide-mediated killing, have traditionally relied on mass release of sterile males that have been produced in facilities through exposure to chemical mutagens or radiation. However, the sterile insect technique [SIT] approaches against human disease vectors have shown limited success, mostly because of the compromised competition of the sterile males that are released [61–64]. To overcome the mosquito fitness impact of classical SIT, genetic engineering has been used to create mosquitoes with a germ line-encoded tetracycline-repressible dominant lethal transgene that results in the death of offspring at the larval stage when tetracycline is absent. This release of insects carrying dominant lethal transgenes [RIDL] is currently used for the suppression of *Aedes aegypti* populations [65–67]. However, these SIT approaches, while potentially powerful under certain transmission conditions, do not fulfill the requirement for self-propagating and self-sustainable malaria control.

Gene drive-based population suppression approaches relying on the spread of genetically engineered mosquitoes that produce only male progeny have recently been developed and tested under laboratory conditions. The *A. gambiae* male-biased sex-distorter gene drive [SDGD] system relies on gene drive-propelled super-Mendelian inheritance of the X-chromosome-shredding I-PpoI nuclease inserted into the sex-determining doublesex [dsx] gene, where it only affects the female splice form. Modeling studies predicted that the use of the SDGD system could result in a male-only population within 10–14 generations, even when starting with an allelic frequency of only 2.5%. Importantly, the selection of resistance is not predicted to occur [68]. A key feature of gene drive-based population suppression strategies for malaria control is that the malaria vector is at least transiently eliminated from the treatment area, thereby attenuating the transmission of any *Plasmodium* species or strain or any other pathogen that might be vectored by the same mosquito species. Another advantage of such strategies is a reduction in the number of mosquito bites and a likely higher public acceptance rate because the genetic modification is only affecting the viability of the female offspring and not some other properties that may have unforeseen consequences in terms of pathogen transmission.

Major challenges in the development and implementation of mosquito population suppression-based strategies for malaria control are [**a**] the transient suppression of the mosquitoes, and therefore the likely need for repeated release of the genetically modified mosquitoes, which may prove logistically complex and costly; and [**b**] possible concerns about the unforeseeable consequences of species elimination from an ecologic niche. However, gene drive-based population suppression strategies that rely on a genetic sex ratio distortion mechanism, resulting in only male offspring from crosses between genetically modified and wild-type individuals, are thought to be capable of longer-term elimination of a mosquito species from a treatment area [68, 69].

While the gene drive-based population modification and suppression technologies have been predicted to have a broad and long-term impact on malaria prevalence, these strategies may not be compatible with all transmission settings, in terms of ecology, mosquito and parasite population biology, and socio-political conditions. Therefore, population modification and suppression technologies cannot achieve disease elimination on their own or be considered "silver bullets." However, if implemented in appropriate combination with other compatible malaria control methods, these gene drive-based strategies can greatly boost the success of malaria control programs through synergistic action [24, 70]. For example, a population suppression strategy preceding the implementation of a population replacement strategy may enhance the speed of transgene fixation in a mosquito population, and as such enhance the impact on malaria prevalence. The implementation of a population modification strategy at the end stages of a malaria eradication campaign, when hypoendemicity may have been achieved through other control methods, may serve as a strategy to eradicate residual transmission and prevent reintroduction of the disease. Establishing a portfolio of diverse malaria control methods, that can be optimally combined and adapted to different transmission settings, to achieve favorable synergistic effects, will be necessary along with trained malaria control program managers with a multidisciplinary understanding of these tools and approaches to fight one of the world's most important diseases.

References

1. Talapko, J., Škrlec, I., Alebić, T., Jukić, M., Včev, A. (2019). Malaria: The past and the present. *Microorganisms.* **7**(6), p. 179.
2. Tan, S.Y., Ahana, A. (2009). Charles Laveran, (1845–1922), Nobel laureate pioneer of malaria. *Singap. Med. J.* **50**, pp. 657–658.
3. Hsu, E. (2006). The history of Qing Hao in the Chinese malaria medica. *Trans. R. Soc. Trop. Med. Hyg.* **100**, pp. 505–508.
4. Guo, Z. (2016). Artemisinin anti-malarial drugs in China. *Acta Pharm. Sin.* **6**, pp. 115–124.
5. Achan, J.O., Talisuna, A., Erhart, A., Yeka, A., Tibenderana, J.K., Baliraine, F.N., Rosenthal, P.J., D'Alessandro, U. (2011). Quinine, an old anti-malarial drug in a modern world: Role in the treatment of malaria. *Malar. J.* **10**, p. 144.
6. Meshnick, S.R., Dobson, M.J. (2001). The history of antimalarial drugs. In antimalarial chemotherapy, mechanisms of action, resistance, and new directions in drug discovery. Rosenthal, P. (ed.). Humana Press: Totowa NJ, USA, p. 396.
7. World Health Organization (2001). Antimalarial drug combination therapy. Report of a WHO Technical Consultation (WHO: Geneva, Switzerland).
8. World Health Organization (2018). World Malaria Report (WHO: Geneva, Switzerland).
9. Mahmoudi, S., Keshavarz, H. (2017). Efficacy of phase 3 trial of RTS, S/AS01 malaria vaccine: The need for an alternative development plan. *Hum. Vaccines Immunother.* **13**, pp. 2098–2101.
10. Draper, S.J., Sack, B.K., King, C.R., Nielsen, C.M., Rayner, J.C., Higgins, M.K., Long, C.A., Seder, R.A. (2018). Malaria vaccines, recent advances and new horizons. *Cell Host Microbe.* **24**, pp. 43–56.
11. Crompton, P.D., Pierce, S.K., Miller, L.H. (2010). Advances and challenges in malaria vaccine development. *J. Clin. Investig.* **120**, pp. 4168–4178.
12. Arama, C., Troye-Blomberg, M. (2014). The path of malaria vaccine development, challenges and perspectives. *J. Intern. Med.*, **275**, pp. 456–466.
13. Cartwright, FF., Biddiss, M. (2014). Disease and history. 3rd Ed. Lume Boks, 30 Great Guilford Street, Borough, SE1 oHS.
14. Cox FE. (2010). History of the discovery of the malaria parasites and their vectors. *Parasit Vectors.* **3**(1), pp. 5.

15. Tangpukdee, N., Duangdee, C., Wilairatana, P., Krudsood, S. (2009). Malaria diagnosis: A brief review. *Korean J. Parasitol.* **47**, pp. 93–102.
16. She, R.C., Rawlins, M.L., Mohl, R., Perkins, S.L., Hill, H.R., Litwin, C.M. (2007). Comparison of immunofluorescence antibody testing and two enzyme immunoassays in the serologic diagnosis of malaria. *J. Travel Med.* **14**, pp. 105–111.
17. Oh, J.S., Kim, J.S., Lee, C.H., Nam, D.H., Kim, S.H., Park, D.W., Lee, C.K., Lim, C.S., Park, G.H. (2008). Evaluation of a malaria antibody enzyme immunoassay for use in blood screening. *Mem. Inst. Oswaldo Cruz.* **103**, pp. 75–78.
18. Gachelin G., Garner P., Ferroni E., Verhave .JP., Opinel A. (2018). Evidence and strategies for malaria prevention and control: A historical analysis. *Malar J.* **17**(1), p. 96.
19. World Health Organization (2015). Guidelines for the treatment of malaria. 3rd Ed. WHO: Geneva, Switzerland.
20. Ray, D. (2010). Organochlorine and pyrethroid insecticides. *Compr. Toxicol.* **13**, pp. 445–457.
21. Flannery, E.L., Chatterjee, A.K., Winzeler, E.A. (2013). Antimalarial drug discovery—Approaches and progress towards new medicines. *Nat. Rev. Microbiol.* **11**, pp. 849–862.
22. Zhao, X., Smith, D.L., Tatem, A.J. (2016). Exploring the spatiotemporal drivers of malaria elimination in Europe. *Malar. J.* **15**, p. 122.
23. White, N.J.N., Pukrittayakamee, S., Hien, T.T.T., Faiz, M.A., Mokuolu, O.A.O., Dondorp, A.A.M. (2014). Malaria. *The Lancet.* **383**, pp. 723–735.
24. Bhatt, S., Weiss, D.J., Cameron, E., Bisanzio, D., Mappin, B., Dalrymple, U., Battle, K., Moyes, C.L., Henry, A., Eckhoff, P.A., Wenger, E.A., Briët, O., Penny, M.A., Smith, T.A., Bennett, A., Yukich, J., Eisele, T.P., Griffin, J.T., Fergus, C.A., Lynch, M., Lindgren, F., Cohen, J.M., Murray, C.L.J., Smith, D.L., Hay, S.I., Cibulskis, R.E., Gething, P.W. (2015). The effect of malaria control on Plasmodium falciparum in Africa between 2000 and 2015. *Nature.* **526**, pp. 207–211.
25. Cibulskis, R.E., Alonso, P., Aponte, J., Aregawi, M., Barrette, A., Bergeron, L., Fergus, C.A., Knox, T., Lynch, M., Patouillard, E., Schwarte, S., Stewart, S., Williams, R. (2016). Malaria: Global progress 2000–2015 and future challenges. *Infect Dis Poverty.* **5**.
26. Tatem, A.J., Gething, P.W., Smith, D.L., Hay, S.I. (2013). Urbanization and the global malaria recession. *Malar. J.* **12**, p. 133.
27. Pan American Health Organization (PAHO); World Health Organization (2018). Epidemiological alert, increase of malaria in the Americas. PAHO: Washington, DC, USA.

28. Dhiman, S. (2019). Are malaria elimination efforts on right track? An analysis of gains achieved and challenges ahead. *Infect. Dis. Poverty.* **8**, p. 14.
29. Hemingway, J., Ranson, H., Magill, A., Kolaczinski, J., Fornadel, C., Gimnig, J., Coetzee, M., Simard, F., Roch, D.K., Hinzoumbe, C.K., Pickett, J., Schellenberg, D., Gething, P., Hoppé, M., Hamon, N. (2016). Averting a malaria disaster: Will insecticide resistance derail malaria control? *The Lancet.* **387**, pp. 1785–1788.
30. Ferguson, N. M., Rodríguez-Barraquer, I., Dorigatti, I., Mier-Y-Teran-Romero, L., Laydon, D.J., Cummings, D.A.B. (2016). Benefits and risks of the Sanofi-Pasteur dengue vaccine: Modeling optimal deployment. *Science.* **353**, pp. 1033–1036.
31. Flasche, S., Jit, M, Rodríguez-Barraquer, I., Coudeville, L., Recker, M., Koelle, K., Milne, G., Hladish, T.J., Perkins, T.A., Cummings, D.A., Dorigatti, I., Laydon, D.J., España, G., Kelso, J., Longini, I., Lourenco, J., Pearson, C.A., Reiner, R.C., Mier-Y-Terán-Romero, L., Vannice, K., Ferguson, N. (2016). The long-term safety, public health impact, and cost-effectiveness of routine vaccination with a recombinant, live-attenuated dengue vaccine (Dengvaxia): A model comparison study. *PLOS Medicine.* **13**, p. e1002181.
32. Olotu, A., Fegan., G., Wambua, J., Nyangweso, G., Leach, A., Lievens, M., Kaslow, D.C., Njuguna, P., Marsh, K., Bejon, P. (2016). Seven-year efficacy of RTS,S/AS01 malaria vaccine among young African children. *New. Engl. J. Med.* **374**, pp. 2519–2529.
33. Penny, M.A., Verity, R., Bever, C.A., Sauboin, C., Galactionova, K., Flasche, S., White, M.T., Wenger, E.A., Van de Velde, N., Pemberton-Ross, P., Griffin, J.T., Smith, T.A., Eckhoff, P.A., Muhib, F., Jit, M., Ghani, A.C. (2016). Public health impact and cost-effectiveness of the RTS,S/AS01 malaria vaccine: a systematic comparison of predictions from four mathematical models. *The Lancet.* **387**, pp. 367–375.
34. RTS S Clinical Trials Partnership. (2015). Efficacy and safety of RTS,S/AS01 malaria vaccine with or without a booster dose in infants and children in Africa: final results of a phase 3, individually randomized, controlled trial. *The Lancet.* **386**, pp. 31–45.
35. Feachem, R.G.A., Chen, I., Akbari, O., Bertozzi-Villa, A., Bhatt, S., Binka, F., Boni, M.F., Buckee, C., Dieleman, J., Dondorp, A., Eapen, A., Sekhri Feachem, N., Filler, S., Gething, P., Gosling, R., Haakenstad, A., Harvard, K., Hatefi, A., Jamison, D., Jones, K.E., Karema, C., Kamwi, R.N., Lal, A., Larson, E., Lees, M., Lobo, N.F., Micah, A.E., Moonen, B., Newby, G., Ning,

X., Pate, M., Quiñones, M., Roh, M., Rolfe, B., Shanks, D., Singh, B., Staley, K., Tulloch, J., Wegbreit, J., Woo, H.J., Mpanju-Shumbusho, W. (2019). Malaria eradication within a generation: ambitious, achievable, and necessary. *The Lancet.* **394**(10203), pp. 1056–1112.

36. Barreaux, P., Barreaux, A.M.G., Sternberg, E.D., Suh, E., Waite, J.L., Whitehead, S.A., Thomas, M.B. (2017). Priorities for broadening the malaria vector control tool kit. *Trends Parasitol.* **33**(10), pp. 763–774.
37. Walker, P.G.T., Griffin, J.T., Ferguson, N.M., Ghani, A.C. (2016). Estimating the most efficient allocation of interventions to achieve reductions in Plasmodium falciparum malaria burden and transmission in Africa: A modeling study. *The Lancet Global Health.* **4**, pp. e474–e484.
38. Ferguson, N.M. (2018). Challenges and opportunities in controlling mosquito-borne infections. *Nature.* **559**(7715), pp. 490–497.
39. Griffin, J.T., Bhatt, S., Sinka, M.E., Gething, P.W., Lynch, M., Patouillard, E., Shutes, E., Newman, R.D., Alonso, P., Cibulskis, R.E., Ghani, A.C. (2016). Potential for reduction of burden and local elimination of malaria by reducing Plasmodium falciparum malaria transmission: A mathematical modeling study. *The Lancet Infectious Diseases.* **16**, pp. 465–472.
40. Eisele, T.P., Bennett, A., Silumbe, K., Finn, T.P., Chalwe, V., Kamuliwo, M., Hamainza, B., Moonga, H., Kooma, E., Chizema, Kawesha, E., Yukich, J., Keating, J., Porter, T., Conner, R.O., Earle, D., Steketee, R.W., Miller, J.M. (2016). Short-term impact of mass drug administration with dihydroartemisinin plus piperaquine on malaria in southern province Zambia: A cluster-randomized controlled trial. *The Journal of Infectious Diseases.* **214**, pp. 1831–1839.
41. Brady, O.J., Slater, H.C., Pemberton-Ross, P., Wenger, E., Maude, R.J., Ghani, A.C., Penny, M.A., Gerardin, J., White, L.J., Chitnis, N., Aguas, R., Hay, S.I., Smith, D.L., Stuckey, E.M., Okiro, E.A., Smith, T.A., Okell, L.C. (2017). Role of mass drug administration in elimination of Plasmodium falciparum malaria: A consensus modeling study. *The Lancet Global Health.* **5**, pp. e680–e687.
42. Bigira, V., Kapisi, J., Clark, T.D., Kinara, S., Mwangwa, F., Muhindo, M.K., Osterbauer, B., Aweeka, F.T., Huang, L., Achan, J., Havlir, D.V., Rosenthal, P.J., Kamya, M.R., Dorsey, G. (2014). Protective efficacy and safety of three antimalarial regimens for the prevention of malaria in young Ugandan children: A randomized controlled trial. *PLOS Medicine.* **11**, p. e1001689.
43. Caragata, E.P., Dong, S., Dong, Y., Simões, M.L., Tikhe, C.V., Dimopoulos, G. (2020). Prospects and pitfalls: Next-generation tools to control

mosquito-transmitted disease. *Annu Rev Microbiol.* **8**(74), pp. 455–475. doi: 10.1146/annurev-micro-011320-025557. PMID: 32905752.

44. Fiorenzano, J. M., Koehler, P. G., Xue, R.-D. (2017). Attractive toxic sugar bait (ATSB) for control of mosquitoes and its impact on non-target organisms: A review. *International Journal of Environmental Research and Public Health.* **14**, p. 398.
45. Qualls, W.A., Müller, G.C., Traore, S.F., Traore, M.M., Arheart, K.L., Doumbia, S., Schlein, Y., Kravchenko, V.D., Xue, R.D., Beier, J.C. (2015). Indoor use of attractive toxic sugar bait (ATSB) to effectively control malaria vectors in Mali, West Africa. *Malar. J.* **14**, p. 301.
46. Slater, H.C., Walker, P.G.T., Bousema, T., Okell, L.C., Ghani, A.C. (2014). The potential impact of adding ivermectin to a mass treatment intervention to reduce malaria transmission: A modeling study. *J. Infect. Dis.* **210**, pp. 1972–1980.
47. Marshall, J.M., Raban, R.R., Kandul, N.P., Edula, J.R., León, T.M., Akbari, O.S. (2019). Winning the tug-of-war between effector gene design and pathogen evolution in vector population replacement strategies. *Front Genet.* **10**, p. 1072.
48. Carballar-Lejarazú, R., James A.A. (2017). Population modification of Anopheline species to control malaria transmission. *Pathog Glob Health.* **111**(8), pp. 424–435.
49. Shaw, W.R., Catteruccia, F. (2019). Vector biology meets disease control: Using basic research to fight vector-borne diseases. *Nat Microbiol.* **4**(1), pp. 20–34.
50. Doudna, J. A., Charpentier, E. (2014). Genome editing: The new frontier of genome engineering with CRISPR-Cas9. *Science.* **346**, p. 1258096.
51. Burt, A., Coulibaly, M., Crisanti, A., Diabate, A., Kayondo, J. K. (2018). Gene drive to reduce malaria transmission in sub-Saharan Africa. *J. Responsible Innovation.* **5**, pp. S66–S80.
52. Burt, A. (2003). Site-specific selfish genes as tools for the control and genetic engineering of natural populations. *Proc. Biol. Sci.* **270**, pp. 921–928.
53. Adelman, Z. (2015). Genetic control of malaria and dengue. Oxford: Elsevier Academic Press.
54. Eckhoff, P.A., Wenger, E.A., Godfray, H.C., Burt A. (2016). Impact of mosquito gene drive on malaria elimination in a computational model with explicit spatial and temporal dynamics. *Proc Nat Acad Sci USA.* **14**(2), pp. E255–E264.

55. Beaghton, A., Hammond, A., Nolan, T., Crisanti, A., Godfray, H.C., Burt, A. (2017). Requirements for driving antipathogen effector genes into populations of disease vectors by Homing. *Genetics*. **205**(4), pp. 1587–1596.
56. Simões, M.L., Caragata, E.P., Dimopoulos, G. (2018). Diverse host and restriction factors regulate mosquito-pathogen interactions. *Trends Parasitol.* **34**(7), pp. 603–616.
57. Carballar-Lejarazu, R., Rodríguez, M.H., de la Cruz Hernández-Hernández, F., Ramos-Castañeda, J., Possani, L.D., Zurita-Ortega, M., Reynaud-Garza, E., Hernández-Rivas, R., Loukeris, T., Lycett, G., Lanz-Mendoza, H. (2008). Recombinant scorpine: A multifunctional antimicrobial peptide with activity against different pathogens. *Cell Mol Life Sci.* **65**(19), pp. 3081–3092.
58. Dong, Y., Simões, M.L., Dimopoulos, G. (2020). Versatile transgenic multistage effector-gene combinations for Plasmodium falciparum suppression in Anopheles. *Sci Adv.* **6**(20), p. e5898.
59. Dong, Y., Simões, M.L., Marois, E., Dimopoulos, G. (2018). CRISPR/Cas9-mediated gene knockout of Anopheles gambiae FREP1 suppresses malaria parasite infection. *PLoS Pathog.* **14**(3), p. e1006898.
60. Simões, M.L., Mlambo, G., Tripathi, A., Dong, Y., Dimopoulos, G. (2017). Immune regulation of plasmodium is anopheles species specific and infection intensity dependent. *mBio.* **8**(5), pp. e01631–17.
61. Oliva, C.F., Jacquet, M., Gilles, J., Lemperiere, G., Maquart, P.O., Quilici, S., Schooneman, F., Vreysen, M.J., Boyer, S. (2012). The sterile insect technique for controlling populations of Aedes albopictus (Diptera: Culicidae) on Reunion Island: Mating vigor of sterilized males. *PLoS ONE.* **7**, p. e49414.
62. Bellini, R., Medici, A., Puggioli, A., Balestrino, F., Carrieri, M. (2013) Pilot field trials with Aedes albopictus irradiated sterile males in Italian urban areas. *J. Med. Entomol.* **50**, pp. 317–325.
63. Vreysen, M.J., Saleh K.M., Ali M.Y., Abdulla A.M., Zhu Z.R., Juma K.G., Dyck V.A., Msangi A.R., Mkonyi P.A., Feldmann H.U. (2000). Glossina austeni (Diptera: Glossinidae) eradicated on the island of Unguja, Zanzibar, using the sterile insect technique. *J. Econ. Entomol.* **93**, pp. 123–135.
64. Dame, D.A., Curtis, C.F., Benedict, M.Q., Robinson, A.S., Knols, B.G. (2009). Historical applications of induced sterilization in field populations of mosquitoes. *Malar. J.* **2**, p. S2.
65. Carvalho, D.O., McKemey, A.R., Garziera, L., Lacroix, R., Donnelly, C.A., Alphey, L., Malavasi, A., Capurro, M.L. (2015). Suppression of a field

population of Aedes aegypti in Brazil by sustained release of transgenic male mosquitoes. *PLoS Negl. Trop. Dis.* **9**, p. e0003864.

66. Harris, A.F., Nimmo, D., McKemey, A.R., Kelly, N., Scaife, S., Donnelly, C.A., Beech, C., Petrie, W.D., Alphey, L. (2012). Field performance of engineered male mosquitoes. *Nat. Biotechnol.* **29**, pp. 1034–1037.
67. Harris A.F., McKemey, A.R., Nimmo, D., Curtis, Z., Black, I., Morgan, S.A., Oviedo, M.N., Lacroix, R., Naish, N., Morrison, N.I., Collado, A., Stevenson, J., Scaife, S., Dafa'alla, T., Fu, G., Phillips, C., Miles, A., Raduan, N., Kelly, N., Beech, C., Donnelly, C.A., Petrie, W.D., Alphey, L. (2012). Successful suppression of a field mosquito population by sustained release of engineered male mosquitoes. *Nat. Biotechnol.* **30**, pp. 828–830.
68. Simoni, A., Hammond, A.M., Beaghton, A.K., Galizi, R., Taxiarchi, C., Kyrou, K., Meacci, D., Gribble, M., Morselli, G., Burt, A., Nolan, T., Crisanti, A. (2020). A male-biased sex-distorter gene drive for the human malaria vector Anopheles gambiae. *Nat Biotechnol.* **38**(9), pp. 1054–1060.
69. Edgington, M.P., Harvey-Samuel, T., Alphey, L. (2020). Split drive killer-rescue provides a novel threshold-dependent gene drive. *Sci Rep.* **10**(1), p. 20520.
70. Bhatt, S., Gething, P.W., Brady, O.J., Messina, J.P., Farlow, A.W., Moyes, C.L., Drake, J.M., Brownstein, J.S., Hoen, A.G., Sankoh, O., Myers, M.F., George, D.B., Jaenisch, T., Wint, G.R., Simmons, C.P., Scott, T.W., Farrar, J.J., Hay, S.I. (2013). The global distribution and burden of dengue. *Nature.* **496**, p. 504.

Section II

Mosquito Genetic Manipulation for Malaria

Chapter 2

Transgenesis and Paratransgenesis for the Control of Malaria

Sibao Wang[a,b] and Marcelo Jacobs-Lorena[c]

[a]*CAS Key Laboratory of Insect Developmental and Evolutionary Biology, CAS Center for Excellence in Molecular Plant Sciences, Shanghai Institute of Plant Physiology and Ecology, Chinese Academy of Sciences, Shanghai, China*
[b]*CAS Center for Excellence in Biotic Interactions, University of Chinese Academy of Sciences, Beijing, China*
[c]*Johns Hopkins Bloomberg School of Public Health and Malaria Research Institute, Department of Molecular Microbiology and Immunology, Baltimore, Maryland, USA*
sbwang@sibs.ac.cn, ljacob13@jhu.edu

2.1 Transgenesis

Genetic engineering of organisms was first demonstrated in an exciting succession of publications in the early 1970s. In 1973, Cohen et al. showed that an in vitro engineered plasmid can be transformed into *Escherichia coli*, conferring it antibiotic resistance [1]. In 1981, by injecting DNA into the pronuclei of mouse eggs, the laboratory of Frank Ruddle created the first genetically engineered animal capable of transmitting the introduced DNA to subsequent generations [2]. However, the mode of DNA integration into the

Mosquito Gene Drives and the Malaria Eradication Agenda
Edited by Rebeca Carballar-Lejarazú

ISBN 978-981-4968-33-1 (Hardcover), 978-1-003-30877-5 (eBook)
www.jennystanford.com

host chromosome was not determined. In 1982, Spradling and Rubin created the first genetically modified insect, using the *P* transposable element [3]. A *P* element carrying the rosy gene complemented with high efficiency, the rosy mutation of the recipient flies. The high efficiency of *P* element integration into *Drosophila* chromosome prompted an intense wave of experimentation that attempted to use this element to transform a variety of organisms, including mammals. For instance, a heroic set of injections of a *P* construct into mosquito embryos resulted in the generation of a single stable transformant. However, one case was not mediated by the *P* element [4]. The *P* element encodes its own transposase, but regulation of its expression is complex. Regulation occurs not at the transcription, but the splicing level, as removal of the third intron occurs exclusively in the *Drosophila* germline, not in somatic cells. The incompletely spliced mRNA (containing the third intron) encodes a 66 kDa protein that acts as a repressor of transposition. Moreover, additional proteins of *Drosophila* origin are involved in the transposition event and this complex regulation probably explains the failure of achieving *P*-mediated transgenesis of non-*Drosophila* organisms.

The *Minos* transposon from *Drosophila hydei* was used for the germline transformation of the medfly, *Ceratitis capitata* [5]. Transposon-mediated transgenesis of a mosquito was first reported in 1998 with the germline transformation of *Aedes aegypti* with the *Hermes* or *mariner* transposons [6, 7]. As was the case for *Drosophila* and *Ceratitis*, identification of successful transformation events was based on the complementation of an eye color mutation in the recipient organism by a gene in the transposon. This strategy imposed a significant limitation, as transgenesis was dependent on the availability of a mutant insect. Attempts of using antibiotic resistance for selecting integration events may be feasible [8], but it was not efficient. Such limitations were overcome by the development of dominant fluorescent markers, such as the green fluorescent protein (GFP), expressed either from a ubiquitous actin promoter [9] or from a synthetic eye 3xP3 promoter [10]. The latter is the most frequently utilized promoter.

Of all possible applications, control of pathogen transmission provides the greatest promise for mosquito transgenesis. This application has two essential requirements: (i) the identification of effective tissue-specific promoters targeting the pathogen in the appropriate mosquito compartment and (ii) the development of effectors, which kill or thwart pathogen development and are harmless to the mosquito.

2.1.1 Tissue-Specific Promoters

The cycle of most pathogens involves three mosquito compartments: the midgut, the hemocoel, and the salivary gland. To decide which is the most favorable compartment for targeting the parasite, one needs to consider that a dramatic bottleneck of pathogen numbers occurs in the mosquito gut. In the case of *Plasmodium*, the causative agent of malaria, a mosquito biting an infected individual may ingest on the order of 1,000 gametocytes, but in the field, typically 10 or fewer oocysts are formed after traversal of the mosquito midgut epithelium [11, 12]. This is followed by the formation of thousands of sporozoites in each oocyst. The sporozoites are then released into the hemocoel, about 20% of which invade the salivary gland [13]. Similarly, infection of mosquitoes by arboviruses is initiated by the infection of just a few midgut epithelial cells followed by replication and steady release of a large number of virions into the hemocoel. The low pathogen numbers in the midgut make this organ a prime target for control. Of course, the simultaneous targeting of more than one compartment should increase blockage efficiency.

In the early days, characterization of organ-specific promoters took advantage of the well-developed and efficient *Drosophila* transgenesis [3] and the finding that regulation of gene expression transcends evolution. For instance, silk moth genes are expressed with correct sex, tissue, and temporal specificity in transgenic *Drosophila*, an evolutionary distant organism [14]. Early work on the characterization of gut-specific genes from the black fly led to the identification and characterization of trypsin and carboxypeptidase genes [15]. Follow-up studies determined that the black fly carboxypeptidase promoter is fully functional and expressed with correct tissue specificity in transgenic *Drosophila* [16]; while using the same approach, the trypsin promoter was not active (MJ-L, unpublished). With the advent of mosquito transgenesis, robust reporter gene expression driven by the *Ae. aegypti* and *Anopheles gambiae* carboxypeptidase promoters was demonstrated in transgenic *Ae. aegypti* mosquitoes [17]. These promoters are now almost universally used to drive gene expression in the gut of transgenic mosquitoes. Carboxypeptidase expression is low in sugar-fed mosquitoes and is strongly induced by blood ingestion, ramping up early after a blood meal (~3-hour peak) in *An. gambiae* [18] and late (~20-hour peak) in *Ae. aegypti* [19]. For some applications, it is desirable to have available a strong and constitutive gut-specific promoter to confer maximal impact of the gene under its control. In anophelines, the expression of genes encoding proteins of the peritrophic matrix (PM, an acellular matrix secreted by the midgut epithelium that surrounds the entire blood meal) is constitutive. Ag-Aper1 is an *An. gambiae* PM protein whose mRNA is constitutively expressed; the Ag-Aper1 protein is stored in midgut cell apical vesicles and secreted immediately upon blood ingestion to be incorporated into the PM [20, 21]. The Ag-Aper1 promoter was used to drive a robust and constitutive expression of a phospholipase effector gene, resulting in a ~80% inhibition of *Plasmodium berghei* development in transgenic *An. stephensi* mosquitoes [22]. Twenty-four hours after the blood meal there is a strong induction of the vitellogenin gene that encodes a major yolk protein precursor secreted from fat body cells into the hemocoel. The Raikhel laboratory demonstrated that the vitellogenin promoter can be used for the robust expression of transgenes in *Ae. aegypti* [23]. Vitellogenin promoter sequences for expression in anophelines were also identified [24]. The vitellogenin promoter is now universally used for transgenic protein secretion into the hemocoel. Early attempts to express transgenes in the salivary gland used promoters derived from the *Maltase-like* and *Apyrase* genes [25]. While luminescence of the luciferase reporter was detected, expression was very weak. More recently, the promoter of the anopheline antiplatelet protein (AAPP) gene was reported to drive robust expression in the salivary glands of *An. stephensi* [26]. The *Ae. aegypti* 30K salivary-specific promoter was used to demonstrate dengue suppression in transgenic mosquitoes [27].

2.1.2 Effector Genes

Genes whose products kill pathogens or inhibit their development are here referred to as effector genes, and their products as effectors. An effector should not impose a fitness cost to the mosquito. There are typically three broad classes of effectors: peptides/proteins with various activities, antibodies that recognize the pathogen and inhibit its development, and mosquito immune genes engineered to be over-expressed. Moreover, small RNAs (sRNAs) have been used mostly to inhibit arbovirus infections of culicine mosquitoes [28, 29].

2.1.2.1 Peptides/proteins

The first effector used to inhibit *Plasmodium* development in a transgenic mosquito was the SM1 peptide. The 12-amino acid SM1 (Salivary gland and Midgut peptide 1) was identified with a screen of a phage peptide-display library for peptides that bind to the *Anopheles* salivary gland and midgut epithelium. Importantly, SM1 strongly inhibited *Plasmodium* invasion of both salivary gland and midgut, presumably by binding to a parasite receptor [30]. Transgenic mosquitoes expressing an SM1 tetramer driven by a carboxypeptidase promoter are strongly impaired for the transmission of a *Plasmodium* parasite [31]. Also, the expression of an SM1-scorpine fusion protein by the *Metarhizium anisopliae* fungus accelerates the killing of human malaria parasites in mosquitoes [32]. Scorpine is a small protein (75 amino acids) with structural similarity to defensins and cecropins that has potent anti-*Plasmodium* activity [33]. In 2001, the Ribeiro laboratory made the interesting observation that snake and bee phospholipase A_2 (PLA2) inhibits *Plasmodium* ookinete association with the mosquito midgut and oocyst formation [34]. The gene encoding a bee venom PLA2 was placed under the control of the carboxypeptidase promoter and inserted into the *An. stephensi* germline. PLA2 expression greatly reduced oocyst formation and parasite transmission to mice [35]. Whereas, the native PLA2 activity imposes a fitness load on mosquitoes [22, 36], hydrolytic activity of the enzyme is not required for antiparasitic action [34]. Indeed, the expression of a mutated inactive PLA2 mutant inhibits *Plasmodium* development without fitness costs to the mosquito [37, 38]. Expression of the C-type lectin CEL-III from the sea cucumber in the mosquito midgut was also shown to inhibit *Plasmodium* development [39]. Several additional effector genes expressed from bacteria (see paratransgenesis below; [40]) have been used to inhibit malaria parasite development: (i) a chitinase pro-peptide (Pro) that inhibits the enzyme and blocks ookinete traversal of the mosquito peritrophic matrix [41]; (ii) a synthetic antiparasitic lytic peptide, Shiva1 [42]; (iii) *Plasmodium* Enolase–Plasminogen Interaction Peptide (EPIP) that inhibits mosquito midgut invasion by preventing plasminogen binding to the ookinete surface [43]; (iv) a fusion peptide composed of Pro and EPIP (Pro:EPIP); and (v) midgut peptide 2 (MP2), a 12-amino acid peptide identified in the same phage display screen that identified SM1 [44]. All effectors were highly efficient, inhibiting *P. falciparum* development in *An. gambiae* by 80–98% without measurable impact on mosquito fitness cost under laboratory conditions. Recently, mosquitoes that express a polypeptide containing an array of five effectors — melittin [45], TP10 [45], Shiva1, EPIP, and scorpine — were shown to efficiently suppress *P. falciparum* development in transgenic *An. stephensi* mosquitoes [46].

2.1.2.2 Antibodies

The use of antibody transgenic effectors to curtail *Plasmodium* development in mosquitoes was pioneered by the James laboratory. They engineered *An. stephensi* to express single-chain antibodies (scFvs) that recognize *P. falciparum* chitinase or the Pfs25 ookinete surface protein in the midgut (both driven by the carboxypeptidase promoter) and the CSP sporozoite surface protein in the hemocoel target (driven by the vitellogenin promoter). Mosquitoes that expressed m4B7, m2A10, Pfs25scFv and CSPscFv, separately fused to the *Anopheles* cecropin antimicrobial peptide, were highly resistant to *P. falciparum* infection as they carried a few or no sporozoites in their salivary glands [47]. Transgenic mosquito lines expressing each of the scFvs, m1C3, m4B7 and m2A10 that are derived from mouse monoclonal antibodies scFv genes, had significantly lower infection of P. falciparum [48]. CSPscFv fused to scorpine was also very effective in reducing sporozoite load [46].

2.1.2.3 Mosquito immune genes

The manipulation of mosquito immune gene expression to inhibit *Plasmodium* development was pioneered by the Dimopoulos laboratory. Over-expression in the midgut and the fat body of Rel2, a key regulator of the IMD immune pathway, conferred significant resistance of transgenic *An. stephensi* to *P. falciparum* infection [46, 49]. Other mosquito immune factors, Cec A [50], TEP1 [51], NF-κB-regulated splicing factors Caper and IRSF1 [53], the fibrinogen domain-containing immunolectin FBN9 [52], the fibrinogen-related protein 1 (FREP1) [53], and the Down syndrome cell-adhesion molecule (AgDscam) splice forms [54], have also been used to enhance mosquito resistance to *P. falciparum* infection.

2.2 Paratransgenesis

An alternative to engineering mosquitoes to express effectors is to engineer gut symbiotic bacteria instead. This approach — commonly referred to as paratransgenesis — has a number of advantages: (i) engineering bacteria is much simpler than engineering mosquitoes; (ii) in the midgut, bacteria with the malaria parasite during its most vulnerable stages of development in the mosquito. In contrast, if secreted from the mosquito midgut epithelium, effectors need to diffuse through the blood bolus to reach the parasites; (iii) bacteria numbers increase dramatically after a mosquito takes a blood meal, resulting in a comparative increase in the effector production; and (iv) logistics of introducing engineered bacteria into malaria-affected areas is simpler than of introducing engineered mosquitoes, especially if the bacteria can self-propagate through mosquito populations.

The concept of paratransgenesis is not new. In 1997, Durvasula et al. showed that a symbiont bacterium of the Chagas disease vector *Rhodnius prolixus* engineered to produce cecropin, curtails transmission of the *Trypanosoma cruzi* parasite [55]. In 2001, Yoshida et al. showed that mosquitoes fed with *E. coli* engineered to express a single-chain antibody targeting the *P. berghei* major ookinete surface protein Pbs21 were strongly inhibited for transmission of the parasite [56]. In 2007, Riehle et al. reported on the construction of *E. coli* bacteria that display SM1 or PLA2 on their surface. These bacteria significantly inhibited *P. berghei* development in mosquitoes. However, inhibition was relatively weak, presumably because the effectors were bound to the bacteria and could not diffuse to reach the parasites. Also, inhibition did not last because the laboratory *E. coli* strain colonized poorly in the mosquito gut [57]. A survey for symbiotic bacteria that are well adapted to the mosquito midgut environment led to the isolation of *Pantoea agglomerans* [57]. In a subsequent step, *P. agglomerans* were engineered to secrete (instead, of being attached to the bacteria surface) a variety of effectors. Mosquitoes carrying these bacteria inhibited *P. falciparum* and *P. berghei* by up to 98% and the proportion of mosquitoes carrying parasites (prevalence) decreased by up to 84% [40]. A further step addressed the issue of how to introduce parasite-inhibiting bacteria in the field. In pioneering work, Favia et al. identified a bacterium of the genus *Asaia* that in addition to the midgut, associates with salivary glands, male and female reproductive organs [58], and is paternally transmitted to females [59]. Shane et al. engineered *Asaia* to express the scorpine effector from a promoter that is activated only in the presence of blood, to achieve improved bacterial fitness [60]. The spread of fluorescent kanamycin-resistant *Asaia* in mosquito populations maintained in large cages in the presence of the antibiotic, which was demonstrated for both *An. stephensi* and *An. gambiae* [61].

As engineered symbiotic bacteria have been shown to render mosquitoes resistant to the malaria parasite, the challenge remained how to introduce such bacteria into mosquito populations in the field. By plating bacteria associated with *An. stephensi* ovaries, Wang et al. identified a bacterium — termed *Serratia* AS1 — that is effectively transmitted transstadially from larva to adult, vertically from female mosquitoes to progeny, and horizontally (sexually) from male to female. In the laboratory, this bacterium rapidly spreads through mosquito populations and remains without selection for at least three generations [62]. *Serratia* AS1 was genetically engineered for the secretion of anti-*Plasmodium* effector proteins, and the recombinant strains strongly inhibit the development of *P. falciparum* in mosquitoes [62].

Mosquito viral symbionts can also be used to express foreign genes in mosquitoes. Ren et al. characterized the first known densovirus (DNV) in *An. gambiae* and developed an AgDNV-based vector to express eGFP in *An. gambiae* midgut, fat body, and ovaries, and was transmitted to next-generation [63]. However, such vectors can carry only a limited amount of foreign genetic material.

Fungi also have the potential to serve as a tool for paratransgenesis in mosquitoes. The yeast *W. anomalus* was found to colonize the mosquito midgut and reproductive organs [64], raising the possibility of its engineering for the production of anti-pathogen effectors. *Saccharomyces cerevisiae* has been engineered to produce shRNAs targeting *An. gambiae* and *Ae. aegypti* larval essential genes. These recombinant yeasts proved to be effective in killing larvae that feed on it [65, 66]. The entomopathogenic fungus *M. anisopliae* has been engineered to produce anti-malaria molecules, which effectively block *Plasmodium* development in *An. gambiae*; in this way enhancing transmission blockage by killing both the mosquito vector and the parasite [32].

2.3 Prospects

After a significant decrease in malaria cases and deaths at the beginning of this decade, in great part thanks to the wide distribution of long-lasting insecticide-impregnated bed nets; in recent years, the number of cases has plateaued and even increased in certain regions [67]. Parasite resistance to drugs, mosquito resistance to insecticides, and mosquito behavior shifting to outdoor biting are largely responsible. These facts strongly argue for the deployment of additional weapons, together with the existing ones, will aid in reducing mortality caused by this devastating disease. Whereas, the effectiveness of transgenesis and paratransgenesis (Fig. 2.1) in a laboratory setting has been robustly demonstrated, challenges remain for field deployment. Among them are the development of efficient drive mechanisms to introduce transgenes into mosquito populations (dealt with in another chapter), satisfying regulatory issues related to the introduction of genetically modified organisms in nature, and social issues such as obtaining consent from local residents. In this respect, one of the laboratories (S.W.) has obtained promising results in the identification of a naturally occurring *Serratia* strain that not only can spread through mosquito populations, but also strongly inhibits *Plasmodium* development without affecting mosquito fitness [68]. This non-recombinant strain would overcome regulatory restrictions related to field introduction.

Another consideration is which of the approaches — transgenesis or paratransgenesis — is the most promising and should be emphasized. We feel these should be implemented concomitantly, as the two approaches are complementary, not mutually exclusive. One of the laboratories (M.J.-L.) has shown data that transgenic mosquitoes and recombinant bacteria, each expressing the same

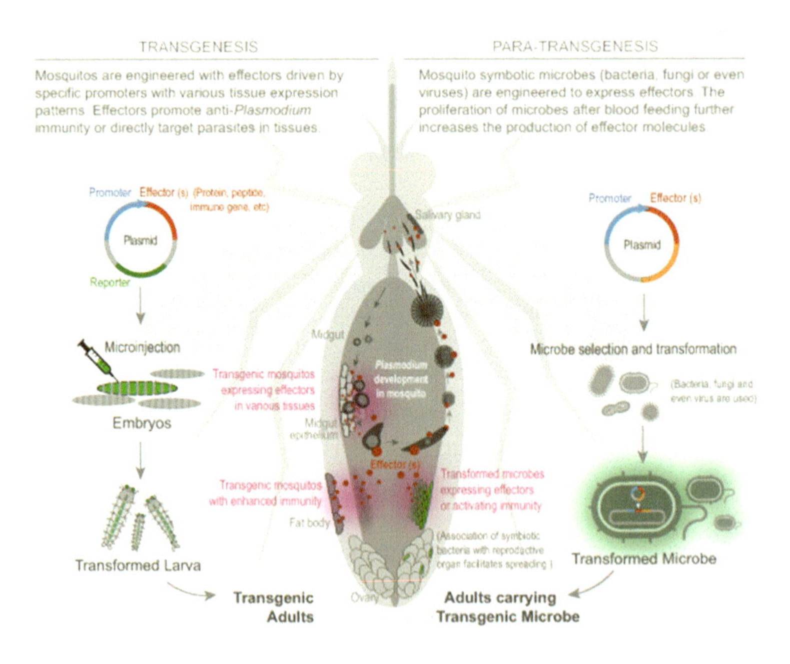

Figure 2.1 Transgenesis and paratransgenesis approaches to contain malaria parasite development in mosquitoes.

effector genes, more effectively inhibit parasite development and transmission when the two are combined than each separately.

Finally, we emphasize the notion that malaria will never be conquered by any single approach and that it is fundamental to combine all available weapons in the fight against this deadly disease.

Acknowledgments

This work was supported by grants from the National Key R&D Program of China (grants 2020YFC1200100 and 2018YFA0900502), the National Natural Science Foundation of China (grants 31830086, 31772534 and 32021001), and NIH grant R01AI031478, and by the Bloomberg Philanthropies. We are grateful to Han Gao for illustration support.

References

1. Cohen SN, Chang AC, Boyer HW, Helling RB (1973). Construction of biologically functional bacterial plasmids in vitro. *Proc. Natl. Acad. Sci. USA*. 70:3240–3244.
2. Gordon JW, Ruddle FH (1981). Mammalian gonadal determination and gametogenesis. *Science* 211:1265–1271.
3. Rubin GM, Spradling AC (1982). Genetic transformation of *Drosophila* with transposable element vectors. *Science* 218:348–353.
4. Miller LH, Sakai RK, Romans P, Gwadz RW, Kantoff P, Coon HG (1987). Stable integration and expression of a bacterial gene in the mosquito *Anopheles gambiae*. *Science*. 237:779–781.
5. Loukeris TG, Livadaras I, Arcà B, Zabalou S, Savakis C (1995). Gene transfer into the medfly, *Ceratitis capitata*, with a *Drosophila hydei* transposable element. *Science*. 270:2002–2005.
6. Jasinskiene N, Coates CJ, Benedict MQ, Cornel AJ, Rafferty CS, James AA, Collins FH (1998). Stable transformation of the yellow fever mosquito, *Aedes aegypti*, with the Hermes element from the housefly. *Proc. Natl. Acad. Sci. USA*. 95:3743–3747.

7. Coates CJ, Jasinskiene N, Miyashiro L, James AA (1998). Mariner transposition and transformation of the yellow fever mosquito, *Aedes aegypti*. *Proc. Natl. Acad. Sci. USA*. 95:3748–3751.
8. Sakai RK, Miller LH (1992). Effects of heat shock on the survival of transgenic *Anopheles gambiae* (*Diptera*: *Culicidae*) under antibiotic selection. *J. Med. Entomol.* 29:374–375.
9. Pinkerton AC, Michel K, O'Brochta DA, Atkinson PW (2000). Green fluorescent protein as a genetic marker in transgenic *Aedes aegypti*. *Insect Mol. Biol.* 9:1–10.
10. Berghammer AJ, Klingler M, Wimmer EA (1999). A universal marker for transgenic insects. *Nature*. 402:370–371.
11. Bompard A, Da DF, Yerbanga SR, Morlais I, Awono-Ambéné PH, Dabiré RK, Ouédraogo JB, Lefèvre T, Churcher TS, Cohuet A (2020). High *Plasmodium* infection intensity in naturally infected malaria vectors in Africa. *Int. J. Parasitol.* 50:985–996.
12. Wang S, Jacobs-Lorena M(2013). Genetic approaches to interfere with malaria transmission by vector mosquitoes. *Trends Biotechnol.* 31:185–193.
13. Hillyer JF, Barreau C, Vernick KD (2007). Efficiency of salivary gland invasion by malaria sporozoites is controlled by rapid sporozoite destruction in the mosquito haemocoel. *Int. J. Parasitol.* 37:673–681.
14. Mitsialis SA, Kafatos FC (1985). Regulatory elements controlling chorion gene expression are conserved between flies and moths. *Nature*. 317:453–456.
15. Ramos A, Mahowald A, Jacobs-Lorena M (1993). Gut-specific genes from the black fly *Simulium vittatum* encoding trypsin-like and carboxypeptidase-like proteins. *Insect Mol. Biol.* 1:149–163.
16. Xiong B, Jacobs-Lorena M (1995). Gut-specific transcriptional regulatory elements of the carboxypeptidase gene are conserved between black flies and *Drosophila*. *Proc. Natl. Acad. Sci. USA*. 92:9313–9317.
17. Moreira LA, Edwards MJ, Adhami F, Jasinskiene N, James AA, Jacobs-Lorena M (2000). Robust gut-specific gene expression in transgenic *Aedes aegypti* mosquitoes. *Proc. Natl. Acad. Sci. USA*. 97:10895–898.
18. Edwards MJ, Lemos FJ, Donnelly-Doman M, Jacobs-Lorena M (1997). Rapid induction by a blood meal of a carboxypeptidase gene in the gut of the mosquito *Anopheles gambiae*. *Insect Biochem Mol. Biol.* 27:1063–1072.

19. Edwards MJ, Moskalyk LA, Donelly-Doman M, Vlaskova M, Noriega FG, Walker VK, Jacobs-Lorena M (2000). Characterization of a carboxypeptidase A gene from the mosquito, *Aedes aegypti*. *Insect Mol. Biol.* 9:33–38.
20. Shen Z, Jacobs-Lorena M (1998). A type I peritrophic matrix protein from the malaria vector *Anopheles gambiae* binds to chitin. Cloning, expression, and characterization. *J. Biol. Chem.* 273:17665–17670.
21. Devenport M, Fujioka H, Jacobs-Lorena M (2004). Storage and secretion of the peritrophic matrix protein Ag-Aper1 and trypsin in the midgut of *Anopheles gambiae*. *Insect Mol. Biol.* 13:349–358.
22. Abraham EG, Donnelly-Doman M, Fujioka H, Ghosh A, Moreira L, Jacobs-Lorena M (2005). Driving midgut-specific expression and secretion of a foreign protein in transgenic mosquitoes with AgAper1 regulatory elements. *Insect Mol. Biol.* 14:271–279.
23. Kokoza V, Ahmed A, Cho WL, Jasinskiene N, James AA, Raikhel A (2000). Engineering blood meal-activated systemic immunity in the yellow fever mosquito, *Aedes aegypti*. *Proc. Natl. Acad. Sci. USA*. 97:9144–9149.
24. Nirmala X, Marinotti O, Sandoval JM, Phin S, Gakhar S, Jasinskiene N, James AA (2006). Functional characterization of the promoter of the vitellogenin gene, AsVg1, of the malaria vector, *Anopheles stephensi*. *Insect. Biochem. Mol. Biol.* 36:694–700.
25. Coates CJ, Jasinskiene N, Pott GB, James AA (1999). Promoter-directed expression of recombinant fire-fly luciferase in the salivary glands of Hermes-transformed *Aedes aegypti*. *Gene*. 226:317–325.
26. Yoshida S, Watanabe H (2006). Robust salivary gland-specific transgene expression in *Anopheles stephensi* mosquito. *Insect Mol. Biol.* 15:403–410.
27. Mathur G, Sanchez-Vargas I, Alvarez D, Olson KE, Marinotti O, James AA (2010). Transgene-mediated suppression of dengue viruses in the salivary glands of the yellow fever mosquito, Aedes aegypti. *Insect Mol. Biol.* 19:753–763.
28. Yen PS, James A, Li JC, Chen CH, Failloux AB (2018). Synthetic miRNAs induce dual arboviral-resistance phenotypes in the vector mosquito *Aedes aegypti*. *Commun. Biol.* 1:11.
29. Buchman A, Gamez S, Li M, Antoshechkin I, Li HH, Wang HW, Chen CH, Klein MJ, Duchemin JB, Paradkar PN, Akbari OS (2019). Engineered resistance to Zika virus in transgenic *Aedes aegypti* expressing a polycistronic cluster of synthetic small RNAs. *Proc. Natl. Acad. Sci. USA*. 116:3656–3661.

30. Ghosh AK, Ribolla PE, Jacobs-Lorena M (2001). Targeting *Plasmodium* ligands on mosquito salivary glands and midgut with a phage display peptide library. *Proc. Natl. Acad. Sci. USA*. 98:13278–13281.
31. Ito J, Ghosh A, Moreira LA, Wimmer EA, Jacobs-Lorena M (2002). Transgenic anopheline mosquitoes impaired in transmission of a malaria parasite. *Nature*. 417:452–455.
32. Fang W, Vega-Rodríguez J, Ghosh AK, Jacobs-Lorena M, Kang A, St Leger RJ (2011). Development of transgenic fungi that kill human malaria parasites in mosquitoes. *Science*. 331:1074–1077.
33. Possani LD, Corona M, Zurita M, Rodríguez MH (2002). From noxiustoxin to scorpine and possible transgenic mosquitoes resistant to malaria. *Arch. Med. Res*. 33:398–404.
34. Zieler H, Keister DB, Dvorak JA, Ribeiro JM (2001). A snake venom phospholipase A(2) blocks malaria parasite development in the mosquito midgut by inhibiting ookinete association with the midgut surface. *J. Exp. Biol*. 204:4157–4167.
35. Moreira LA, Ito J, Ghosh A, Devenport M, Zieler H, Abraham EG, Crisanti A, Nolan T, Catteruccia F, Jacobs-Lorena M (2002). Bee venom phospholipase inhibits malaria parasite development in transgenic mosquitoes. *J. Biol. Chem*. 277:40839–40843.
36. Moreira LA, Wang J, Collins FH, Jacobs-Lorena M (2004). Fitness of anopheline mosquitoes expressing transgenes that inhibit *Plasmodium* development. *Genetics*. 166:1337–1341.
37. Rodrigues FG, Santos MN, de Carvalho TX, Rocha BC, Riehle MA, Pimenta PF, Abraham EG, Jacobs-Lorena M, Alves de Brito CF, Moreira LA (2008). Expression of a mutated phospholipase A2 in transgenic *Aedes fluviatilis* mosquitoes impacts *Plasmodium gallinaceum* development. *Insect Mol. Biol*. 17:175–183.
38. Smith RC, Kizito C, Rasgon JL, Jacobs-Lorena M (2013). Transgenic mosquitoes expressing a phospholipase A(2) gene have a fitness advantage when fed *Plasmodium falciparum*-infected blood. *PLoS One*. 8(10):e76097.
39. Yoshida S, Shimada Y, Kondoh D, Kouzuma Y, Ghosh AK, Jacobs-Lorena M, Sinden RE (2007). Hemolytic C-type lectin CEL-III from sea cucumber expressed in transgenic mosquitoes impairs malaria parasite development. *PLoS Pathog*. 3:e192.
40. Wang S, Ghosh AK, Bongio N, Stebbings KA, Lampe DJ, Jacobs-Lorena M (2012). Fighting malaria with engineered symbiotic bacteria from vector mosquitoes. *Proc. Natl. Acad. Sci. USA*. 109:12734–12739.

41. Bhatnagar RK, Arora N, Sachidanand S, Shahabuddin M, Keister D, Chauhan VS (2003). Synthetic propeptide inhibits mosquito midgut chitinase and blocks sporogonic development of malaria parasite. *Biochem. Biophys. Res. Commun.* 304:783–787.

42. Jaynes JM, Burton CA, Barr SB, Jeffers GW, Julian GR, White KL, Enright FM, Klei TR, Laine RA (1988). In vitro cytocidal effect of novel lytic peptides on *Plasmodium falciparum* and *Trypanosoma cruzi. FASEB. J.* 2:2878–2883.

43. Ghosh AK, Coppens I, Gårdsvoll H, Ploug M, Jacobs-Lorena M (2011). *Plasmodium* ookinetes coopt mammalian plasminogen to invade the mosquito midgut. *Proc. Natl. Acad. Sci. USA.* 108:17153–17158.

44. Vega-Rodríguez J, Ghosh AK, Kanzok SM, Dinglasan RR, Wang S, Bongio NJ, Kalume DE, Miura K, Long CA, Pandey A, Jacobs-Lorena M (2014). Multiple pathways for *Plasmodium* ookinete invasion of the mosquito midgut. *Proc. Natl. Acad. Sci. USA.* 111:E492–500.

45. Carter V, Underhill A, Baber I, Sylla L, Baby M, Larget-Thiery I, Zettor A, Bourgouin C, Langel U, Faye I, Otvos L, Wade JD, Coulibaly MB, Traore SF, Tripet F, Eggleston P, Hurd H (2013). Killer bee molecules: antimicrobial peptides as effector molecules to target sporogonic stages of *Plasmodium. PLoS Pathog.* 9(11):e1003790.

46. Dong Y, Simões ML, Dimopoulos G (2020). Versatile transgenic multistage effector-gene combinations for *Plasmodium falciparum* suppression in *Anopheles. Sci. Adv.* 6(20):eaay589.

47. Isaacs AT, Jasinskiene N, Tretiakov M, Thiery I, Zettor A, Bourgouin C, James AA (2012). Transgenic *Anopheles stephensi* coexpressing single-chain antibodies resist *Plasmodium falciparum* development. Proc. *Natl. Acad. Sci. USA.* 109:E1922–1930.

48. Isaacs AT, Li F, Jasinskiene N, Chen X, Nirmala X, Marinotti O, Vinetz JM, James AA (2011). Engineered resistance to *Plasmodium falciparum* development in transgenic *Anopheles stephensi. PLoS Pathog.* 7(4):e1002017.

49. Dong Y, Das S, Cirimotich C, Souza-Neto JA, McLean KJ, Dimopoulos G (2011). Engineered *Anopheles* immunity to *Plasmodium* infection. *PLoS Pathog.* 7(12):e1002458.

50. Kim W, Koo H, Richman AM, Seeley D, Vizioli J, Klocko AD, O'Brochta DA (2004). Ectopic expression of a cecropin transgene in the human malaria vector mosquito *Anopheles gambiae* (Diptera: Culicidae): effects on susceptibility to *Plasmodium. J. Med. Entomol.* 41(3):447–55.

51. Volohonsky G, Hopp AK, Saenger M, Soichot J, Scholze H, Boch J, Blandin SA, Marois E (2017). Transgenic expression of the anti-parasitic factor TEP1 in the malaria mosquito *Anopheles gambiae*. *PLoS Pathog.* 13(1):e1006113.

52. Simões ML, Dong Y, Hammond A, Hall A, Crisanti A, Nolan T, Dimopoulos G (2017). The *Anopheles* FBN9 immune factor mediates *Plasmodium* species-specific defense through transgenic fat body expression. *Dev. Comp. Immunol.* 67:257–265.

53. Dong Y, Simões ML, Marois E, Dimopoulos G (2018). CRISPR/Cas9 - mediated gene knockout of *Anopheles gambiae FREP1* suppresses malaria parasite infection. *PLoS Pathog.* 14(3): e1006898.

54. Dong Y, Cirimotich CM, Pike A, Chandra R, Dimopoulos G (2012). *Anopheles* NF-κB-regulated splicing factors direct pathogen-specific repertoires of the hypervariable pattern recognition receptor AgDscam. *Cell Host Microbe.* 12(4):521–530.

55. Durvasula RV, Gumbs A, Panackal A, Kruglov O, Aksoy S, Merrifield RB, Richards FF, Beard CB (1997). Prevention of insect-borne disease: an approach using transgenic symbiotic bacteria. Proc. Natl. Acad. Sci. USA. 94:3274–3278.

56. Yoshida S, Ioka D, Matsuoka H, Endo H, Ishii A (2001). Bacteria expressing single-chain immunotoxin inhibit malaria parasite development in mosquitoes. *Mol. Biochem. Parasitol.* 113:89–96.

57. Riehle MA, Moreira CK, Lampe D, Lauzon C, Jacobs-Lorena M (2007). Using bacteria to express and display anti-*Plasmodium* molecules in the mosquito midgut. *Int. J. Parasitol.* 37:595–603.

58. Favia G, Ricci I, Damiani C, Raddadi N, Crotti E, Marzorati M, Rizzi A, Urso R, Brusetti L, Borin S, Mora D, Scuppa P, Pasqualini L, Clementi E, Genchi M, Corona S, Negri I, Grandi G, Alma A, Kramer L, Esposito F, Bandi C, Sacchi L, Daffonchio D (2007). Bacteria of the genus *Asaia* stably associate with *Anopheles stephensi*, an Asian malarial mosquito vector. *Proc. Natl. Acad. Sci. USA.* 104:9047–9051.

59. Damiani C, Ricci I, Crotti E, Rossi P, Rizzi A, Scuppa P, Esposito F, Bandi C, Daffonchio D, Favia G (2008). Paternal transmission of symbiotic bacteria in malaria vectors. *Curr. Biol.* 18:R1087–1088.

60. Shane JL, Grogan CL, Cwalina C, Lampe DJ (2018). Blood meal-induced inhibition of vector-borne disease by transgenic microbiota. *Nat. Commun.* 9(1):4127.

61. Mancini MV, Spaccapelo R, Damiani C, Accoti A, Tallarita M, Petraglia E, Rossi P, Cappelli A, Capone A, Peruzzi G, Valzano M, Picciolini M, Diabaté

A, Facchinelli L, Ricci I, Favia G (2016). Paratransgenesis to control malaria vectors: a semi-field pilot study. *Parasit. Vectors.* 10;9:140.

62. Wang S, Dos-Santos ALA, Huang W, Liu KC, Oshaghi MA, Wei G, Agre P, Jacobs-Lorena M (2017). Driving mosquito refractoriness to *Plasmodium falciparum* with engineered symbiotic bacteria. *Science.* 357:1399–1402.
63. Ren X, Hoiczyk E, Rasgon JL (2008). Viral paratransgenesis in the malaria vector *Anopheles gambiae. PLoS Pathog.* 4(8):e1000135.
64. Ricci I, Damiani C, Scuppa P, Mosca M, Crotti E, Rossi P, Rizzi A, Capone A, Gonella E, Ballarini P, Chouaia B, Sagnon N, Esposito F, Alma A, Mandrioli M, Sacchi L, Bandi C, Daffonchio D, Favia G (2011). The yeast *Wickerhamomyces anomalus* (*Pichia anomala*) inhabits the midgut and reproductive system of the Asian malaria vector *Anopheles stephensi. Environ. Microbiol.* 13:911–921.
65. Mysore K, Hapairai LK, Sun L, Harper EI, Chen Y, Eggleson KK, Realey JS, Scheel ND, Severson DW, Wei N, Duman-Scheel M (2017). Yeast interfering RNA larvicides targeting neural genes induce high rates of *Anopheles* larval mortality. *Malar. J.* 16(1):461.
66. Hapairai LK, Mysore K, Chen Y, Harper EI, Scheel MP, Lesnik AM, Sun L, Severson DW, Wei N, Duman-Scheel M (2017). Lure-and-kill yeast interfering RNA larvicides targeting neural genes in the human disease vector mosquito *Aedes aegypti. Sci. Rep.* 7(1):13223.
67. WHO World Malaria Report 2020. https://www.who.int/publications/i/item/9789240015791
68. Gao H, Bai L, Jiang Y, Huang W, Wang L, Li S, Zhu G, Wang D, Huang Z, Li X, Cao J, Jiang L, Jacobs-Lorena M, Zhan S, Wang S (2021). A natural symbiotic bacterium drives mosquito refractoriness to *Plasmodium* infection via secretion of an antimalarial lipase. *Nat. Microbiol.* 6: 806–817.

Chapter 3

Gene Drives for *Anopheles* Mosquitoes

Jackson Champer
Center for Bioinformatics, School of Life Sciences,
Peking-Tsinghua Center for Life Sciences, Peking University, Beijing, China
jchamper@pku.edu.cn

3.1 Introduction to Gene Drives and Their Characteristics

3.1.1 The Concept of Gene Drive and Its Applications

Gene drives could potentially provide an inexpensive, easily implemented, and powerful solution to malaria and many other vector-borne diseases. These engineered alleles act to bias their inheritance, allowing them to quickly increase in frequency and spread through a mosquito population within only a few years [1–10]. These drives could carry an antimalaria cargo gene, disrupting the ability of a mosquito to transmit the disease. Alternatively, the gene drive could directly suppress the population of the mosquito vector.

Though the idea of gene drive was first proposed several decades ago, mosquito genetic tools have only recently enabled

Mosquito Gene Drives and the Malaria Eradication Agenda
Edited by Rebeca Carballar-Lejarazú

ISBN 978-981-4968-33-1 (Hardcover), 978-1-003-30877-5 (eBook)
www.jennystanford.com

intensive study on this topic. CRISPR nucleases in particular have created a recent explosion of gene drive studies, several of which have demonstrated successful laboratory results. Nevertheless, gene drives remain unproven in the field, and they may face substantial obstacles to field deployment based on public opinion. Several technical challenges also remain, even for *Anopheles* species where certain types of gene drives are well-developed (Table 3.1).

3.1.2 Outcomes of a Gene Drive Strategy

Proposed gene drive strategies against *Anopheles* generally fall under two categories. In the modification strategy, the gene drive is designed to either spread a cargo gene or disrupt one or more native genes such that malaria transmission is impaired or eliminated (Fig. 3.1). In the suppression strategy, the gene drive disrupts an essential but haplosufficient gene without rescuing the function of the target gene, thus eventually resulting in population reduction or elimination (Fig. 3.1). Both these strategies have been attempted with promising results.

One key component for modification drives is the effector element, which ideally should completely block transmission of malaria while having a low fitness cost and remaining compact. Several possibilities for this already exist, including antimicrobial factors and modifications to the mosquitos' immune system [27–29]. A couple of studies have also reduced malaria transmission by disrupting a native gene, a strategy that is less vulnerable to inactivating mutations in the cargo gene [30, 31]. These mutations would be expected to reduce the fitness cost of the allele and thus outcompete the complete drive, particularly if the effector carries a high-fitness cost. In either case, it is possible that the malaria parasite itself may be able to evolve resistance to the gene drive effector, which could necessitate the release of the second-generation drives with a new effector. Combination strategies with multiple effectors could potentially circumvent this.

Suppression drives can be simpler to design due to the lack of need for a cargo gene, though in this case, the target gene becomes more important and controls the mechanism of suppression. If the suppression drive is 100% efficient, then in simple models,

Table 3.1 Major classes of gene drives in *Anopheles* mosquitoes

Drive class	Drive type	Speed	Introduction threshold	Construction difficulty	Status
Chromosomal rearrangement	Modification	Slow	High	High	Developed but not tested as a drive [11–13]
Transposon	Modification	Fast	None	High	Demonstrated [14, 15]
Homing	Modification or suppression	Fast	None	Low	Successful for suppression [16], nearly so for modification [17, 18]
X-shredder	Suppression	Fast	None	High	Autosomal demonstrated [19, 20], successful with homing drive [21]
RNAi toxin-antidote	Modification	Slow- medium	Low-high	High	Successful in flies [22, 23]
CRISPR toxin-antidote	Modification or suppression	Slow-medium	Low-high	Low	Successful in flies [24–26] for modification

it will always eliminate the population. However, an imperfect suppression drive may lack the power to eliminate the population, instead of reaching an equilibrium allele frequency. In general, the suppressive power of the drive is specified by its genetic load when it reaches this equilibrium frequency. Genetic load refers to the fractional reduction in reproductive output compared to a complete wild-type population with otherwise similar characteristics, with "1" referring to complete lack of successful reproduction and "0" referring to no effect on reproduction. Note, however, that the exact degree of population suppression is dependent not just on genetic load, but on species-specific and ecological factors such as the density dependence of larval viability [32]. A high genetic load can be imposed without targeting an essential gene via cargo genes with high-fitness costs, but such fitness costs would greatly slow the spread of the drive, and this option has not been preferred. Instead, engineered suppression drives have targeted haplosufficient but essential genes, allowing the rapid spread of the drive-in heterozygotes. Furthermore, the targets have been female-specific fertility genes, which allows for higher genetic loads than gene targets affecting both sexes. In most situations, targeting female-specific genes will likely result in stronger suppression since males are usually not the limiting factor in a species reproductive capacity. Aside from lacking the power to impose a sufficiently high genetic load, suppression drives have a number of other potential issues that could affect the success of a gene drive program. These include a wide variety of factors such as the importance of the species to the local ecosystem or replacement of the disease vector by other species with similar vectorial capacity. Several computational models have also indicated that even drives with enough power to suppress the population in a simple model may fail to do so when the population has a complex spatial structure [33–36]. Instead, both drive and wild-type alleles can persist in a chaotic manner. The population size in this situation is still generally reduced compared to the initial population.

The persistence of the gene drive is another important aspect of its outcome. Most gene drive designs are expected to persist indefinitely in their idealized forms for as long as the population exists. However, some designs exist for self-limiting drives that

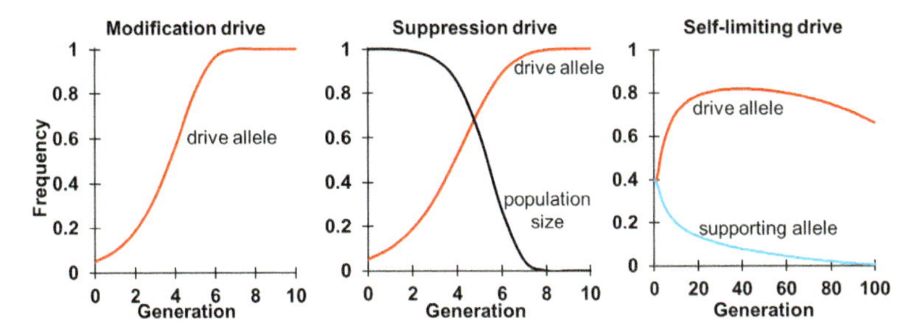

Figure 3.1 Types of gene drives. A modification drive (ideal homing drive in the figure) is designed to alter the entire target population on a long-term basis while keeping the population intact. A suppression drive (ideal homing suppression drive targeting a female fertility gene) is designed to reduce or even eliminate the population as the drive reaches high frequency. A self-limiting drive (a killer-rescue drive in the figure) can cause temporary population modification, but the population should revert to wild type after a moderate interval.

are intended to increase in frequency only temporarily before declining to extinction (Fig. 3.1). The best example of this is the killer-rescue drive [37], which has been demonstrated in flies [38]. In this system, the killer element will result in death unless the unlinked rescue element is present. This initially results in the rescue element increasing in frequency while the killer element immediately decreases. When the killer frequency is low, the rescue element will lose its advantage over wild-type alleles and decline in frequency over time due to fitness costs. Another method for limiting a drive system is called the "split drive" method where the drive is broken into two components. Only one of them is capable of increasing in frequency, and this requires the other component, the supporting element (often a lone Cas9 allele), to be present. Since the supporting element will decline in frequency over time, the system will eventually lose its ability to drive, though the rate that this happens depends entirely on the fitness costs involved, potentially making it less controllable. Split drive systems are common for safe laboratory testing, with several experimental demonstrations [39–43]. They could potentially be used in the wild for population modification, though the necessary release sizes are often substantially higher than needed for the standard drive systems. A variant for the split drive system is the daisy chain, where several drive elements are supported by the next in turn, with only the final supporting element being unable to drive [44]. These allow for smaller release sizes and the potential for suppression, but this comes at the cost of greater engineering complexity and less predictable population dynamics [45].

3.1.3 Confinement of a Gene Drive to Target Populations

While eliminating malaria may necessarily encompass continental-scale deployments of gene drive insects, gene drives that could be confined to only target populations may be needed in a variety of situations. For many gene drive applications, it would be highly undesirable for a gene drive to spread globally through a species, for example, when targeted eradication of invasive populations of some species is called for. Even for malaria control, the use of a confined drive may prove beneficial for building the necessary public support, or avoiding the need for international regulation of a drive predicted to cross state borders. In these cases, a confined drive could still prove highly effective, as well as pave the way for subsequent releases of more powerful drive forms. However, the confinement of a drive is a potentially complex issue.

There are many types of confined drives with a variety of molecular mechanisms and performance characteristics. However, all of these share the characteristic of not being able to directly increase their allele count in the population. Instead, these drives function by removing wild-type alleles, thus increasing the relative frequency of the drive. Drives that show underdominance characteristics (where drive heterozygotes are less fit than wild-type or drive homozygous individuals) can also tend to remove drive alleles, sometimes at greater rates than wild-type alleles are removed, further reducing drive power and increasing the level of confinement. The simplest example of this is species-like incompatibilities, where no viable offspring are produced between crosses of individuals with drive alleles and individuals with wild-type alleles [46]. Confinement is usually measured by the drive's introduction threshold frequency (Fig. 3.2). This refers to the critical frequency of the drive in a population, below which the drive will decrease in frequency and

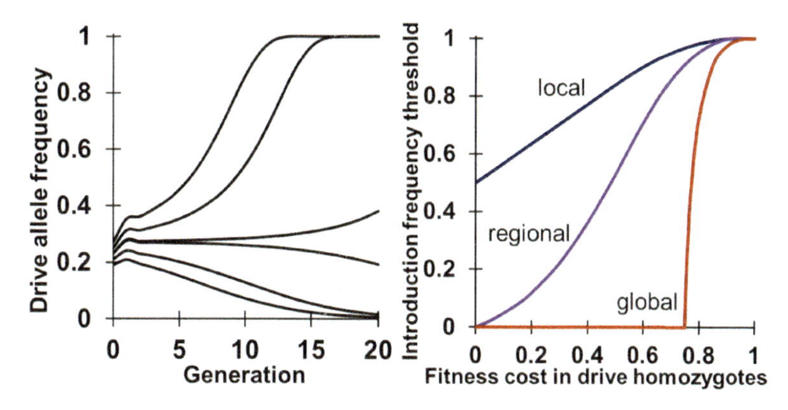

Figure 3.2 Introduction thresholds of gene drives. Confined gene drives have an introduction frequency threshold, above which the drive increases in frequency (either to fixation or a higher equilibrium frequency) and below which the drive is eliminated from the population (an ideal CRISPR toxin-antidote TADE suppression drive [54, 55] with 50% embryo cut rate is shown in the figure). The graph shows drive frequency trajectories after releases of a variable number of drive-carrying individuals, demonstrating its drive allele (in heterozygotes) introduction threshold of 23.2%. The introduction frequency threshold is dependent on several factors, but the most important are usually the drive variant and the drive fitness. Example thresholds are shown for local (single-locus underdominance), regional (CISPR toxin-antidote drive targeting an essential but haplosufficient gene), and global (homing) type drives.

above which the drive will increase in frequency. If the drive is not able to surpass the introduction threshold in a population, it will not be able to spread, thus confining the drive. Drives generally fall into two broad categories for confinement, "regional" drives that have a zero-introduction threshold frequency but gain a nonzero threshold if they carry any fitness cost (likely a realistic assumption), and "local" underdominance drives that have an introduction threshold frequency even without fitness costs. Among local drives, those with frequency thresholds over 50% have particularly stringent confinement, requiring high release numbers over broad areas to be able to establish in populations [47].

Most modeling of drive confinement has involved panmictic demes, sometimes connected by fixed migration to another deme [45, 48–59]. Such modeling provides a good starting point, particularly for comparing the qualitative levels of confinement between drive systems or between populations where panmixia and long-distance migration may be a good representation, such as birds on island chains. However, real landscapes can have complex spatial structure, potentially limiting the utility of such basic models. Some groups have thus used networks of connected demes to analyze the confinement of various types of drives [60–63]. At the cost of greater computing power, models can be brought even closer to many real-world situations by incorporating continuous space where individuals can move over a landscape, thus potentially providing the best estimates of drive confinement within connected areas [47, 64–69].

One way to limit the ultimate spread of a gene drive without the need to use an intrinsically confined drive system is to use "locally fixed" or "private" alleles [70, 71]. If the gene drive can only spread among individuals with particular alleles, and these target alleles are fixed in only particular populations, then the gene drive will only be able to spread effectively in these populations. If the gene drive can only spread among individuals with particular alleles, and these target alleles are fixed in only target populations, then the gene drive will only be able to spread effectively in these populations. Even if some target alleles exist outside the target population, gene drive spread in these areas may be short-term due to fitness effects of the drive compared to non-target alleles, particularly in the case of a suppression drive, where the drive would have substantially reduced fitness after replacing target alleles.

3.2 Types of Gene Drive

3.2.1 Chromosomal Rearrangements

Some of the first engineered gene drives involved exchanging fragments of two chromosomes. Among these large fragments, each contains many essential genes, so individuals must have two copies of each gene for viability. Thus, insects must possess all wild-type chromosomes, all rearranged chromosomes, or one of each type to be viable. This results in a high introduction threshold of 50%, which would be somewhat higher with additional fitness costs. In early experiments, substantial fitness costs were indeed present because the reciprocal translocations were radiation-induced [11–13], damaging the genome. This drive system also can only be used for modification, and when originally constructed it could not be combined with a suitable cargo gene. Modern genome engineering techniques could address this issue, but the high threshold would potentially be difficult to achieve in malarial regions.

3.2.2 Transposons

Common in nature, transposons are selfish genetic elements that can copy themselves into random locations of a genome, thus increasing in frequency in the population as a whole. This is often a slow process, in part due to effective natural piRNA-based defenses against such systems. It was initially thought that transposons could be engineered or ported between species to copy themselves at higher rates, allowing them to serve as effective modification gene drives with a zero-introduction threshold. However, while a transposon-based gene drive was successfully created in *Anopheles stephensi*, these systems have not yet demonstrated sufficiently high effectiveness to act as a useful gene drive [14, 15], since at low mobilization efficiency, even a small fitness cost from a linked cargo gene could prevent effective spread through a population.

3.2.3 Homing Drives

The first homing drive in mosquitoes was based on the I-SceI nuclease, targeting a synthetic sequence inserted into *Anopheles gambiae* [72]. It was able to successfully bias its inheritance using its unique "copy-and-paste" mechanism. This occurs when the homing drive cuts an allele (usually a wild-type target sequence) on the homologous chromosome and is then inserted into the sequence during the process of homology-directed repair, which uses the drive allele as the template to repair the DNA break (Fig. 3.3). This ability to be inherited potentially by all progeny of a heterozygote makes homing drive one of the more powerful forms of drive, able to rapidly modify or suppress populations. However, as initial

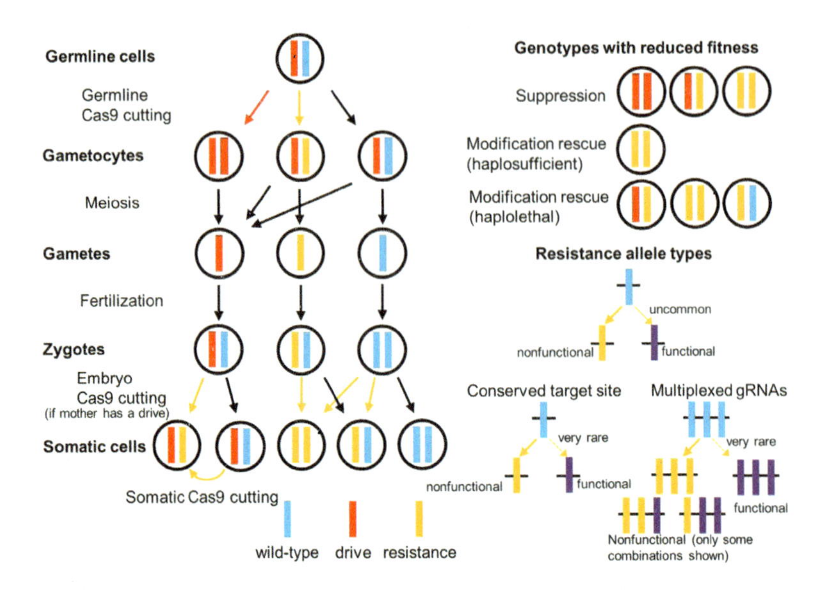

Figure 3.3 Homing drive designs for modification and suppression. Homing drives work by cutting wild-type alleles and then being copied to the cut site by homology-directed repair. If end-joining repair occurs instead, a resistance allele forms, which can be functional or nonfunctional. For the most advanced modification drives and suppression drives, the rarer functional resistance alleles are the most problematic, but they can potentially be avoided by using multiplexed gRNAs and/or highly conserved target sites. Note that CRISPR toxin-antidote drives have similar mechanisms, except that germline drive conversion does not take place.

experiments indicated, DNA repair could also take place by end-joining, mutating the target sequence, and creating a "resistance allele" that can no longer be cut by the drive and converted to a drive allele (Fig. 3.3) [73–76]. These resistance alleles represent the main initial challenges that researchers faced when developing homing drives.

With the advent of flexible CRISPR/Cas9 nucleases that can be directed by a guide RNA (gRNA) to target wild-type sequences, homing drives became far more viable, quickly resulting in demonstrations of modification drive in *A. stephensi* [74] and suppression drive in *A. gambiae* [73]. The modification drive carried single-chain antibodies that could inhibit malaria [74], while the suppression drive targeted an essential but haplosufficient female fertility gene [73]. This theoretically would allow the suppression drive to spread rapidly in heterozygotes, eventually forming sterile females when it reaches high frequency, resulting in population suppression. Both drives had high drive conversion efficiencies of over 90% (the rate at which wild-type alleles are converted to drive alleles in the germline prior to inheritance by offspring), but both also suffered from resistance allele formation [73, 74]. Efforts to study these resistance alleles in both *Anopheles* and *Drosophila melanogaster* indicated that they could be formed in the early germline and inherited [73, 77, 78], but a larger problem was high persistence of maternally deposited Cas9 into the embryo, resulting in cleavage and resistance allele formation in offspring with drive mothers, even if a drive allele was available as a template for homology-directed repair [73, 74, 76, 77]. Somatic expression of Cas9 was also seen with the *vasa* promoter [73, 76, 77], which rendered drive carrying female mosquitoes with the suppression drive largely infertile [73].

When the *A. stephensi* modification drive mosquitoes were released into a wild-type cage, they initially spread rapidly and sometimes succeeded in modifying the entire population if released at high frequency [79]. However, most cages saw the formation of resistance alleles that preserved the function of the *kh* target gene. These rarer functional resistance alleles had a substantial fitness advantage, since *kh* is important for *A. stephensi* female fecundity, allowing them to outcompete drive alleles that disrupted *kh*. To overcome this issue in *A. gambiae*, a gene was targeted where disruption conferred lower fitness costs, and the *nanos* promoter was used in place of *vasa*, forming far fewer resistance alleles in the early embryo while preserving high drive conversion efficiency [17]. This drive had far higher success, reaching and remaining at high frequency despite some forms of resistance alleles, though such alleles may be an issue in natural populations with larger sizes. Another study took the original modification drive and added a recoded rescue version of *kh* (in place of the antimalarial cargo genes) to the drive that could not be cut by the drive's gRNA [18]. This drive also reached and remained at high frequency, despite the formation of functional resistance alleles, because the drive did not carry a substantial fitness cost. Though nonfunctional resistance alleles were formed at high rates, these disrupted *kh* and thus were largely eliminated when two copies were present in females. However, more efficient systems could be needed, since antimalarial effectors will likely carry at least a small fitness cost. An earlier study in *Drosophila melanogaster* suggests one possible method for achieving this. Therein, an essential gene was targeted with two gRNAs, with the drive carrying a recoded version of the target gene [41]. By using multiple gRNAs, functional resistance alleles were avoided because each cut site had the potential to independently disrupt the gene (with deletions after cutting at both sites assured to disrupt the target gene). Because the target gene was haplolethal, nonfunctional resistance alleles were immediately eliminated. One issue in general with the recoded essential gene strategy is that recoded regions could serve as a template for homology-directed repair (potentially partial homology-directed repair followed by end-joining [76, 77]), thus forming functional resistance alleles [80]. The rate at which this occurs could be increased by multiplexed gRNAs, though it remains unclear how large these rates would be on an absolute scale. Nevertheless, this issue could be circumvented by targeting native genes that are nonessential for the mosquito, but essential for malaria transmission, thus potentially avoiding issues with functional resistance allele formation with multiplexed gRNAs [80].

When the *A. gambiae* suppression drive mosquitoes were released into a cage, they also increased in frequency at first [81]. However, the drive frequency quickly declined due to the rapid formation of functional resistance alleles. Such alleles have a large advantage over a suppression drive since their carriers will always be fertile. Like the modification drive, increased success was seen when improved promoters (*zpg* in particular) were used that minimized maternal deposition of Cas9, while also reducing (though not eliminating) female fecundity reduction from somatic expression [82]. In cage populations, such a drive reached much higher frequency, but it still ended up rapidly declining due to the eventual formation of functional resistance alleles with a large fitness advantage. To overcome this issue, another essential but haplosufficient female development/fertility exon (of the *doublesex* gene) was targeted at a site that was highly conserved [16]. This prevented the formation of functional resistance alleles, resulting in successful population suppression in multigeneration cage populations. Though small numbers of nonfunctional resistance alleles were generated, these did not impede suppression. Indeed, such resistance alleles would only slightly reduce the equilibrium genetic load of a suppression drive [83] (and a peak genetic load could be even higher if such alleles are slow to form). Genetic load is mainly determined by the drive conversion efficiency and fitness costs. This could potentially present an issue if additional measures are needed to eliminate the formation of functional resistance alleles in much larger natural populations. While two gRNAs are predicted to increase the drive conversion efficiency of *Anopheles* homing suppression drives, a greater number of gRNAs would likely reduce drive conversion efficiency [80]. This results in an intermediate optimal number of gRNAs to balance the need for high drive conversion efficiency/genetic load and low functional resistance allele formation [80].

It should be noted that while homing drives represent rapid and powerful tools for population modification or suppression, methods for mitigating their effect on natural populations are potentially available in the event of an unanticipated need to revert the gene drive. A second-generation homing drive could, for example, disrupt and replace an existing drive, which has been demonstrated in fruit flies [84].

3.2.4 X-Shredders

An X-shredder is composed of a nuclease that is designed to shred the X-chromosome in male gametes, thus allowing only sperm with Y-chromosomes to fertilize eggs and thereby, increasing the male to female sex ratio in the next generation (Fig. 3.4) [85]. If this X-shredder is located on the Y chromosome, it will become a gene drive that can increase its frequency in a population. This potentially allows a small release of driving Y alleles to induce population suppression by eliminating females from the population, or at least reducing their number in an imperfect drive. However, driving Y chromosomes have not yet been experimentally demonstrated. This is due to difficulties in engineering transgenics onto the Y

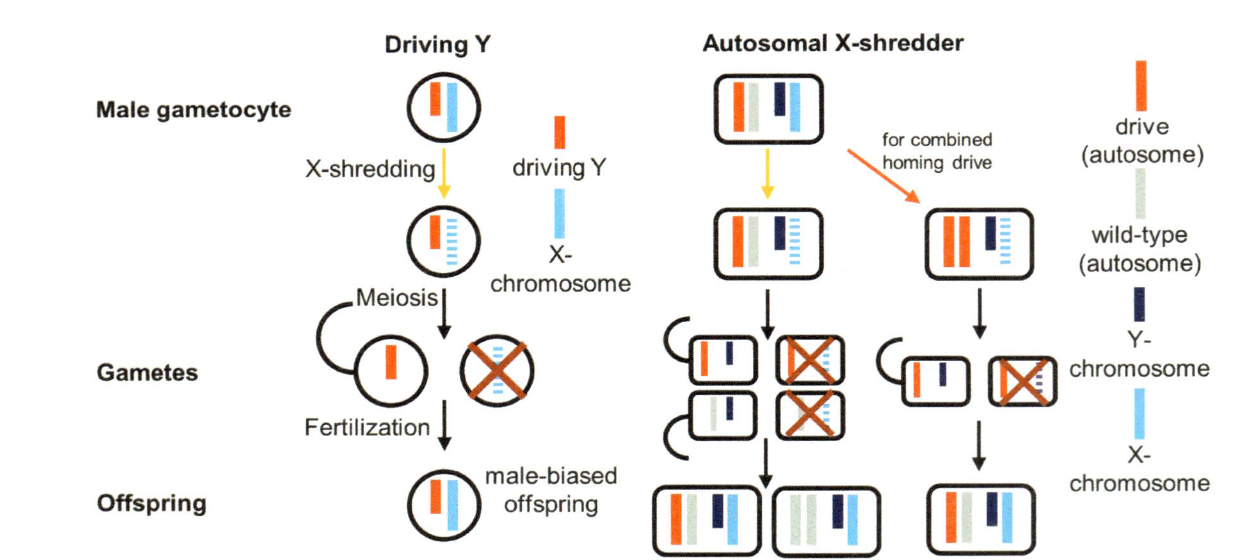

Figure 3.4 X-shredder drives. In the male germline, the X-shredder will eliminate the X-chromosome. This results in only sperm with the Y chromosome remaining viable, biasing the sex ratio of offspring towards males.

chromosome, as well as expressing genes on the Y chromosome at high levels, which is needed to obtain sufficient nuclease activity for X-chromosome shredding. Nevertheless, research on autosomal X-shredders has progressed rapidly over the past decade.

The earliest X-shredder was constructed in *A. gambiae* using the I-PpoI nuclease [86–88], which fortuitously has a target sequence that is highly repeated in the *A. gambiae* X-chromosome. Expressed under control of a male germline-specific promoter, X-shredding activity was high. However, paternally deposited nuclease also induced X-chromosome shredding in all embryos, resulting in no viable progeny from fathers with the X-shredder [86, 87]. This was corrected by using a destabilized form of I-PpoI, which successfully resulted in viable, male-biased offspring [19]. A follow-up study successfully transferred this system to *Anopheles arabiensis* [89], and another study similarly utilized CRISPR/Cas9 for effective X-shredding in *A. gambiae* [20].

Though not able to function as a gene drive on its own, the I-PpoI was combined with the *A. gambiae* homing suppression drive targeting *doublesex* to bias the drive population toward males and thus avoid the loss in fertility seen in drive carrying-females [21]. This also essentially changed the drive's suppression mechanism from one dominated by reduction of female fertility to one dominated by male-bias. This drive was successful in rapidly suppressing cage populations, but since the X-shredder mechanism may have lower performance than the female fertility suppression mechanism in models involving continuous space [33], it is unclear if this variant would perform better in natural populations.

3.2.5 RNAi-Based Toxin-Antidote Drives

A wide variety of additional gene drives have been developed in flies but not yet in mosquitoes. These are noteworthy because they may be more controllable in terms of confinement than the homing drives that have thus far been the most successful in *Anopheles*.

Medea drive was engineered in *D. melanogaster* by using toxin-antidote principles. It contains an RNAi toxin targeting *myd88*, a gene that is expressed in mothers and provided maternally to offspring, where it plays an essential role in development.

Medea alleles also contain a recoded version of *myd88* that is expressed zygotically, which provides the antidote. Offspring with a *Medea*-carrying mother will thus perish unless they are rescued by inheriting a *Medea* allele from either parent. This drive has no introduction threshold frequency in the absence of fitness costs, but any fitness cost will result in a nonzero threshold. Though successful in spreading through multigeneration caged *D. melanogaster* populations, efforts to bring *Medea* to mosquitoes have stalled due to the specific timing requirements for successful expression of the RNAi toxin and the antidote elements.

A newer form of RNAi-based underdominance drive was also developed in fruit flies, targeting a haploinsufficient gene with RNAi and providing a recoded rescue copy of the target gene [22]. With both wild-type copies knocked down in heterozygotes and only one rescue copy present, these individuals suffered from very low fitness due to the haploinsufficiency of the target gene. Such a drive would have an introduction threshold of 50% in the absence of fitness costs to drive homozygotes. This drive was also successful in spreading through the caged fly population when released at sufficiently high frequency.

Other forms of RNAi-based drive designs have been proposed that also work via underdominance principles [53]. These involve two separate drive alleles that each carry a toxin (generally an RNAi targeting an essential gene), as well as an antidote for the other type of allele (generally a recoded version of the target gene). If both alleles must occupy the same genomic site, only individuals with each type of drive allele are viable, resulting in an introduction threshold frequency of 67%. If each type of drive allele is placed at a separate locus, then more drive-carrier genotypes are viable, resulting in a lower threshold of 27%.

3.2.6 CRISPR-Based Toxin-Antidote Drives

Because RNAi systems can be difficult to successfully engineer, there has been increasing interest in using CRISPR/Cas9 and gRNAs as the "toxin", while still retaining recoded forms of the target gene as the antidote. This has the major advantage over RNAi systems of being easier to construct, potentially easing their deployment into mosquitoes.

The only experimental demonstrations of such systems thus far have been in flies with essential but haplosufficient target genes [24, 26]. Any individuals with a drive would always be viable, but individuals receiving two disrupted target genes without a drive would be eliminated. This allowed even embryo cutting due to maternally deposited Cas9 and gRNAs to contribute to the overall effectiveness of the gene drive. These experimental demonstrations were successful in cage populations, avoiding functional resistance alleles easily by use of multiplexed gRNAs [24, 26], without the downsides of this method as found in homing drives [80]. Such drives are similar to *Medea* in that they would have a zero-introduction threshold in the absence of fitness costs, but any fitness cost would result in at least a small threshold, potentially allowing them to be confined to a target population.

While the demonstrated forms of CRISPR toxin-antidote drives are only suitable for modification, other feasible designs targeting haplolethal genes could be used for suppression [55], and still more designs exist for underdominance forms with greater introduction thresholds (for both modification and suppression) [54]. Confined CRISPR toxin-antidote drives could also work well as a "tether" for homing drives, in which they provide the Cas9 needed for the homing drive to function. This allows the power of a homing drive (useful for suppression and modification with costly cargo genes) to be used with the confinement of a toxin-antidote system [90].

3.2.7 Wolbachia

Wolbachia is a common type of intracellular bacteria that infect a wide variety of insects. They achieve this success due to a gene drive-like mechanism involving cytoplasmic incompatibility. This refers to a bacteriophage-mediated phenomenon where wild-type females cannot produce viable eggs when mated to *Wolbachia*-carrying males. Coupled with the maternal inheritance of *Wolbachia*, this allows *Wolbachia* to increase in frequency in a population with no introduction threshold frequency in the absence of fitness costs. However, in practice, fitness costs from the *Wolbachia* infection result in a higher threshold, which is highly variable at around 15–40% [91]. In addition to acting as a modification drive, *Wolbachia*-infected males can be released for short-term population suppression.

Though not an engineered gene drive with a cargo, *Wolbachia* can still act as an effective modification type gene drive because the presence of *Wolbachia* itself often effectively prevents transmission of disease, at least in *Aedes aegypti* [92]. This has been successfully used for combatting dengue in several regions [93]. However, though *Wolbachia* was recently discovered in *Anopheles*, it remains unclear what effect it might have on malaria transmission [94, 95].

Another way to harness *Wolbachia* would be to use its genetic elements for cytoplasmic incompatibility as a conventional gene drive. Such a system has been engineered in *Drosophila* [96] and would have a moderately high introduction frequency threshold of 37% if it did not induce additional fitness costs. It is possible that its actual fitness costs would be far smaller than the whole *Wolbachia* parasite, though expression timing would need to be well-controlled to achieve sufficiently high drive efficiency.

References

1. Burt A, Crisanti A. Gene drive: Evolved and synthetic. *ACS Chem Biol*, 13, 343–346, 2018.
2. Macias VM, Ohm JR, Rasgon JL. Gene drive for mosquito control: Where did it come from and where are we headed? *Int J Environ Res Public Health*, 14, 1006, 2017.
3. Gantz VM, Bier E. The dawn of active genetics. *Bioessays*, 38, 50–63, 2016.
4. Champer J, Buchman A, Akbari OS. Cheating evolution: Engineering gene drives to manipulate the fate of wild populations. *Nat Rev Genet*, 17, 146–159, 2016.
5. Hay BA, Oberhofer G, Guo M. Engineering the composition and fate of wild populations with gene drive. *Annu Rev Entomol*, 66, annurev-ento-020117-043154, 2021.
6. Bull JJ. Evolutionary decay and the prospects for long-term disease intervention using engineered insect vectors. *Evol Med Public Heal*, 2015, 152–166, 2015.

7. Esvelt KM, Smidler AL, Catteruccia F, Church GM. Concerning RNA-guided gene drives for the alteration of wild populations. *Elife*, e03401, 2014.
8. Carballar-Lejarazú R, James AA. Population modification of Anopheline species to control malaria transmission. *Pathog Glob Health*, 111, 424–435, 2017.
9. Quinn CM, Nolan T. Nuclease-based gene drives, an innovative tool for insect vector control: Advantages and challenges of the technology. *Curr. Opin. Insect Sci.* 39, 77–83, 2020.
10. Leftwich PT, Edgington MP, Harvey-Samuel T, Carabajal Paladino LZ, Norman VC, Alphey L. Recent advances in threshold-dependent gene drives for mosquitoes. *Biochem Soc Trans*, 46, 1203–1212, 2018.
11. Baker RH, Sakai RK, Perveen A, Raana K. Isolation of a homozygous translocation in Anopheles culicifacies: Selection utilizing adult polytene chromosomes and insecticide resistance. *J Hered*, 71, 25–28, 1980.
12. Kaiser PE, Seawright JA, Benedict MQ, Narang S. Homozygous translocations in Anopheles albimanus. *Theor Appl Genet*, 65, 207–211, 1983.
13. Kaiser PE, Seawright JA, Benedict MQ, Narang S, Suguna SG. Radiation induced reciprocal translocations and inversions in Anopheles albimanus. *Can J Genet Cytol*, 24, 177–188, 1982.
14. O'Brochta DA, Alford RT, Pilitt KL, Aluvihare CU, Harrell RA. PiggyBac transposon remobilization and enhancer detection in Anopheles mosquitoes. *Proc Natl Acad Sci USA*, 108, 16339–16344, 2011.
15. Macias VM, Jimenez AJ, Burini-Kojin B, Pledger D, Jasinskiene N, Phong CH, Chu K, Fazekas A, Martin K, Marinotti O, James AA. Nanos-driven expression of piggyBac transposase induces mobilization of a synthetic autonomous transposon in the malaria vector mosquito, Anopheles stephensi. *Insect Biochem Mol Biol*, 87, 81–89, 2017.
16. Kyrou K, Hammond AM, Galizi R, Kranjc N, Burt A, Beaghton AK, Nolan T, Crisanti A. A CRISPR-Cas9 gene drive targeting doublesex causes complete population suppression in caged Anopheles gambiae mosquitoes. *Nat Biotechnol*, 2018.
17. Carballar-Lejarazú R, Ogaugwu C, Tushar T, Kelsey A, Pham TB, Murphy J, Schmidt H, Lee Y, Lanzaro GC, James AA. Next-generation gene drive for population modification of the malaria vector mosquito, Anopheles gambiae. *Proc Natl Acad Sci USA*, 117, 22805–22814, 2020.
18. Adolfi A, Gantz VM, Jasinskiene N, Lee HF, Hwang K, Terradas G, Bulger EA, Ramaiah A, Bennett JB, Emerson JJ, Marshall JM, Bier E, James AA. Efficient population modification gene-drive rescue system in the malaria mosquito Anopheles stephensi. *Nat Commun*, 11, 1–13, 2020.

19. Galizi R, Doyle LA, Menichelli M, Bernardini F, Deredec A, Burt A, Stoddard BL, Windbichler N, Crisanti A. A synthetic sex ratio distortion system for the control of the human malaria mosquito. *Nat Commun*, 5, 3977, 2014.
20. Galizi R, Hammond A, Kyrou K, Taxiarchi C, Bernardini F, O'Loughlin SM, Papathanos PA, Nolan T, Windbichler N, Crisanti A. A CRISPR-Cas9 sex-ratio distortion system for genetic control. *Sci Rep*, 6, 31139, 2016.
21. Simoni A, Hammond AM, Beaghton AK, Galizi R, Taxiarchi C, Kyrou K, Meacci D, Gribble M, Morselli G, Burt A, Nolan T, Crisanti A. A male-biased sex-distorter gene drive for the human malaria vector Anopheles gambiae. *Nat Biotechnol*, 1–7, 2020.
22. Reeves RG, Bryk J, Altrock PM, Denton JA, Reed FA. First steps towards underdominant genetic transformation of insect populations. *PLoS One*, 9, e97557, 2014.
23. Akbari OS, Matzen KD, Marshall JM, Huang H, Ward CM, Hay BA. A synthetic gene drive system for local, reversible modification and suppression of insect populations. *Curr Biol*, 23, 671–677, 2013.
24. Champer J, Lee E, Yang E, Liu C, Clark AG, Messer PW. A toxin-antidote CRISPR gene drive system for regional population modification. *Nat Commun*, 11, 1082, 2020.
25. Oberhofer G, Ivy T, Hay BA. Gene drive and resilience through renewal with next generation Cleave and Rescue selfish genetic elements. *Proc Natl Acad Sci USA*, 117, 9013–9021, 2020.
26. Oberhofer G, Ivy T, Hay BA. Cleave and Rescue, a novel selfish genetic element and general strategy for gene drive. *Proc Natl Acad Sci*, 201816928, 2019.
27. Fuchs S, Nolan T, Crisanti A. Mosquito transgenic technologies to reduce plasmodium transmission. *Methods Mol Biol*, 923, 601–622, 2013.
28. Wang S, Jacobs-Lorena M. Genetic approaches to interfere with malaria transmission by vector mosquitoes. *Trends Biotechnol*, 31, 185–193, 2013.
29. Adelman ZN, Kojin BB. Malaria-resistant mosquitoes (Diptera: Culicidae): The principle is proven, but will the effectors be effective? *J Med Entomol*, 2021.
30. Dong Y, Simões ML, Marois E, Dimopoulos G. CRISPR/Cas9 -mediated gene knockout of Anopheles gambiae FREP1 suppresses malaria parasite infection. *PLoS Pathog*, 14, e1006898, 2018.
31. Garver LS, Dong Y, Dimopoulos G. Caspar controls resistance to Plasmodium falciparum in diverse Anopheline species. *PLoS Pathog*, 5, e1000335, 2009.

32. Dhole S, Lloyd AL, Gould F. Gene drive dynamics in natural populations: The importance of density dependence, space, and sex. *Annu Rev Ecol Evol Syst*, 51, 505–531, 2020.
33. Champer J, Kim IK, Champer SE, Clark AG, Messer PW. Suppression gene drive in continuous space can result in unstable persistence of both drive and wild-type alleles. *Mol Ecol*, 30, 1086–1101, 2021.
34. North AR, Burt A, Godfray HCJ. Modeling the suppression of a malaria vector using a CRISPR-Cas9 gene drive to reduce female fertility. *BMC Biol*, 18, 98, 2020.
35. Bull JJ, Remien CH, Krone SM. Gene-drive-mediated extinction is thwarted by population structure and evolution of sib mating. *Evol Med public Heal*, 2019, 66–81, 2019.
36. North AR, Burt A, Godfray HCJ. Modeling the potential of genetic control of malaria mosquitoes at national scale. *BMC Biol*, 17, 26, 2019.
37. Gould F, Huang Y, Legros M, Lloyd AL. A killer-rescue system for self-limiting gene drive of anti-pathogen constructs. *Proc Biol Sci*, 275, 2823–2829, 2008.
38. Webster SH, Vella MR, Scott MJ. Development and testing of a novel killer–rescue self-limiting gene drive system in Drosophila melanogaster. *Proc R Soc B Biol Sci*, 287, 20192994, 2020.
39. Champer J, Chung J, Lee YL, Liu C, Yang E, Wen Z, Clark AG, Messer PW. Molecular safeguarding of CRISPR gene drive experiments. *Elife*, 8, 2019.
40. Guichard A, Haque T, Bobik M, Xu X-RS, Klanseck C, Kushwah RBS, Berni M, Kaduskar B, Gantz VM, Bier E. Efficient allelic-drive in Drosophila. *Nat Commun*, 10, 1640, 2019.
41. Champer J, Yang E, Lee E, Liu J, Clark AG, Messer PW. A CRISPR homing gene drive targeting a haplolethal gene removes resistance alleles and successfully spreads through a cage population. *Proc Natl Acad Sci*, 2020.
42. Oberhofer G, Ivy T, Hay BA. Split versions of Cleave and Rescue selfish genetic elements for measured self limiting gene drive. *PLOS Genet*, 17, e1009385, 2021.
43. DiCarlo JE, Chavez A, Dietz SL, Esvelt KM, Church GM. Safeguarding CRISPR-Cas9 gene drives in yeast. *Nat Biotechnol*, 33, 1250–1255, 2015.
44. Noble C, Min J, Olejarz J, Buchthal J, Chavez A, Smidler AL, DeBenedictis EA, Church GM, Nowak MA, Esvelt KM. Daisy-chain gene drives for the alteration of local populations. *Proc Natl Acad Sci*, 201716358, 2019.

45. Dhole S, Vella MR, Lloyd AL, Gould F. Invasion and migration of spatially self-limiting gene drives: A comparative analysis. *Evol Appl*, 11, 794–808, 2018.
46. Maselko M, Feltman N, Upadhyay A, Hayward A, Das S, Myslicki N, Peterson AJ, O'Connor MB, Smanski MJ. Engineering multiple species-like genetic incompatibilities in insects. *Nat Commun*, 11, 1–7, 2020.
47. Champer J, Zhao J, Champer SE, Liu J, Messer PW. Population dynamics of underdominance gene drive systems in continuous space. *ACS Synth Biol*, acssynbio.9b00452, 2020.
48. Edgington MP, Alphey LS. Conditions for success of engineered underdominance gene drive systems. *J Theor Biol*, 430, 128–140, 2017.
49. Edgington MP, Alphey LS. Population dynamics of engineered underdominance and killer-rescue gene drives in the control of disease vectors. *PLoS Comput Biol*, 14, e1006059, 2018.
50. Altrock PM, Traulsen A, Reeves RG, Reed FA. Using underdominance to bi-stably transform local populations. *J Theor Biol*, 267, 62–75, 2010.
51. Huang Y, Lloyd AL, Legros M, Gould F. Gene-drive in age-structured insect populations. *Evol Appl*, 2, 143–159, 2009.
52. Khamis D, El Mouden C, Kura K, Bonsall MB. Ecological effects on underdominance threshold drives for vector control. *J Theor Biol*, 456, 1–15, 2018.
53. Davis S, Bax N, Grewe P. Engineered underdominance allows efficient and economical introgression of traits into pest populations. *J Theor Biol*, 212, 83–98, 2001.
54. Champer J, Champer SE, Kim IK, Clark AG, Messer PW. Design and analysis of CRISPR-based underdominance toxin-antidote gene drives. *Evol Appl*, eva.13180, 2020.
55. Champer J, Kim IK, Champer SE, Clark AG, Messer PW. Performance analysis of novel toxin-antidote CRISPR gene drive systems. *BMC Biol*, 18, 27, 2020.
56. Marshall JM, Hay BA. General principles of single-construct chromosomal gene drive. *Evolution (N Y)*, 66, 2150–2166, 2012.
57. Marshall JM, Hay BA. Confinement of gene drive systems to local populations: A comparative analysis. *J Theor Biol*, 294, 153–171, 2012.
58. Marshall JM. The toxin and antidote puzzle: New ways to control insect pest populations through manipulating inheritance. *Bioeng Bugs*, 2, 235–240, 2011.

59. Huang Y, Magori K, Lloyd AL, Gould F. Introducing transgenes into insect populations using combined gene-drive strategies: Modeling and analysis. *Insect Biochem Mol Biol*, 37, 1054–1063, 2007.
60. Láruson ÁJ, Reed FA. Stability of underdominant genetic polymorphisms in population networks. *J Theor Biol*, 390, 156–63, 2016.
61. Altrock PM, Traulsen A, Reed FA. Stability properties of underdominance in finite subdivided populations. *PLoS Comput Biol*, 7, e1002260, 2011.
62. Sánchez C. HM, Wu SL, Bennett JB, Marshall JM. MGD <scp>riv</scp> E: A modular simulation framework for the spread of gene drives through spatially explicit mosquito populations. *Methods Ecol Evol*, 2041–210X.13318, 2019.
63. Wu SL, Bennett JB, Sánchez C. HM, Dolgert AJ, León TM, Marshall JM. MGDrivE 2: A simulation framework for gene drive systems incorporating seasonality and epidemiological dynamics. *PLOS Comput Biol*, 17, e1009030, 2021.
64. Piálek J, Barton NH. The spread of an advantageous allele across a barrier: The effects of random drift and selection against heterozygotes. *Genetics*, 145, 1997.
65. Faber NR, McFarlane GR, Gaynor RC, Pocrnic I, Whitelaw CBA, Gorjanc G. Novel combination of CRISPR-based gene drives eliminates resistance and localizes spread. *Sci Rep*, 11, 3719, 2021.
66. Huang Y, Lloyd AL, Legros M, Gould F. Gene-drive into insect populations with age and spatial structure: A theoretical assessment. *Evol Appl*, 4, 415–428, 2011.
67. Barton NH, Turelli M. Spatial waves of advance with bistable dynamics: Cytoplasmic and genetic analogs of allee effects. *Am Nat*, 178, E48–E75, 2011.
68. Barton NH, Hewitt GM. Adaptation, speciation and hybrid zones. *Nature*, 341, 497–503, 1989.
69. Tanaka H, Stone HA, Nelson DR. Spatial gene drives and pushed genetic waves. *Proc Natl Acad Sci*, 201705868, 2017.
70. Willis K, Burt A. Double drives and private alleles for localized population genetic control. *PLOS Genet*, 17, e1009333, 2021.
71. Sudweeks J, Hollingsworth B, Blondel D V, Campbell KJ, Dhole S, Eisemann JD, Edwards O, Godwin J, Howald GR, Oh KP, Piaggio AJ, Prowse TAA, Ross J V, Saah JR, Shiels AB, Thomas PQ, Threadgill DW, Vella MR, Gould F, Lloyd AL. Locally fixed alleles: A method to localize gene drive to island populations. *Sci Rep*, 9, 15821, 2019.

72. Windbichler N, Menichelli M, Papathanos PA, Thyme SB, Li H, Ulge UY, Hovde BT, Baker D, Monnat Jr. RJ, Burt A, Crisanti A. A synthetic homing endonuclease-based gene drive system in the human malaria mosquito. *Nature*, 473, 212–215, 2011.

73. Hammond A, Galizi R, Kyrou K, Simoni A, Siniscalchi C, Katsanos D, Gribble M, Baker D, Marois E, Russell S, Burt A, Windbichler N, Crisanti A, Nolan T. A CRISPR-Cas9 gene drive system targeting female reproduction in the malaria mosquito vector Anopheles gambiae. *Nat Biotechnol*, 34, 78–83, 2015.

74. Gantz VM, Jasinskiene N, Tatarenkova O, Fazekas A, Macias VM, Bier E, James AA. Highly efficient Cas9-mediated gene drive for population modification of the malaria vector mosquito Anopheles stephensi. *Proc Natl Acad Sci USA*, 112, E6736–E6743, 2015.

75. Gantz VM, Bier E. Genome editing. The mutagenic chain reaction: A method for converting heterozygous to homozygous mutations. *Science (80)*, 348, 442–444, 2015.

76. Champer J, Reeves R, Oh SY, Liu C, Liu J, Clark AG, Messer PW. Novel CRISPR/Cas9 gene drive constructs reveal insights into mechanisms of resistance allele formation and drive efficiency in genetically diverse populations. *PLoS Genet*, 13, e1006796, 2017.

77. Champer J, Liu J, Oh SY, Reeves R, Luthra A, Oakes N, Clark AG, Messer PW. Reducing resistance allele formation in CRISPR gene drive. *Proc Natl Acad Sci*, 115, 5522–5527, 2018.

78. KaramiNejadRanjbar M, Eckermann KN, Ahmed HMM, Sánchez C. HM, Dippel S, Marshall JM, Wimmer EA. Consequences of resistance evolution in a Cas9-based sex-conversion suppression gene drive for insect pest management. *Proc Natl Acad Sci*, 201713825, 2018.

79. Pham TB, Phong CH, Bennett JB, Hwang K, Jasinskiene N, Parker K, Stillinger D, Marshall JM, Carballar-Lejarazú R, James AA. Experimental population modification of the malaria vector mosquito, Anopheles stephensi. *PLOS Genet*, 15, e1008440, 2019.

80. Champer SE, Oh SY, Liu C, Wen Z, Clark AG, Messer PW, Champer J. Computational and experimental performance of CRISPR homing gene drive strategies with multiplexed gRNAs. *Sci Adv*, 6, eaaz0525, 2020.

81. Hammond AM, Kyrou K, Bruttini M, North A, Galizi R, Karlsson X, Kranjc N, Carpi FM, D'Aurizio R, Crisanti A, Nolan T. The creation and selection of mutations resistant to a gene drive over multiple generations in the malaria mosquito. *PLOS Genet*, 13, e1007039, 2017.

82. Hammond A, Karlsson X, Morianou I, Kyrou K, Beaghton A, Gribble M, Kranjc N, Galizi R, Burt A, Crisanti A, Nolan T. Regulating the expression of gene drives is key to increasing their invasive potential and the mitigation of resistance. *PLOS Genet*, 17, e1009321, 2021.
83. Beaghton AK, Hammond A, Nolan T, Crisanti A, Burt A. Gene drive for population genetic control: Non-functional resistance and parental effects. *Proceedings Biol Sci*, 286, 20191586, 2019.
84. Xu X-RS, Bulger EA, Gantz VM, Klanseck C, Heimler SR, Auradkar A, Bennett JB, Miller LA, Leahy S, Juste SS, Buchman A, Akbari OS, Marshall JM, Bier E. Active genetic neutralizing elements for halting or deleting gene drives. *Mol Cell*, 2020.
85. Haghighat-Khah RE, Sharma A, Wunderlich MR, Morselli G, Marston LA, Bamikole C, Hall A, Kranjc N, Taxiarchi C, Sharakhov I, Galizi R. Cellular mechanisms regulating synthetic sex ratio distortion in the Anopheles gambiae germline. *Pathog Glob Health*, 114, 370–378, 2020.
86. Windbichler N, Papathanos PA, Crisanti A. Targeting the X chromosome during spermatogenesis induces Y chromosome transmission ratio distortion and early dominant embryo lethality in Anopheles gambiae. *PLoS Genet*, 4, e1000291, 2008.
87. Klein TA, Windbichler N, Deredec A, Burt A, Benedict MQ. Infertility resulting from transgenic I-PpoI male Anopheles gambiae in large cage trials. *Pathog Glob Health*, 106, 20–31, 2012.
88. Windbichler N, Papathanos PA, Catteruccia F, Ranson H, Burt A, Crisanti A. Homing endonuclease mediated gene targeting in Anopheles gambiae cells and embryos. *Nucleic Acids Res*, 35, 5922–5933, 2007.
89. Bernardini F, Kriezis A, Galizi R, Nolan T, Crisanti A. Introgression of a synthetic sex ratio distortion system from Anopheles gambiae into Anopheles arabiensis. *Sci Rep*, 9, 5158, 2019.
90. Dhole S, Lloyd AL, Gould F. Tethered homing gene drives: A new design for spatially restricted population replacement and suppression. *Evol Appl*, eva.12827, 2019.
91. Hancock PA, Ritchie SA, Koenraadt CJM, Scott TW, Hoffmann AA, Godfray HCJ. Predicting the spatial dynamics of Wolbachia infections in Aedes aegypti arbovirus vector populations in heterogeneous landscapes. *J Appl Ecol*, 56, 1674–1686, 2019.
92. Blagrove MS, Arias-Goeta C, Failloux AB, Sinkins SP. Wolbachia strain wMel induces cytoplasmic incompatibility and blocks dengue transmission in Aedes albopictus. *Proc Natl Acad Sci USA*, 109, 255–260, 2012.

93. Ross PA, Turelli M, Hoffmann AA. Evolutionary ecology of Wolbachia releases for disease control. *Annu. Rev. Genet.* 53, 93–116, 2019.

94. Ross PA, Hoffmann AA. Vector control: Discovery of Wolbachia in malaria vectors. *Curr Biol*, 31, R738–R740, 2021.

95. Bian G, Joshi D, Dong Y, Lu P, Zhou G, Pan X, Xu Y, Dimopoulos G, Xi Z. Wolbachia invades Anopheles stephensi populations and induces refractoriness to Plasmodium infection. *Science (80)*, 340, 748–751, 2013.

96. Shropshire JD, Bordenstein SR. Two-By-One model of cytoplasmic incompatibility: Synthetic recapitulation by transgenic expression of cifA and cifB in Drosophila. *PLoS Genet*, 15, e1008221, 2019.

Chapter 4

Gene Drive Applications for Malaria Control

Vanessa Macias[a] and Anthony James[b]
[a]*Department of Biological Sciences, University of North Texas, Denton, Texas, USA*
[b]*Departments of Microbiology & Molecular Genetics and Molecular Biology & Biochemistry, University of California, Irvine, California, USA*
aajames@uci.edu

4.1 Introduction

Genetic approaches that focus on the vector mosquitoes to prevent malaria parasite transmission have been considered for many decades (Chapter 1, Section I). Some efforts, mostly using variations of sterile insect technologies (SITs), have progressed to open field trials [19, 77]. However, genetic control strategies were given a significant boost with the successful development of gene-drive systems, genetic methods for rapidly spreading beneficial genes and phenotypes through mosquito populations. We review in this chapter some concepts of gene drive systems and describe pioneering applications to control mosquito populations and prevent parasite transmission.

What is gene drive? Gene drive most commonly refers to genetic phenomena in which one of a pair of alternate forms of a gene in

Mosquito Gene Drives and the Malaria Eradication Agenda
Edited by Rebeca Carballar-Lejarazú

ISBN 978-981-4968-33-1 (Hardcover), 978-1-003-30877-5 (eBook)
www.jennystanford.com

a heterozygous organism is inherited preferentially by the offspring of the parent carrying the gene drive system. Gene drive is a term that may also be applied broadly to a number of approaches that do not involve germ-line inheritance (Subchapter 2.1, Chapter 2, Section II) [5]. Gene drive can be assayed simply in what is formally a test cross, a mating between a gene-drive heterozygous parent and a parent homozygous for the lack of the gene drive system (Fig. 4.1). Inheritance patterns consistent with those described originally by Mendel should result in half the progeny carrying one of the copies of the heterozygous chromosomes as a result of random segregation during the first division of meiosis in the parental germ cells [11]. These are the '50:50' inheritance and segregation ratios, respectively. In contrast, gene drive phenotypes result in a significant deviation in either segregation or inheritance from a 50:50 ratio in favor of the chromosome carrying the gene drive system. As we shall see, this can have profound effects on the genetic structure of a population if the gene drive system is stable and heritable.

This genetic phenomenon has been observed for a long time and was designated 'meiotic drive' when it was described in the fruit fly, *Drosophila melanogaster* [114]. Similar phenomena in single-cell eukaryotes were attributed to 'gene conversion' and were later found in a wide variety of organisms and in both meiotically- and mitotically-active cells. We now know much more about the mechanisms involved in both uni- and multi-cellular organisms to the point that these phenomena have been adapted for practical purposes and form the basis for what has been termed 'active genetics' [59]. We focus here on gene drive discovery and applications in the mosquito vectors of malaria parasites.

Gene drive concepts: It is important first to distinguish gene drive from the larger field of gene editing and classical transformation (transgenesis) (Fig. 4.2). Gene editing refers to a number of techniques that allow precise changes, often single nucleotide (nt), in a DNA molecule to achieve the desired phenotype. A number of powerful techniques have been developed over the last 20 years based on the site-specific cleavage properties of an array of DNA endonucleases and the fundamental biochemistry

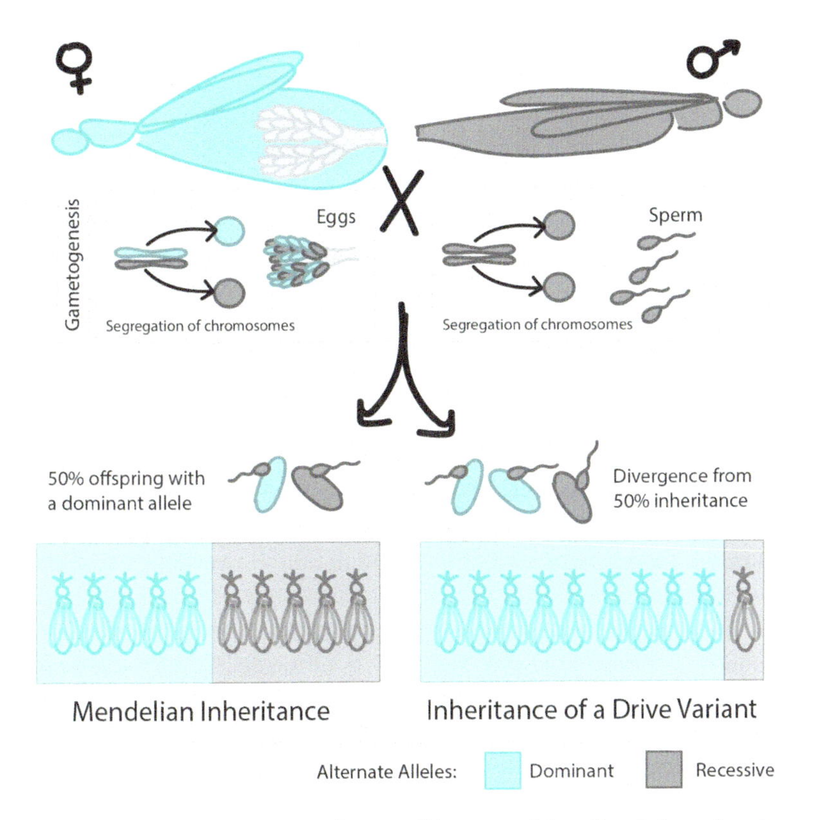

Figure 4.1 Test-cross assay for possible gene drive. Top left: a female parent (♀) heterozygous for the dominantly-marked potential gene-drive chromosome (blue) produces eggs in the ovaries that contain either the gene drive (blue) or wild-type (gray) chromosomes. The male parent (♂) is homozygous for the lack of the gene drive chromosomes (top right) and all sperm are wild-type (gray). Following a cross between these two, a 50:50 segregation indicates no gene drive (bottom left) while deviation from this inheritance ratio is indicative of gene drive (bottom left).

of DNA repair mechanisms. Gene- and genome-editing systems based on transcription activator-like effector nucleases (TALENs) and zinc-finger nucleases (ZFNs) contributed to advance basic and biomedical research, and more recently, the application of technology based on nucleases from the bacterially-derived CRISPR biology (clustered regularly interspaced short palindromic repeats) has set off a revolution in basic and applied gene editing research.

Classical transformation techniques seek to achieve the introduction of exogenous DNA into an organism to modify its phenotype and may be heritable or transient[1]. The site of integration of the DNA is determined by the means used to introduce the DNA. For example, modified transposable elements have served as the work-horses of insect transgenesis, and while they often have a high degree of target specificity (specific nucleotide sequences of 4–6 base pairs in length), these sites are so abundant in large genomes that their insertion appears random (see Chapter 1, Section I; Subchapter 2.1, Chapter 2, Section II). Gene-editing technologies may confer more precision to the target insertion site and this has been exploited significantly with these tools. Modifications made using these techniques can be somatic, affecting unselected or selected subsets of cells of a single organism, or can occur in the germline cells, where modifications made to DNA can be inherited through the gametes to modify a population. However, if heritable, modifications are usually designed so that they are transmitted from a heterozygous parent to its progeny in the canonical Mendelian inheritance ratio of 50:50. The success of the spread of the altered trait through a population will depend on a number of factors including whether or not the edited gene confers a genetic load that will affect its ability to compete with wild-type genotypes. Genetic loads are defined as the presence of unfavorable genetic material in a population that is subject to negative selection to remove it [66]. Loads can be defined numerically to range from 0 (most fit) to 1 (most unfit), although specific mathematical models may adopt different ranges of values [83].

An experimental gene drive is the result of the specific activity of the DNA encoded on a transgene. The activity of a nuclease-based gene drive begins with a gene-editing step, but is designed to target the germline to ensure heritability (Fig. 4.2). This targeting is coupled with some mechanism for achieving an inheritance bias that results in a genotypic frequency and phenotypic change at the population level as a result of the rapid increase in the frequency of

[1]'Exogenous' refers to a genetic or epigenetic element *not* originating from or common to the wild-type of the species of interest. 'Endogenous' refers to a genetic or epigenetic element originating from or common to the wild-type of the species of interest.

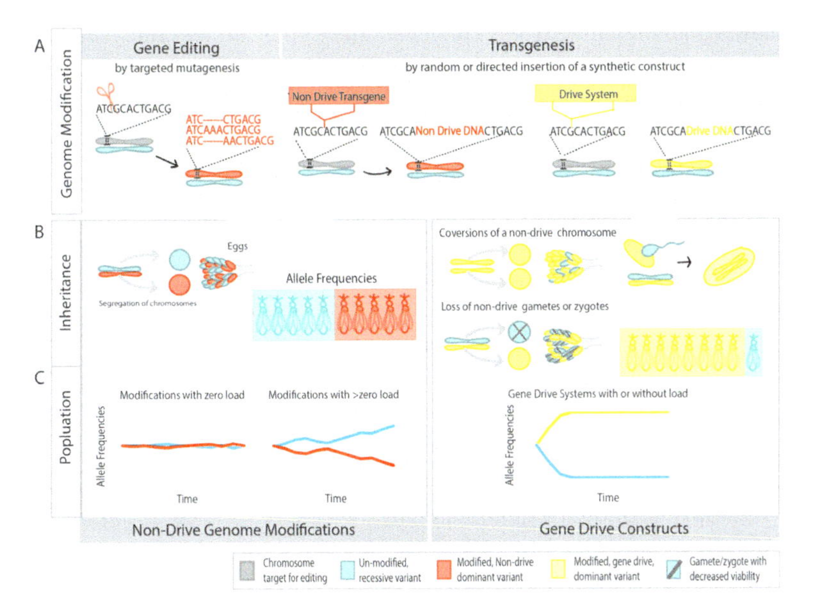

Figure 4.2 Gene editing and transgenesis are distinct from gene-drive. **A.** Gene-editing results from targeted cleavage of the genome that is repaired by the cell leaving modifications at the target site (top-left panel). Transgenesis can integrate exogenous DNA into the genome at a target site (top middle and left panels). Note that the target site can be sufficiently abundant so that the integration appears random or highly specific. The integrated DNA may include a drive system (left panel). **B.** Inheritance of non-drive edits or transgenes segregates normally during meiosis and so are inherited according to Mendelian frequencies. Transgenes encoding a drive system bias inheritance resulting in a deviation from Mendelian inheritance of alleles. **C.** Alleles frequencies in populations of mosquitoes with non-drive genome will change as a result of fitness effects, drift, shift, and natural selection, whose variants with a drive system are expected to increase in frequency despite these effects.

the gene drive elements in the population. As such, gene drives are expected to work best in those organisms with large brood sizes and relatively short life cycles. The mosquito vectors of malaria parasites easily meet these criteria.

The genetic conditions for observing gene drive require allelic differences and/or chromosomal variants because the biased inheritance of one of two alternate variants is inherent in the definition of gene drive. However, the types of variants can be complex. While it is straightforward to consider single nucleotide polymorphisms ([SNPs], alleles) as the simplest variants, they are not expected to be able to bias inheritance unless they cause a change in a gene that can impact normal chromosome segregation during meiosis [81]. Here again, genetic loads conferring fitness costs could affect the variant frequency at the population level, but this is not recognized generally as gene drive [5]. More common are variants that result from the integration of large DNA segments that encode the necessary components for gene drive. These large insertions are paired in a heterozygous organism with an alternate allele, which is often the unaltered wild-type version. A dominant, no-load marker gene encoded on the insertion helps by producing an easily-distinguishable phenotype, but this is not absolutely necessary as modern genome analysis technologies can be used to directly survey the DNA. We have discussed already the need for heterozygosity to define a gene drive effect, and this can be set up in a straightforward manner in diploid and haplo/diploid organisms. Gene drive can be developed to work in polyploid organisms and convert the multiple copies of a target allele that are present, but there is no evidence yet of whole-genome polyploidy in common strains of malaria vector mosquitoes.

We defined a 'drive mechanism' as the underlying biological phenomenon that results in the observed biased inheritance, and the 'drive system' as the final synthetic gene drive product [71]. A recent publication stresses the need for the adoption of a common language to describe gene-drive concepts and the uses herein are generally consistent with these suggestions [5]. The biological basis for many naturally-occurring drive mechanisms is in many cases still unknown (Subchapter 2.1, Chapter 2, Section II). However, synthetic mechanisms have been developed that mimic those occurring naturally, and the lack of understanding of the biological basis of the natural drive mechanism has not hindered their development (Subchapter 2.1, Chapter 2, Section II). Gene drive mechanisms have been reviewed elsewhere and in this book (Subchapter 2.2, Chapter 2, Section II). These include genetic phenomena that can be adapted to manipulating the DNA and/or whole chromosomes to achieve biased inheritance. These approaches result in competitive displacement, reduced heterozygous fitness, under-dominance, hybrid sterility, meiotic drive, segregation distortion (SD), or gene conversion, the latter of which involves nuclease-mediated DNA cleavage and DNA-break induced repair [24, 28, 47, 53]. The latter include homing endonuclease genes (HEGs) [27]. These DNA endonucleases recognize and cleave DNA at highly-specific sites that are sufficiently long enough in length to be rare in the genome. Cleavage in their native species is followed by a recombination event that copies the cleavage machinery into the target chromosome as a result of the DNA repair processes. However, by far those exploiting on nuclease-mediated DNA cleavage and DNA-break induced repair based on CRISPR biology have seen the most success. Strictly speaking, these systems actually exploit the powers of the CRISPR-associated proteins (Cas) and a single guide RNA (sgRNA or gRNA) (Cas9/gRNA). The Cas component is the specific endonuclease that cleaves the DNA, and the protein from *Streptococcus pyogenes*, Cas9, has been used extensively in mosquito gene-drive systems. The gRNA is designed based on the two RNAs (CRISPR-RNA, crRNA, and trans-activating crRNA, tracrRNA) used by the bacterial Cas9, such that the single RNA version contains both a stem-looped region that scaffolds the sgRNA to Cas9 and a ~21 nt domain complementary to a specific genomic DNA target [74]. Interaction of Cas9/gRNA complex with the target region in the genome results in sequence-specific cleavage (Fig. 4.3). Because the ~21 nt target region can be easily encoded to be complementary to almost any sequence on the genome, this system is more flexible in terms of homing than analogous systems where the endonuclease itself confers targeting and cleavage, such as those based on HEGs. Cas proteins from other bacterial species have different specificity features and some of these may be adapted to the mosquito systems [79]. Other proposed approaches use mechanisms based on infectious and infectious-like agents including extracellular and intracellular symbiotic microorganisms (for example, *Wolbachia* sp.), paratransgenesis (engineered symbiotic micro-organisms), viruses, and replicative transposable elements [3, 15, 43, 76, 90] (Chapter 1, Section I; Subchapter 2.1, Chapter 2, Section II).

Successful gene-drive systems can be adapted to a number of purposes [59]. For example, they can be designed to facilitate

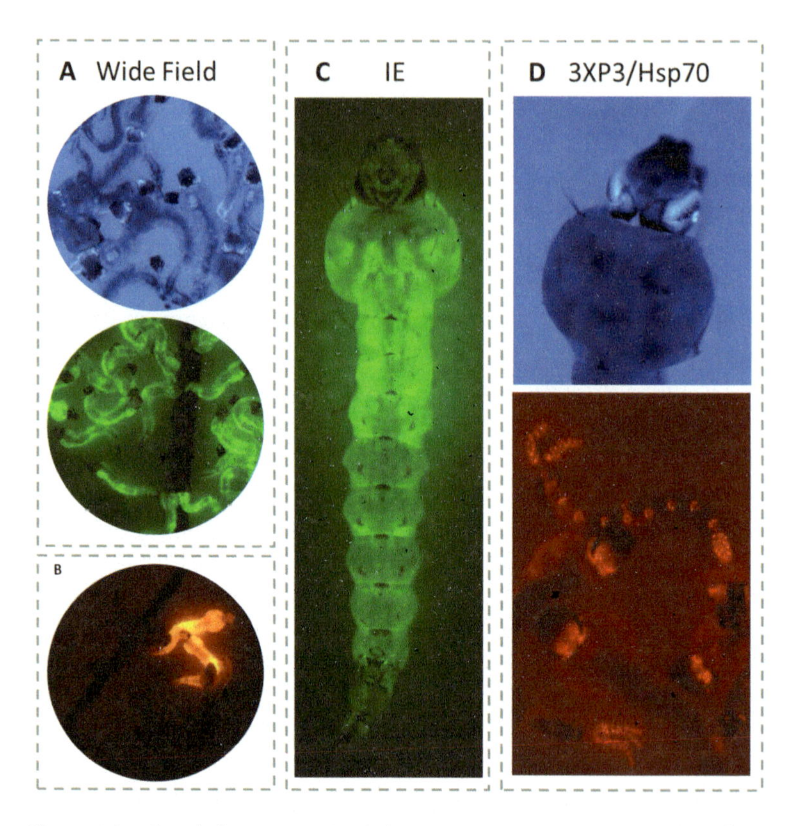

Figure 4.3 *Anopheles* sp. transgenic larvae expressing genes encoding dominant fluorescent proteins. A, B: Wide-field fluorescence microscope images of larvae. A. Larvae expressing enhanced Cyan fluorescent protein using the 3XP3-Hsp70 promoter (top); enhanced green fluorescent protein using an IE-promoter (bottom). B. Transgenic larvae expressing DsRed from an IE promoter with Hr5 enhancer can be distinguished from non-transgenic larvae with no fluorescence phenotype. C. EGFP expressed in the body using IE promoter. D. Larvae expressing ECFP (top) and DsRed (bottom), visible in the eyes and ventral nerve ganglia. Abbreviations: IE, immediate early baculoviruses gene. Images courtesy of Kiona Parker/Vanessa Macias.

drive and bias inheritance, while at the same time conferring a phenotype with a beneficial effect on human health. These beneficial phenotypes for malaria control generally fall into two categories, those that result in vector population suppression and those that elicit reduced pathogen transmission without reducing vector abundance (population replacement/modification/alteration) (Chapter 1, Section I). Examples of these will be explored in detail in the following sections.

Gene drive systems for population suppression or modification can be made with distinct design features. One of the most significant is whether or not they are autonomous (also known as 'autocatalytic') [5, 59]. Autonomous systems carry all the genetic information needed to self-mobilize or cause an inheritance bias tightly linked in a *cis* configuration as part of a single construct (Fig. 4.4). Other systems are designed as 'split' in which the necessary components are at separate loci on homologous or heterologous chromosomes. They also can be temporally distinct, acting at different developmental stages or generational times. Split systems function only when all components of the drive system are active in the same cell at the same time. These alternate configurations, autonomous or split, can affect population-level drive performance and produce systems with different properties and drive dynamics. Autonomous systems are generally designed to have low thresholds for increasing their frequency in populations ('no' and 'low' threshold designations are used interchangeably [5, 72] (Subchapter 2.2, Chapter 2, Section II). Single releases of small numbers of gene drive organisms are expected to result eventually in every organism in the population carrying the drive system. Split and variations of autonomous systems can have more stringent threshold dynamics in which mosquitoes carrying the drive system have to be released above a minimal frequency in relation to the target population (either by one-time releases of larger numbers of mosquitoes or by a succession of serial releases) before the system moves throughout the whole population. Both low and threshold-dependent systems can be designed to be non-limiting or self-limiting. These properties can influence how long the gene drive system remains in the population, the geographical extent to which it can spread, and importantly, how long the beneficial phenotype will be available to exert its effects (Subchapter 2.2, Chapter 2, Section II).

Other features of gene drive systems include the efficacy of their drive. This is measured most directly in Cas9/gRNA-derived systems as to how frequently the heterozygote converts to a

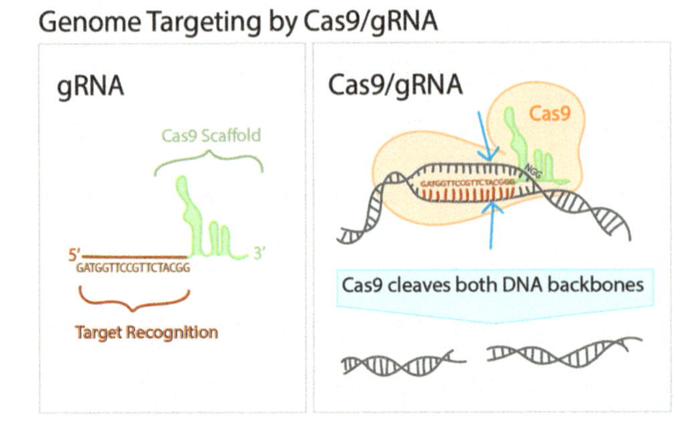

Figure 4.4 The CRISPR-associated (Cas) protein 9 (Cas9) and engineered single guide RNAs (sgRNA, gRNA) derived from those used by Streptococcus pyogenes have been used extensively in the mosquito systems for target specified cleavage of the genome. The gRNA has a stem-loop region that interacts with the Cas protein and a short sequence of ~21 nucleotides that dictates the genomic target for specific cleavage.

homozygote. As we shall see, a number of systems easily achieve 100% efficacy, meaning that all offspring from an outcross of the gene-drive line inherits the drive system. Stability, defined as how long the drive system stays intact and continues to drive, also is important. There are a number of ways a system can cease to function, including mutations that inactivate system components and recombination that may disrupt drive system architecture so that critical components become unlinked or segregated [107].

Gene drive systems for population level effects are designed to mobilize the diploid cells of the germline. Having this occur prior to meiosis facilitates the use of the homologous chromosome as the template for DNA repair following Cas9/gRNA cleavage at the target site. The control sequences ('5- and 3'-ends and promoters) of a number of genes expressed during early development of the mosquito germline have been used to drive the expression of the Cas9 nuclease. Mosquito homologs of the *vasa*, *nanos* (*nos*), and *zero population growth* (*zpg*) genes have provided control sequences in a number of gene-drive systems [96, 105]. The most versatile promoters for expressing the guide RNAs are derived from members of the conserved *U6* gene family, RNA polymerase III (pol III) promoters that normally drive the expression of small nuclear RNAs [48]. The control DNA sequences that have worked the best are constitutive, expressed in many tissues, and produce an abundant product. At least one useful gene family member has been identified in both *An. gambiae and An. stephensi*. Finally, the selection of the genomic target gene is critical to the success of a population suppression gene drive system. Appropriate gene disruptions can alter the proportions of males and females (conferring a sex bias), male or female fertility, or female survival. Controlling any one of these phenotypes can lead to successful population suppression. As discussed previously, a dominant effect is most desirable and this could require a feature for conditional control. Selecting sites with minimally-occurring natural variation also is a good design feature [116].

A number of 'accidental' gene drive outcomes that have been extensively discussed include off- and non-target effects. Off-target effects include those in systems dependent on site-specific activity. Any direct action at a genomic site in the organism other than that intended by the drive design is an off-target effect. These are important because off-target effects may confer a genetic load or cause a phenotype that not only impacts the function of the beneficial genes but may have a novel and potentially undesirable phenotype. Non-target effects are those that impact any organism other than the designated target and the potential for these is one of the most common concerns expressed about gene drive systems for malaria control. Most people, scientists included, know little of the place of mosquitoes in ecosystems. While an analysis of available data has shown that mosquitoes do not appear to represent a 'keystone' species in any known ecosystem, we cannot equate the absence of evidence as definitive [112]. This supports the need for understanding the target organism in its environment and applying cost-benefit analysis.

Horizontal transfer is another major concern and it is worth considering what it would take to make that happen [112][2].

[2]'Horizontal' transmission is the transfer of a genetic or epigenetic element from one eukaryotic organism to another (same or different species) in the absence of sexual

Elucidating mechanisms of horizontal transfer have increased as efforts are being made to reconcile the presence of common DNA sequences in genomes with different phylogenetic histories determined by whole-genome sequencing. For example, such an analysis provides evidence of an evolutionarily remote transfer of genetic material from *Wolbachia* to *Aedes* species [77]. This also can be seen most clearly by those looking at the distribution of mobile DNA elements (all classes of transposons) in eukaryotic organisms. DNA transposons in insects such as the well-studied *P* and *mariner* elements clearly have managed horizontal transfer at some level, but the precise mechanism is still speculative [8, 113, 143]. Perhaps a clue may be in the discovery of the *piggyBac* transposable element. It was found first as an insert in a baculovirus that was capable of infecting a number of different, reproductively-isolated lepidoptera species [54]. Such infections could in principle allow horizontal transfer of a gene drive system into a novel, unintended target species. *piggyBac* was adapted easily as a transformation vector for a wide diversity of organisms from insects to human cells and is the preferred element for mosquito transgenesis [49] (Chapter 1, Section I; Subchapter 2.1, Chapter 2, Section II).

Discussions of genetic loads and fitness have long been a part of the insect transgenesis literature. Researchers with different perspectives (for example, ecologists, geneticists, physiologists, and molecular biologists) have contributed opinions and in some cases, made efforts at empirical measurements [6, 70, 106]. Contributions to the load and impact on fitness are likely to result from two effects, the consequences of the integration of the gene drive system into the genome (insertion/position effects) and those that result from the expression of the components of the system. The insertion event is essentially mutagenic, it disrupts the DNA at the site of integration, and depending upon where it inserts, may affect the gene into which it inserts or genes linked closely enough to be affected by the change in the DNA architecture or the introduction of enhancer DNA sequences. We will see an example of this in the next section. Expression of novel gene products or ectopic expression of host-

exchange. 'Vertical' transmission describes a genetic or epigenetic element being passed from parent to progeny (for example, by incorporation into germ cells).

derived gene products also may impact fitness and load. Regardless of their origin, reliable tests for measuring loads and fitness impacts vary. The ultimate impact is a potential for a reduction in the mating competitiveness of the engineered mosquito so that natural or directed selection processes lead to a loss of the introduced genes from the population. Here gene-drive systems have an interesting competitive advantage. It was shown early in models that gene-drive systems could have significant negative effects on the fitness of an organism, but the inheritance bias could be strong enough to overcome this disadvantage [92, 111]. Simply stated, fitness costs can be outweighed by the strength of the drive. Balancing these two characteristics (fitness/drive efficiency) is the key to the development of successful gene drive systems for both mosquito population suppression and modification.

We discussed already the potential complexity of the genetic variants that may be subject to or components of gene drive systems. One issue receiving considerable attention in the literature is the potential impact of allelic variation on the efficacy of the drive system [29, 107]. This is important particularly with those systems that seek to deploy a DNA endonuclease with a high degree of target-site specificity for a cleavage site. The larger the minimally-required target site, the higher the probability that variants exist. These variants may represent drive-resistant alleles and mosquitoes carrying them may exhibit reduced drive efficacy or a complete loss of the drive system. Genomic target sites resistant to cleavage may be part of natural standing variation (for example, one or more SNPs) or induced by the mechanics of the gene-drive system. The latter has been implicated particularly in Cas9/gRNA-derived gene-drive systems where the repair process after DNA cleavage could result in insertions and deletions (indels) of one or more nucleotides near the cut site followed by non-homologous end-joining (NHEJ) [41, 64, 106]. However, a number of successful efforts in developing gene-drive systems and large population surveys show that these potential challenges can be mitigated or may not be as problematic as anticipated [2, 34, 107, 115] (discussed below).

Gene-drive systems can have different effects on drive transmission dynamics depending upon whether the system is propagated through the paternal or maternal lineage. This may be an explicit feature of the design characteristics of systems for population suppression of malaria vector mosquitoes. Here drive characteristics may feature differential impacts on males and females to maximize drive spread while at the same time disabling one of the sexes, usually females. An interesting phenomenon was observed in which drive dynamics in male-derived lineages remained near 100% through continuous outcrosses of drive males to wild-type females, but drive efficacy was suppressed in the third generation of outcrosses in lineages derived from drive-system females with progeny phenotypes nearing standard Mendelian-predicted 50:50 segregation ratio [58]. This was explained to result from known aspects of mosquito gametogenesis and the impact of females producing eggs loaded with gene-drive nucleoprotein complexes that act on incoming male chromosomes at fertilization. This model is supported by DNA sequencing of paternally-derived alleles and somatic mosaicism seen in progeny carrying visible markers. As discussed below, the choice of promoters for expressing the gene-drive components can significantly affect this phenomenon.

4.2 Gene-Drive Applications

We begin this discussion of applications of gene-drive technologies for malaria control by briefly reviewing some of the features and approaches used to generate and test the systems. We now will look at examples of these in successful drive systems for population suppression and modification. All efforts to adapt various gene-drive mechanisms or produce synthetic analogs of natural phenomena depend on a good working knowledge of basic principles of DNA manipulation and cloning, and utilization of the significant online resources available for whole genomes. For the latter, databases exemplified by Vectorbase provide a portal through which to access a large amount of useful information [128]. In particular, we shall see how the judicious choice of promoter DNA sequences for expression of gene-drive system components and identification of genomic target sites for integration can enhance greatly the probability of success.

Most gene-drive systems comprise a mixture of components that are derived from endogenous (native) DNAs combined with purpose-built synthetic fragments. One of the most important components is a dominant marker gene that is easy to screen. Indeed, the rate of discovery in the field of insect transgenesis accelerated significantly in terms of productive output following the adaptation of genes expressing fluorescent proteins that appear to confer no significant impacts on fitness when expressed in vivo. The three main classes of proteins most frequently used are green fluorescent protein (GFP and enhanced versions, EGFP), red fluorescent protein (*Discosoma* species [Ds] Red and variants), and cyan fluorescent protein (CFP and enhanced versions, ECFP). The array of promoters used to drive the expression of these dominant marker genes depends on the purpose but the most useful has been a synthetic construct comprising three copies of the P3 binding elements and minimal promoter from the *heat shock protein 70* (*hsp70*) gene of *D. melanogaster* [20]. This gene construct is expressed strongly in the nervous tissues of mosquitoes and is most evident in larval and pupal eyes, and in the segmented ventral nerve cord (Fig. 4.5). The signal in wild-type adult eyes can be occluded partially by the dark pigmentation present there and so this marker is often scored best in the sub-adult stages [58]. The fluorescent marker genes also have been used with great success as reporters for promoter function in tissue-specific gene characterization studies [95, 101].

Additional analytic tools include the suite of bioinformatic packages available online and simple methods for gene amplification. The polymerase chain reaction (PCR) is an extremely powerful tool for editing, manipulating, and screening specific fragments of DNA and RNA, coupled with DNA sequencing, is necessary for validating the structure, insertion-site specificity, and component function of any gene-drive system. DNA sequencing also is important for validating the target sites in vivo for natural and induced allelic variants. Techniques are now available to examine whole genome sequences of individual mosquitoes [7].

All gene-drive experiments begin with the introduction into chromosomes in the germline progenitor cells of DNA carrying genes encoding the various components. This has been termed a

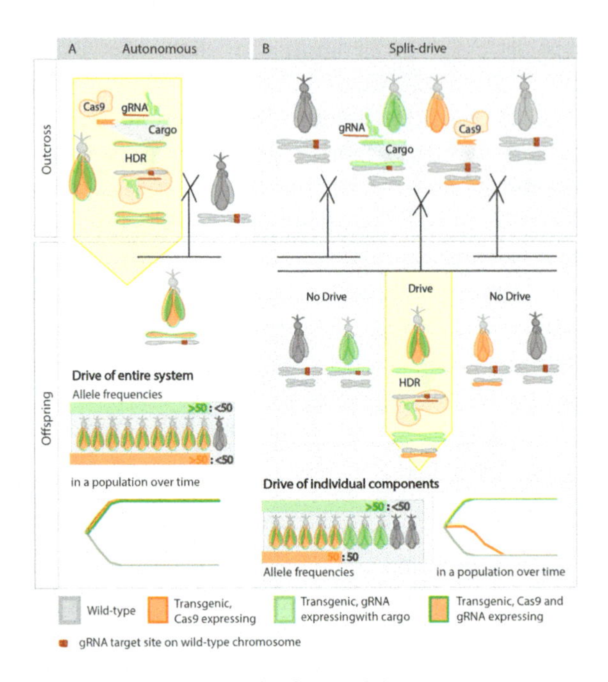

Figure 4.5 Autonomous vs split drives. (A) Autonomous systems carry all the genetic information needed to self-mobilize or cause an inheritance bias tightly linked in a cis configuration as part of a single construct. Top, colocalized expression (yellow banner) of Cas9 (orange) and guideRNA (gRNA, green) in the germline can induce cleavage at the chromosomal target site (red) on the homologous wild-type chromosome (gray) following an outcross, resulting in biased inheritance of the gene drive system and associated cargo to greater than 50% of offspring (bottom). All parts of the gene drive system are expected to increase in frequency together in a population (graph). (B) Split-drive systems with the drive system components (Cas9 [orange] and gRNA [green]) on different chromosomes (top). Mating between gRNA and cargo-bearing transgenic mosquitoes (green) or Cas9 transgenic mosquitoes with wild-type results in Mendelian inheritance of the transgene. Only when all components of the split drive are expressed concurrently in the germline (orange and green together) does Cas9/gRNA targeting of the target site (red) on the homologous wild-type chromosome (gray) occur (yellow banner). Homology directed repair (HDR) following target site cleavage leads to integration of the gRNA and cargo into the wild-type chromosome (conversion) resulting in biased inheritance of the gRNA/cargo transgene (bottom left). A biased inheritance leads to an increase of the gRNA/cargo transgene at the population level, which Cas9 alone (orange) is expected to decrease due to a >0 fitness load (bottom right).

'primary' transformation event as it involves the first introduction of novel DNA into the mosquito [1]. As we shall see in our review of specific experiments, the complement of components depends on the intended outcome, population suppression or modification, and the desired dynamic features of the drive and function of the beneficial genes. Microinjection techniques permit the introduction of DNA plasmids carrying drive-system components into early mosquito embryos at the posterior pole with the intent of maximizing the probability of the exogenous DNA being inserted into the germline [124]. A recent adaptation to the Indo-Pakistan malaria mosquito, *Anopheles stephensi*, of a technique that allows germline modifications to the DNA by injecting females with an ovary-targeting nucleoprotein complex has proven effective for editing of target genes but has yet to achieve DNA insertions [91].

It is not possible to visually distinguish male from female embryos, so putative transgenic animals (designated generation zero, G_0) of both sexes are recovered following injection. Surviving adult mosquitoes are then outcrossed individually or in pools with wild-type members of the opposite sex. The progeny (designated G_1) are screened for the dominant marker phenotypes and the transgenic insects again outcrossed, or in some circumstances intercrossed with recovered transgenic siblings, to establish a colony. The inserted DNA in the mosquitoes is characterized molecularly for the integrity of the introduced DNA and insertion site in the target chromosome. Once colonies are established, outcrosses of subsequent generation mosquitoes and screening phenotypes of their progeny can give a first look at the efficacy of the drive system. It is advisable to initially keep two lines, one derived from males and one from females, to monitor for the lineage effects described previously. Life table analyses comparing transgenic with wild-types mosquitoes provide a preliminary indication of genetic loads, if any, that could affect fitness [2, 34, 106].

Small laboratory cage trials have emerged as an effective way to get information on drive dynamics and quantitative estimates for values used in modeling parameters. Competitive mating studies of a gene-drive strain with wild-type mosquitoes can test the capacity of the system to reach full introduction (every individual carrying at least one copy of the drive system) and also may reveal evidence of loads that may affect reproductive fitness. The sizes of the cages are generally small (0.005–0.2 m^3), although larger ones have been used [2, 34, 106]. At least three replicates of each condition are advised to provide the opportunity to see possible stochastic events. Mosquito densities in the cages are necessarily higher than what would be found in nature to allow robust statistical analyzes. Trial design variables include the origin of gene-drive mosquitoes (hemizygous or homozygous), using target populations with or without age-structure, overlapping or non-overlapping generations, sources of blood meals (human or other animals, and/or artificial feeders), single or multiple releases of gene-drive mosquitoes, release ratios of gene-drive to wild-type mosquitoes and male-only, and female-only or mixed-sex releases [106]. Arguments have been made for male-only releases to mitigate possible future community concerns over releasing female mosquitoes, but modeling and other considerations support female releases being more operationally and cost-effective [32, 108, 123].

The number of generations scored in cage trials depends on the gene-drive dynamics but also should allow sufficient breeding and population sizes to identify low-frequency exceptional phenotypes. These phenotypes generally result from drive-resistant alleles or the lineage-specific maternal effects described previously. Drive-resistant allele frequencies in population suppression trials could start low, but due to the load imposed by the drive system be positively selected and eventually achieve fixation [13]. Trials with three replicates carried out over 15–20 generations can yield total population surveys of several hundred thousand mosquitoes and identify exceptional events with frequencies at $\leq 1 \times 10^{-5}$ [106].

Phenotypic screening of cage trial mosquitoes can be laborious due to the large sample sizes and manipulation required. As described previously, dominant marker genes expressing fluorescent proteins are used as indicators of the presence of gene-drive systems. These are screened efficiently in larvae where wide-field dissecting microscopes fitted with UV lamps and appropriate filters allow surveying fields of 100–300 mosquitoes at a time. In addition, larvae represent a low risk for escape and are easy to manage because they cannot fly. Sex assignment can be made visually under microscopy at the pupal stage and mosaic phenotypes involving the eye also can be seen at this stage [2, 34, 106]. Adults also can be scored for sex, and mosaic and other exceptional phenotypes.

4.2.1 Population Suppression

Mosquito population suppression technologies have been a core feature of malaria control strategies ever since Ronald Ross discovered that these insects were the vectors of the parasites that cause the disease. Indeed, the first modeling of parasite transmission dynamics was formulated by Ross and featured a significant role for mosquito-specific parameters [120, 121]. In addition to personal protection measures, nets, and other strategies for preventing access by mosquitoes to human hosts, approaches were developed to reduce the size of mosquito populations such that the risk of an encounter with infectious mosquitoes became negligible. Considerable success was achieved following the use of insecticides targeting immature and adult insects. Following success with SIT in insects of agricultural significance, similar and genetics-based strategies were proposed, developed, and in some cases, tested for malaria vector species (Table 4.1) [77]. These mostly involved what are termed 'inundative releases' of reproductively-compromised males to achieve sterile mating and subsequent population reduction. Although some field trials showed promise, male mating competitiveness was an issue, and ultimately the gains were not sustainable resulting in a lapse in efforts to further develop the technologies. This changed as scientists began to consider how new molecular biological tools, specifically DNA manipulation and transgenesis, might be used to develop next-generation mosquito population suppression technologies [16].

Genetic population suppression technologies seek to locally eliminate vector mosquito populations and thereby reduce the risk of an infectious bite. They are essentially a genetic analog of insecticides and other methods to reduce mosquito numbers and densities. However, they face a challenge when designed to exploit gene-drive systems. Consider that their intent is to reduce mosquito numbers. Therefore, the system by definition must impose a significant genetic load, as close to 1 as possible, on the target population. Ideally, the system would be able to do this dominantly whereby only one

Table 4.1 Summary of release trials with sterile or semi-sterile male *Anopheles* species mosquitoes[3]

Target species	Location	Sterilization/sex-separation method	Outcome	Reference
Anopheles quadrimaculatus Say	Lakes in Florida, USA	Pupal irradiation, adult release, sex separation by pupal size	Poor competitiveness of colonized males for wild females, which may have been mismatched for sibling species	[44, 132, 133]
Anopheles gambiae s.s. Giles	Pala, Burkina Faso	*An. melas* Theobald x *An. arabiensis* Patton cross yielding sterile hybrid males and few females	Poor competitiveness of hybrid males	[46]
Anopheles albimanus	Lake Apastapeque, El Salvador	Chemosterilization of pupae with bisazir, inaccurate sex separation based on pupal size	100% sterility induced in a wild population, which fell below detection level after 5 months	[85]
Anopheles albimanus	Pacific coast of El Salvador	Bisazir sterilization, sex separation originally by pupal size + feeding on malathion-treated blood, later by a Y chromosome propoxur-resistance translocation inversion (MACHO Strain)	Eventually, 1 million MACHO released per day and found competitive, a natural population increase was suppressed, but eradication prevented by immigration in spite of a barrier one	[45, 116]
Anopheles culicifacies Giles	Village near Lahore, Pakistan	Bisazir sterilization, sex separation by Y chromosome dieldrin translocation	Released males behaved normally in the field but showed subnormal mating competitiveness	[9, 10, 110]

[3]Klassen and Curtis, 2005 [77]

copy of the complete system would be required to see the effect. With these criteria, it is difficult to make and maintain such a strain without the phenotype being conditional. Conditionality could be achieved by having the system be inducible, not active until triggered by an endogenous or exogenous stimulus, suppressible, exogenous, or endogenous component prevents system activity, or sex-specific where continuous out-crossing to wild-type mosquito allows the production of the unaffected sex. Split design, where the system is active only when the full complement of components is present in the same organism, is essentially a formal variant of an inducible/suppressible system.

An example of a non-drive genetic population suppression system in a malaria vector mosquito exploits conditional gene expression and sex specificity. RIDL (release of insects carrying a dominant lethal) was first demonstrated in *D. melanogaster* and further refinement, fsRIDL (female-specific RIDL), applied to culicine mosquitoes and later adapted for *An. stephensi* [55, 67, 93, 126]. *piggyBac* transposon- and $\varphi C31$ recombinase-based technologies were used to generate *An. stephensi* lines that produce a flight-inhibited female phenotype by combining a tetracycline-repressible lethal factor with the control sequences of a gene expressed preferentially in the flight muscles of the adult female. Flightless females do not mate and this can lead to population suppression [52, 138]. The flightless phenotype is repressed by the addition of tetracycline to the larval diet allowing rearing of the mosquitoes under routine laboratory conditions. Importantly, the minimal level of tetracycline needed to rescue the flightless phenotype was higher than that found as an environmental contaminant in agricultural locales subject to intensive antibiotic use [93].

Another non-drive system was developed to affect sex ratios. The males in *Anopheles* species are heterogametic with the sex chromosomes comprising one X and one Y. Practical suppression technologies were developed in the African malaria mosquito, *Anopheles gambiae*, based on an 'X-shredder' phenotype in which nuclease-mediated damage to the single copy of the X chromosome in males biases inheritance in favor of the Y chromosome resulting in 'extreme reproductive sex ratios' in outcrosses [50, 56, 57]. Both

HEG- (*I-PpoI*) and Cas9/gRNA-based systems targeting a repetitive DNA sequence in the X-linked ribosomal RNA genes resulted in fully fertile mosquito strains that produce high frequencies (>95%) of male offspring with no discernable fitness effects in males. The male mosquitoes suppressed wild-type mosquito populations in cage trials. These systems would be expected to work best if they were located on the Y chromosome, but this has been difficult so far to achieve.

The first demonstration of a synthetic gene drive system in mosquitoes was based on a HEG. The activity of two of these enzymes was demonstrated in mosquito cells and embryos and provided the impetus to use them in a gene-drive system [136]. Windbichler and colleagues working with the HEG, *I-SceI*, derived from a mitochondrial intron of the yeast, *Saccharomyces cerevisiae*, were able to design a gene drive system that propagated through the male germline of *An. gambiae* [137]. The male-specific drive was facilitated by using the control DNA of the *β2-tubulin* gene, which is active only in the testes, to drive the expression of the nuclease. In two separate small cage trials, insertion of the *I-SceI*-based system into its 18 base pairs recognition sequence disrupted expression of a dominant marker gene, GFP, as the drive system went from 10% of the starting population to 40–60% over 11–12 generations, and from 50% of the starting percentage to 75—89% at generations 8–10. The observed drive dynamics fit well to both deterministic and stochastic population genetic models in conditions in which no loads or fitness issues were attributed to the presence of the system.

The application of Cas9/gRNA-based technology to population suppression gene-drive systems has resulted in strains whose properties are far superior to those based in HEGs. Specifically, the ability to target any gene or chromosomal locus has eliminated the need for a fortuitous target site (only one *I-SceI* site was found in the *An. gambiae* genome; [137]. The ability to control the expression of the system components (Cas9 and gRNA) and provide alternate, engineered, DNA templates for DNA repair following site-specific cleavage is a key to their success.

The first Cas9/gRNA-based population suppression strain was developed for *An. gambiae* and was designed to affect female fertility [63]. The Cas9 nuclease was expressed using *vasa* control

DNA sequences and a U6 gene promoter from *An. gambiae* [80], and targeted three genes that confer recessive female reduced fertility phenotypes. Conditionality was achieved because males are fertile and can propagate the line by continuous outcrossing. Intercrossing creates homozygous females that have significantly reduced numbers of viable offspring. Drive dynamic analysis showed a gene drive efficacy of 91–99% in the progeny of presumed heterozygous gene-drive parents. Dual replicates of cage experiments for one of the targeted genes, AGAP007280, a second chromosome locus encoding the putative ortholog of the *nudel* gene, showed 70–75% introduction at the fourth generation after a 1:1 gene-drive:wild-type male release ratio. The observed drive dynamics were consistent with modeling in which homozygous drive females had only ~9% reproductive fitness compared to wild-type counterparts and males had no load at all.

Genotypes that confer severe loads and resulting phenotypic fitness costs are expected to be selected negatively and disappear from a population. If the gene-drive mechanics facilitate the production of cleavage-resistant but functional mutant alleles of the target locus, it can be expected that with time these will increase in frequency and ultimately become fixed. For a population suppression strain, this means that there will be an initial reduction in the population size followed by a rebound as the drive-resistant alleles increase in frequency. The initial gene-drive experiments targeting AGAP007280 were carried out for only four generations [63]. Follow-up cage experiments carried out for 25 generations verified the impact of the accumulation of drive-resistant alleles [64]. There was an initial increase in the frequency of the gene drive system followed by a gradual decrease consistent with the accumulation of functional mutations at the target site. These mutations were selected positively by restoring female fertility and highlighting one of the major challenges of population suppression strategies.

Two efficient systems that target sex ratios to achieve population suppression also have been developed. In the first, the drive system targets the *An. gambiae doublesex* ortholog. Insect *dsx* genes encode two alternatively-spliced, sex-specific transcripts that function as 'downstream' effectors of sexual differentiation [23]. The transcript produced in females has an additional exon (E5) not present in male transcripts as a result of alternate splicing, and its primary sequence is highly-conserved in six *An. gambiae* complex species analyzed [82, 83]. A gene-drive system using the control sequences of the *An. gambiae zpg* gene and a gRNA targeting the intron 4-exon 5-boundary ablates the formation of a functional female transcript and the resulting mosquito homozygous for the mutant allele develop with an intersex phenotype and are sterile [82]. Male development and fertility are not affected, thus providing the conditional feature of the system that allows its propagation. Two replicates of cage trials showed the drive system reaching 100% full introduction within 7–11 generations following a 1:1 gene-drive:wild-type male ratio release. The cages showed a concurrent reduction in egg production and the populations went extinct within one generation (generations 8 and 12) of achieving homozygosity of the disrupted gene. The judicious choice of the target site and promoter sequences driving the Cas9 expression did not allow the emergence of functional drive-resistant alleles, and therefore unlike the previous efforts with the female fertility genes, there was no population recovery.

A compound system designated 'male-biased sex-distorter gene drive' (SDGD) was developed for *An. gambiae* and combines *dsx*-targeting with a HEG, *I-PpoI* [119]. As discussed previously, the recognition site for the *I-PpoI* endonuclease is located on the X chromosome in the rRNA cistrons and was used as the basis of the X-shredder population suppression strategy [56, 57]. The SDGD system uses a modified *β2 tubulin* promoter to drive the expression of a modified *I-PpoI* nuclease in males and the *zpg* gene control elements to drive Cas9 expression. Males carrying the SDGD system outcrossed to wild-type females produce few females due to HEG X-shredding and exhibit a high frequency of drive of the compound construct for the *dsx* targeting. Females homozygous for the system are converted to intersexes due to the lack of the female-specific *dsx* transcript and therefore do not contribute to subsequent generations. Comparative modeling of the invasion dynamics predicted that the SDGD system would reduce the females in the population at a rate faster than the previous gene drives targeting female fertility [64]. Laboratory cage experiments supported this prediction and male-only populations were achieved within 10–14 generations following single releases of 1:9 SDGD:wild-type males. As with the previous *dsx* work by Kyrou et al. [82], populations crashed and there was no evidence of selection for functional resistant alleles. The sequence similarity in the target genes allowed the introgression of the SDGD system from *An. gambiae* into *An. arabiensis* [21]. Introgression is the movement and fixation of genetic material from one species to another. It requires overcoming mating barriers (pre- and post-copulatory) and the production of fertile hybrids. A series of crossing and selection schemes were developed to circumvent F1 hybrid male sterility-producing drive dynamics in *An. arabiensis* similar to those observed in *An. gambiae*.

In summary, the Cas9/gRNA-based population suppression systems performed well and appear to be ready for field trials. In addition to being simpler to encode and having a high degree of flexibility in design than other systems, this work validated a number of endogenous mosquito genes whose control sequences can be used for further system development. While some success has been achieved using other mechanisms as the basis of gene drive, the Cas9/gRNA-based systems far outperform these and appear to be the solution for genetics-based population suppression technology development.

4.2.2 Population Modification (Replacement/Alteration)

Population modification (also called replacement or alteration) refers to a class of disease-control strategies that use genetic technologies to make vectors incapable of transmitting pathogens. The approaches discussed here involve the introduction of exogenous genes or modification of endogenous genes in populations of mosquitoes to confer resistance to the malaria parasites with the expectation that this will result in decreased transmission of the pathogen. Implicit is the outcome that fewer transmission results in less human morbidity and mortality. Other approaches using wild-type or genetically-engineered microorganisms are described in Subchapter 1.1 and 1.2, Section I, Chapter 1 (see also [22, 68, 129, 130]). The benefits of population modification approaches were reviewed recently [32]. These include a predicted sustainable impact on malaria elimination throughout the World Health Organization (WHO) defined stages of control, pre-elimination, elimination, and prevention of reintroduction [134]. Furthermore, these approaches are not expected to result in an 'empty' ecological niche that can be occupied by another vector species or re-invaded by migrating wild-type animals of the same species, as might be expected to occur with population suppression. These features in addition to preserving parasite-resistant mosquitoes and concomitant reductions in transmission also contribute to environmental safety for species that feed on them. Population modification strains are modeled to be stable in low population densities and capable of expanding with their parasite-resistance phenotypes following seasonal reductions in numbers and densities or impacts of other vector control technologies. These features are expected to support a serial approach in resource-constrained control, allowing local elimination with confidence that the area made free of malaria will remain so. This then should result in a consolidation of gains as elimination efforts move on to new areas targeted for elimination.

At their simplest, gene drive population modification strains of mosquitoes comprise two components, a drive system, and DNA sequences that encode the parasite resistance phenotypes, so-called 'effector' genes. The specific phenotype conferred by the effector genes prevents transmission of parasites in amounts sufficient to infect and cause disease in the human hosts. This phenotype can be achieved by gene editing of agonists, mosquito host factors that enable or facilitate parasite infection, or antagonists, factors that block or inhibit mosquito infection by the parasites [118]. A large number of candidate effectors of both types have been identified through comprehensive characterization of active and passive immune responses of mosquitoes to parasites and the literature has many reviews available describing potential genes and genes products that whether by induction or suppression affect one or more mosquito stages of the parasite (Chapter 1, Section I; [32, 103, 118, 121]). The expression of some of these molecules may be modulated by the microbiome, but there are enough that are independent to give a wealth of targets [142].

Another class of effectors is exogenous gene constructs often containing synthetic elements or repurposed antimicrobial components [32, 118]. These potentially would impose less of a load and fitness impact on mosquitoes because they are not expected to disrupt physiologic or metabolic balances in which endogenous target genes would be involved. We have pursued single-chain antibody constructs (single-chain fragment variable, scFvs) driven by mosquito tissue-specific gene promoters and have had some success with those that target model system malaria parasite and the human pathogen, *Plasmodium falciparum* [31, 69, 70].

A key issue for the effector molecules is to determine how well they must perform in order to have an impact on malaria transmission. This can be evaluated in the context of a target product profile (TPP) that lists all the desirable features of a coupled effector-gene drive system, their ideal performance characteristics, and those that would be 'minimally acceptable' [32]. Early work drawing from a human infection protocol and animal model system supported the conclusion that the 'zero parasites in the salivary glands' phenotype were both the ideal and minimally acceptable [73, 127]. However, a recent comprehensive review of animal model parasite challenge experiments supports the conclusion that there may be a minimum threshold inoculum from the mosquito to the vertebrate host required to establish an infection and cause subsequent disease [62]. It is important to not extrapolate too much from the animal systems to humans and human parasites, but further work should help resolve these divergent conclusions.

Other features that will affect the efficiency of gene drive population modification systems are whether or not a single copy is sufficient to confer the malaria-resistance phenotype, that is, it is dominant, and what is the required prevalence (percentage of mosquitoes carrying at least one copy). While the single-copy efficacy can be determined empirically by parasite challenge assays with engineered mosquitoes, mathematical modeling is required to make a preliminary determination of the impact of prevalence on controlling parasite transmission and the expected time to see an epidemiological impact (Chapter 3, Section II; [12, 51]).

The first coupling of a Cas9/gRNA-based gene drive system with anti-malarial effector genes was achieved in *An. stephensi* [60]. This study used an autonomous system comprising the *vasa* control DNA to drive Cas9 endonuclease expression, an orthologous *U6* gene promoter to express the gRNA, a dominant fluorescent marker gene (DsRed), and targeted the gene encoding kynurenine hydroxylase (kh^w, also called kynurenine mono-oxygenase, *kmo*), homozygous mutations of which give a white-eye phenotype. The system also carried two anti-*P. falciparum* scFv effector genes controlled by tissue-specific gene promoters. Progeny of males and females derived from transgenic males exhibited a high frequency of germ-line gene conversion consistent with homology-directed repair (HDR). In addition, the system copied the large (~17 kilo-base pairs in length) construct from the site of its primary insertion to the homologous chromosome. The system was introduced into ~99.5% of the progeny following outcrosses of transgenic lines to wild-type mosquitoes. The effector genes remained transcriptionally inducible upon blood feeding but were not tested with parasite challenge assays. While efficient gene drive was seen in mosquitoes derived from male lineages, later generation female-derived insects produced progeny that had a high frequency of mutations in the targeted kh^w genome sequence and exhibited ~50:50 inheritance ratios of the drive system. These mutations most likely resulted from NHEJ events prior to the segregation in early embryos of somatic and germ-line cell lineages. Additionally, in contrast to other work showing no significant loads associated with homozygous knock-outs of the kh^w locus in this and other mosquito species, decreases in survival and progeny output were observed in females homozygous for the drive system or heteroallelic for the drive system and a nonfunctional NHEJ allele [65, 141].

Extended generation small-cage trials of this prototype gene drive system assessed the endpoint stages of the drive dynamics [106]. Experiments were carried out in triplicate and had two different design features. The first series looked at the initial drive dynamics and used overlapping generations of mosquitoes. The second series used discrete, non-overlapping generations. Both designs involved only a single release of male gene-drive mosquitoes. Two of three cages in which the gene drive:wild-type male release ratio was 1:1 reached full introduction at 6–8 generations while the third cage achieved ~80% introduction within the same time. Release ratios of 1:10 gene drive:wild-type males failed to introduce the drive system. The single-release trials of the same gene-drive strain in a non-overlapping generation design resulted in two of three cages in which 1:1 gene drive:wild-type male were released reaching 100% introduction within 6–12 generations. Two of three cages with 1:3 gene drive:wild-type male single releases achieved full introduction in 13–16 generations. As expected from the severe load conferred on females by homozygosity for the drive system or a heteroallelic condition of the drive system and an NHEJ allele in the *kh* locus, all populations exhibiting full introduction went extinct. All cages with a 1:10 gene drive:wild-type male release ratio failed to reach full introduction. The frequency of gene drive mosquitoes reached a maximum of ~92% in one cage by generation 8, while the other two achieved maxima of 83% at generations 9 and 16. The cages that did not go extinct began to accumulate functional gene-drive resistant alleles that increased in frequency until the termination of the experiment.

The *An. stephensi* gene-drive system was designed originally for population modification and was to introduce dual anti-*P. falciparum* effector molecules [60]. However, the unanticipated and significant genetic load imposed on females by the system targeting the kh^w gene resulted in a strain that could elicit population suppression in small cage trials [106]. Furthermore, the selection for functional drive-resistant target alleles could cause the system to fail and give rise to a drive-resistant population capable of transmitting malaria parasites. A series of experiments were conducted to see if the potentially compromised system could be rescued [2]. A Cas9/gRNA exchange system, designated 'swap', used two specific gRNAs to cleave the integrated gene-drive system to effect direct replacement of the marker gene and included a recoded portion of the kh^w gene that would restore the wild-type function of the gene product while at the same time making the gene resistant to Cas9/gRNA cleavage. The new strain, designated 'recoded', retained the elements of the original autonomous drive system but lacked the effector molecules. The modifications relieved the load in females caused by the integration of the drive into the kh^w gene and non-functional resistant alleles were eliminated by a maternal effect called 'lethal-mosaicism' in conjunction with standard negative selection. Lethal mosaicism is hypothesized to occur in mosquitoes that have a sufficiently large portion of their cells homozygous for non-functional copies of the *kh*^{w} target gene. If the patch of tissue is large enough, it will cause the death of the mosquito following a blood meal, similar to a whole animal homozygous mutant for non-functional *kh*^{w} alleles. Small cage trials of single releases of gene-drive males resulted in efficient population modification with $\geq$95% of mosquitoes carrying the drive within 5–11 generations over a range of initial release ratios (1:1, 1:3, and 1:10 gene-drive: wild-type males). Furthermore, rare functional resistant alleles did not prevent drive invasion. This strategy of inserting a gene drive into a gene essential for viability or fertility and providing at the same time a functional gene that rescues the loss of viability and fertility provides a general solution to drive resistance through females as was demonstrated previously by its application in *D. melanogaster* [61].

An alternate approach to attaining efficient population modification would be to target a gene that confers no-load and therefore has a minimal or negligible impact on fitness. Such a system was developed for *An. gambiae* [34]. The system, designated AgNosCd-1, targets the *cardinal* gene ortholog that encodes a heme peroxidase that catalyzes the last step in ommochrome biosynthesis, homozygous mutations of which produce a red-eye phenotype. The Cas9 endonuclease is driven by the *nanos* gene control elements [96]. Drive achieved 98–100% efficiencies in both sexes with full introduction observed in small cage trials within 6–10 generations following a single release of 1:1 gene-drive:wild-type males. No genetic load or fitness effect was evident that affected drive performance in the cage trials. Potential drive-resistant target-site alleles were found at frequencies $<$0.1 and 5 of the most prevalent polymorphic gRNA target sites in colonized and wild-derived African mosquitoes did not prevent Cas9/guide RNA complex cleavage in vitro, indicating that they may not represent resistant alleles. One predicted off-target site was cleavable in vitro but a negligible number of mutations was observed in vivo. AgNosCd-1 was shown to meet key performance criteria of a TPP and is being used to make a strain field ready for mosquito population modification.

Strategies to target endogenous mosquito genes that act as agonists and antagonists also were proposed as the basis for population modification [100, 118]. The theoretical basis for a concept, 'integral gene drive' (IGD), comprises a small (minimal) number of molecular components, drive-system, and anti-pathogen effector elements introduced into endogenous genes [100]. The design features are intended to target conserved gene coding sequences without disrupting their function. Both autonomous and non-autonomous (split) IGD strains can be envisioned. Modeling comparisons of IGDs with other modification gene-drive systems support significant reductions in the emergence of drive-resistant alleles. Drive systems that target multiple genomic sites also may contribute to phenotypic stability.

The efficiency of Cas9/gRNA-based gene drive systems for quickly converting a population to one that cannot transmit parasites is unmatched at this time by any other mechanism. However, arguments have been made for population modification strategies based on the inundative release of mosquitoes carrying effector genes [17]. The approach would require serial multiple releases of mosquitoes carrying one or more anti-parasite effector genes that in addition to blocking parasite transmission would confer low or negligible loads that would affect fitness. Such a trial was carried out in small cages with *An. stephensi* [106]. The genetically-engineered strain had dual anti-malarial scFvs genes based on the m1C3 and m2A10 monoclonal antibodies linked in a 'tail-to-tail' orientation flanked by *gypsy* insulator sequences and a dominant DsRed marker gene [33, 70]. Triplicate cages were seeded with wild-type larvae over three successive weeks to create age-structured populations [106]. Once established, wild-type male and female pupae in equal numbers were added weekly to replenish the populations following die-offs of adults in the cages. After eight weeks of stabilization, two series of cages also were provided weekly with 1:1 and 10:1 engineered:wild-type male mosquitoes, and the third series was kept as controls. The 1:1 release-ratio cages showed an increase of the engineered mosquitoes to a maximum percentage of 76% at week 21, although one cage fell to ~48%. Since the engineered males were being introduced into cage populations with pre-existing wild-type males, at least one generation after introduction would have to elapse before virgin wild-type females were available for mating with engineered males. Two of three 10:1 cages achieved 95–99% engineered mosquitoes during weeks 16–17, eight weeks (~3 generations) after the first introduction. The levels achieved by the 10:1 releases are high enough to have an impact on transmission dynamics and malaria epidemiology, but the anticipated costs of the serial releases make those approaches that couple the effector gene with a gene drive system more attractive.

4.2.3 Considerations

A considerable amount of work has been done since the first reports of adapting Cas9/gRNA gene-drive systems to insects. A wealth of information is now available, most notably from work done with *D. melanogaster* and the culicine mosquito, *Aedes aegypti*, a major vector of many of the medically-significant arboviruses, including dengue and Zika. This chapter focuses on the Anopheline vectors of human malaria, but what is being done with these and other species is expected to inform future design features for vector strains for malaria control. A list of significant achievements in those other species relevant to gene-drive systems is provided in Table 4.2. Note that there are many applications to studying insect gene function using *gene editing* but these are not listed.

The applications of Cas9/gRNA based gene-drive systems have stimulated a large amount of discussion both within and outside of the research community working with it. Specific chapters in this book address much of this discussion and we will only mention a few points of consideration here and leave the in-depth discussion to others. Researchers developing gene drive systems have to acknowledge spatial and temporal components. Most gene-drive systems are designed with the expectation of a regional impact. However, a more global effect may result from insect migration or human involvement (accidental or intentional). These technologies need to work on a human time-scale, not an evolutionary one. There will be high expectations for an epidemiological impact on morbidity and mortality with only a few years of their initial deployment.

Table 4.2 Sample of experimental gene drive systems for population suppression or modification in insects

Species	Comment	Application[4]	Reference
Aedes aegypti	Radiation-induced reciprocal chromosome translocations in mosquitoes	PM/PS	[88]
Rhodnius prolixis	First paratransgenic system using bacteria to deliver anti-parasite effector molecules	PM	[14]
Drosophila melanogaster	First gene drive system based on synthetic MEDEA	PS	[42]
Ae. Aegupti	Exogenous microbial agent conferring pathogen resistance	PM	[97]
D. melanogaster	First HEG-based synthetic homing drive	PM/PS	[35]
Ae. albopictus	Exogenous microbial agent conferring population suppression	PS	[30]
D. melanogaster	First publication of a synthetic threshold-dependent gene drive system, UD(MEL)	PM	[4]
D. melanogaster	High threshold UD system based on targeting a haploinsufficient gene with rescue	PM/PS	[109]
D. melanogaster	First demonstration of Cas9/gRNA based gene drive systems (MCR) in insects	PM	[58]
D. melanogaster	Cas9-activated 'brake' gene drive system, CATCHA	PM	[139]
D. melanogaster	Analysis of resistance allele formation and improved promoters	PM	[41]
D. melanogaster	Multiplexed gRNAs to reduce resistant allele formation	PM	[40]
D. melanogaster	Cas9/gRNA engineered reciprocal chromosome translocations reversible, high-threshold population modification	PM	[26]
D. melanogaster	Gene drive system utilizing multiplexed guide RNAs	PS	[102]
D. melanogaster	Cas9/gRNA-based system for 'allelic drive' (biased inheritance of one allele over another)		[61]
D. melanogaster	Application of a 'cleave and rescue' system to drive linked beneficial genes	PM	[103]
D. melanogaster	*Wolbachia* gene-based synthetic gene drive	PM	[117]
D. melanogaster	Cas9/gRNA-based, synthetic target sites and split drive system	PM	[38]
D. melanogaster	Cas9/gRNA-based drive performance in genetically diverse backgrounds	PM	[29]

(*Continued*)

Table 4.2 (*Continued*)

Species	Comment	Application[4]	Reference
D. melanogaster	'Killer-rescue' gene drive	PM	[131]
Ae. Aegypti	Cas9/gRNA-based split-drive systems for population modification	PM	[84]
D. melanogaster	Cas9/gRNA-based drive that reduces the prevalence of resistance alleles by targeting a haplo-lethal gene with two gRNAs and providing a rescue allele	PM	[37]
D. melanogaster	CRISPR-Cas9-based gene drive systems to inactivate (e-CHACRs) or delete and replace (ERACRs) pre-existing gene drive	PM	[140]
D. melanogaster	CRISPR-Cas9-based split gene drive system, GME, enables assessment of homing and knockout efficiencies of two target genes simultaneously, and timing and tissue specificity of cleavage and homing rates	PM/PS	[75]
D. melanogaster	Blocking Cas9/gRNA-based gene drives with small molecules	PM/PS	[86]
D. melanogaster	Threshold-dependent Cas9/gRNA-based toxin-antidote, TARE, drive system, similar to ClvR	PM	[36]
D. melanogaster	Self-limiting Cas9/gRNA-based sGD gene drive system clears potential resistance alleles in vital gene targets	PM	[125]
D. melanogaster	Species and strain isolation with EGIs	PM	[92]
D. melanogaster	Cas9/gRNA-based split gene drives, tGD, promote super-Mendelian inheritance of separate transgenes	PM	[87]

[4]Potential or actual application.

Abbreviations: Cas: CRISPR-associated protein; CATCHA: Cas9-triggered chain ablation; e-CHACR: construct hitchhiking on the autocatalytic chain reaction; ClvR: Cleaver and Rescue; ERACR: elements for reversing the autocatalytic chain reaction; EGIs: engineered genetic incompatibilities; GME: gRNA-mediated effector; gRNA: guide RNA; HEG: homing endonuclease gene; MCR: mutagenic chain reaction; MEDEA: maternal effect dominant embryonic arrest; PM: population modification; PS: population suppression; sGD: split gene-drive; TARE: toxin-antidote recessive embryo; tGD: trans-complementing split-gene-drive; UD: underdominance; UD(MEL): underdominance, maternal-effect lethal

Issues of safety and efficacy of gene drive systems are paramount. Detailed comments on these are found in Section I (Chapter 1). The discovery and description of the consequences of off-target and non-target need evaluation. The consequence of drive or effector molecule failures must be anticipated and plans for mitigation in place. All of these issues require robust interactions between the scientist and the society of which they are part. These are fostered by interactions with national and international regulatory realms and explicit criteria for individual (where needed) and community consent (Chapter 9, Section IV).

4.2.4 Pathways to Deployment

The pathway for moving gene-drive technologies out of the laboratory to the field is being defined as the work progresses. Unlike other technologies for controlling malaria such as vaccines, therapeutic and preventative drugs, and insecticides, no industry-vetted pathways exist for genetically-engineered mosquitoes [32, 89]. As a consequence, investigators and scientific advisory groups have offered analyzes of challenges and issued guidelines for moving the science forward. These include a number of studies reviewed in Section V (Chapter 10). An overriding guiding principle is that the work should be done in a stepwise fashion in specific phases and safety and efficacy criteria met before moving from one phase to the next [18, 73, 99, 135].

Acknowledgments

The authors are grateful for the contributions of Rhodell Valdez, Kiona Parker, Omar Akbari, Ethan Bier, and Jackson Champer. AAJ is a Donald Bren Professor at the University of California, Irvine.

References

1. Adelman, Z.N., Jasinskiene, N., Onal, S., Juhn, J., Ashikyan, A., Salampessy, M., MacCauley, T., James, A.A. (2007). Nanos gene control DNA

mediates developmentally-regulated transposition in the yellow fever mosquito. *Aedes aegypti. Proc. Natl. Acad. Sci. USA,* 104:9970–9975.

2. Adolfi, A., Gantz, V.M., Jasinskiene, N., Lee, H.S., Hwang, K., Bulger, E.A., Ramaiah, A., Bennett, J.B., Terradas, G., Emerson, J.J., Marshall, J.M., Bier, E., James, A.A. (2020). A population modification gene-drive rescue system dominantly eliminates resistance alleles in the malaria mosquito, *Anopheles stephensi. Nature Comm.*, 11:5553.
3. Afanasiev, B., Carlson, J. (2000). Densovirinae as gene transfer vehicles. *Contrib. Microbiol.*, 4:33–58.
4. Akbari, O.S., Matzen, K.D., Marshall, J.M., Huang, H., Ward, C.M., Hay, B.A. (2013). A synthetic gene drive system for local, reversible modification and suppression of insect populations. *Curr Biol.,* 23(8):671–677.
5. Alphey, L.S., Crisanti, A., Randazzo, F.F., Akbari, O.S. (2020). Opinion: Standardizing the definition of gene drive. *Proc Natl Acad Sci USA,* 18:202020417. doi: 10.1073/PNAS.2020417117.
6. Amenya, D.A., Bonizzoni, M., Isaacs, A.T., Jasinskienc, N., Chen, H., Marinotti, O., Yan, G., James, A.A (2010). Comparative fitness assessment of *Anopheles stephensi* transgenic lines receptive to site-specific integration. *Insect Molec. Biol.,* 19:263–269.
7. Anopheles gambiae 1000 Genomes Consortium; Data analysis group (2017). Genetic diversity of the African malaria vector *Anopheles gambiae. Nature,* 552(7683):96–100.
8. Anxolabéhère, D., Kidwell, M. G., Periquet, G. (1988). Molecular characteristics of diverse populations are consistent with the hypothesis of a recent invasion of *Drosophila melanogaster* by mobile P elements. *Mol. Biol. Evol.*, 5:252–269.
9. Baker, R.H., Reisen, W.K., Sakai, R.K., Rathor, H.R., Raana, K., Azra, K., Niaz, S. (1980). *Anopheles culicifacies*: Mating behavior and competitiveness in nature of males carrying a complex chromosomal aberration. *Annals of the Entomological Society of America,* 73:581–588.
10. Baker, R.H., Sakai, R.K., Raana, K. (1981). Genetic sexing for a mosquito sterile male release. *J. Heredity,* 72: 216–218.
11. Bateson, W., Mendel, G. (1913). Mendel's principles of heredity. Cambridge University Press. ISBN-13:978-0-486-47701-5, ISBN-10: 0-486-47701-0. Dover edition 2010, Dover Publications Inc., Mineola, New York.
12. Beaghton, A.K., Hammond, A., Nolan, T., Crisanti, A., Godfray, C.H., Burt, A. (2017). Requirements for driving antipathogen effector genes

into populations of disease vectors by homing. *Genetics,* 205(4):1587–1596.

13. Beaghton, A.K., Hammond, A., Nolan, T., Crisanti, A., Burt, A. (2019). Gene drive for population genetic control: Non-functional resistance and parental effects. *Proc Biol Sci.,* 286(1914):20191586.
14. Beard, C.B., Mason, P.W., Aksoy, S., Tesh, R.B., Richards, F.F. (1992). Transformation of an insect symbiont and expression of a foreign gene in the Chagas' disease vector Rhodnius prolixus. *Am J Trop Med Hyg.,* 46(2):195–200.
15. Beard, C.B. et al. (2002). Bacterial symbionts of the triatominae and their potential use in control of Chagas disease transmission. *Annu. Rev. Entomol.,* 47:123–141.
16. Beaty, B.J., Prager, D.J., James, A.A., Jacobs-Lorena, M., Miller, L.H., Law, J.H., Collins, F.C., Kafatos, F.C. (2009). From Tucson to genomics and transgenics: The vector biology network and the emergence of modern vector biology. *PLoS Negl Trop Dis.,* 3:e343.
17. Benedict, M.Q. (2011). Let it snow: Field-testing malaria-refractory strains by inundation. *Malaria World* (blog). https:// malaria-world.org/search/site/let%20it%20snowden
18. Benedict, M., Bonsall, M., James, A.A., James, S., Lavery, J., Mumford, J., Quemada, H., Rose, R., Thompson, P., Toure, Y., Yan. G. (2014). Guidance framework for testing of genetically modified mosquitoes. WHO/TDR publications, ISBN 978 92 4 150748 6.
19. Benedict, M.Q., Robinson, A.S. (2003). The first releases of transgenic mosquitoes: An argument for the sterile insect technique. *Trends in Parasitology,* 19:349–355.
20. Berghammer, A.J., Klingler, M., Wimmer, E.A. (1999). A universal marker for transgenic insects. *Nature,* 402:370–371.
21. Bernardini, F., Kriezis, A., Galizi, R., Nolan, T., Crisanti, A. (2019. Introgression of a synthetic sex ratio distortion system from Anopheles gambiae into Anopheles arabiensis. *Sci. Rep.,* 9: 5158.
22. Bian, G., Joshi, D., Dong, Y., Lu, P., Zhou, G., Pan, X., Xu, Y., Dimopoulos, G., Xi, Z. (2013). Wolbachia invades Anopheles stephensi populations and induces refractoriness to Plasmodium infection. 340(6133):748–51.
23. Bopp, D., Saccone, G., Beye, M. (2014). Sex determination in insects: Variations on a common theme. *Sex Dev.,* 8(1–3):20–8. doi: 10.1159/000356458. Epub 2013 Dec 6.
24. Braig, H.R., Yan, G. (2001). The spread of genetic constructs in natural insect populations. In genetically engineered organisms:

Assessing environmental and human health effects. Letourneau, D.K. and Burrows, B.E. (eds.), 251–314, CRC Press.

25. Buchman, A., Akbari, O.S. (2019). Site-specific transgenesis of the Drosophila melanogaster Y-chromosome using CRISPR/Cas9. *Insect Mol Biol.*, 28:65–73.
26. Buchman, A.B., Ivy, T., Marshall, J.M., Akbari, O.S., Hay, B.A. (2018). Engineered reciprocal chromosome translocations drive high threshold, reversible population replacement in drosophila. *ACS Synth Biol.*, 7(5):1359–1370. doi: 10.1021/acssynbio.7b00451. Epub 2018 Apr 13.
27. Burt, A. (2003). Site-specific selfish genes as tools for the control and genetic engineering of natural populations. *Proc. Biol. Sci.*, 270:921–928.
28. Burt, A., Crisanti, C. (2018). Gene drive: Evolved and synthetic. *ACS Chem Biol.*,13(2):343–346.
29. Callaway, E. (2017). Gene drives meet the resistance. *Nature*, 542:15.
30. Calvitti, M., Moretti, R., Skidmore, A.R., Dobson, S.L. (2012). Wolbachia strain wPip yields a pattern of cytoplasmic incompatibility enhancing a Wolbachia-based suppression strategy against the disease vector Aedes albopictus. *Parasit Vectors*, 5:254. doi: 10.1186/1756-3305-5-254.
31. Capurro, M de L., Coleman, J., Beerntsen, B.T, Myles, K.M., Olson, K.E., Rocha, E., Krettli, A.U., James, A.A. (2000). Virus-expressed, recombinant single-chain antibody blocks sporozoite infection of salivary glands in *Plasmodium gallinaceum*-infected *Aedes aegypti*. *Am. J. Trop. Med. Hygiene*, 62:427–433.
32. Carballar-Lejarazú, R., James, A.A. (2017). Population modification of Anopheline species to control malaria transmission. *Pathog Glob Health*, 111:424–435.
33. Carballar-Lejarazú, R., Jasinskiene, N., James, A.A. (2013). Exogenous gypsy insulator sequences modulate transgene expression in the malaria vector mosquito, *Anopheles stephensi*. *Proc. Natl. Acad. Sci. USA*, 110:7176–7181.
34. Carballar-Lejarazú, R., Ogaugwu, C., Tushar, T., Kelsey, K., Pham, T.B., Murphy, J., Schmidt, H., Lee, Y., Lanzaro, G., James, A.A. (2020). Next-generation gene drive for population modification of the malaria vector mosquito, *Anopheles gambiae*. *Proc. Natl. Acad. Sci. USA*, 117: 22805–22814.

35. Chan, Y.S., Naujoks, D.A., Huen, D.S., Russell, S. (2011). Insect population control by homing endonuclease-based gene drive: An evaluation in Drosophila melanogaster. *Genetics*, 33–44.

36. Champer, J., Lee, E., Yang, E., Liu, C., Clark, A.G., Messer, P.W. (2020). A toxin-antidote CRISPR gene drive system for regional population modification. *Nat Commun.*, 1(1):1082.

37. Champer, J., Yang, E., Lee, E., Liu, J., Clark, A.G., Messer, P.W. (2020). A CRISPR homing gene drive targeting a haplolethal gene removes resistance alleles and successfully spreads through a cage population. *Proc Natl Acad Sci USA*, 24377–24383.

38. Champer, J., Chung, J., Lee, Y.L., Liu, C., Yang, E., Wen, Z., Clark, A.G., Messer, P.W. (2019). Molecular safeguarding of CRISPR gene drive experiments. *Elife.*, 8:e41439.

39. Champer, J., Wen, Z., Luthra, A., Reeves, R., Chung, J., Liu, C., Lee, Y.L., Liu, J., Yang, E., Messer, P.W., Clark, A.G. (2019). CRISPR gene drive efficiency and resistance rate is highly heritable with no common genetic loci of large effect. *Genetics*, 333–341.

40. Champer, J., Liu, J., Oh, S.Y., Reeves, R., Luthra, A., Oakes, N., Clark, A.G., Messer, P.W. (2018). Reducing resistance allele formation in CRISPR gene drive. *Proc Natl Acad Sci USA*, 5522–5527.

41. Champer, J., Reeves, R., Oh, S.Y., Liu, C., Liu, J., Clark, A.G., Messer, P.W. (2017). Novel CRISPR/Cas9 gene drive constructs reveal insights into mechanisms of resistance allele formation and drive efficiency in genetically diverse populations. *PLoS Genet*, 3(7):e1006796.

42. Chen, C.H., Huang, H., Ward, C.M., Su, J.T., Schaeffer, L.V., Guo, M., Hay, B.A. (2007). A synthetic maternal-effect selfish genetic element drives population replacement in Drosophila. *Science*, 316(5824):597–600. Epub 2007 Mar 29.

43. Curtis, C.F., Sinkins, S.P. (1998). Wolbachia as a possible means of driving genes into populations. *Parasitology 116(Suppl.)*, S111–S115.

44. Dame, D.A., Woodward, D.B., Ford, H.R., Weidhaas, D.E. (1964). Field behavior of sexually sterile *Anopheles quadrimaculatus* males. *Mosquito News*, 24:6–16.

45. Dame, D.A., Lowe, R.E., Williamson, D.W. (1981). Assessment of released sterile *Anopheles albimanus* and *Glossina morsitans*, 231–248. *In* R. Pal, J. B. Kitzmiller and T. Kanda (eds.), Cytogenetics and genetics of vectors. Proceedings of XVI International Congress of Entomology, Kyoto, Japan. Elsevier Science Publishers, Amsterdam, The Netherlands.

46. Davidson, G., Odetoyinbo, J.A., Colussa, B., Coz, J. (1970). A field attempt to assess the mating competitiveness of sterile males produced by crossing two member species of the *Anopheles gambiae* complex. Bulletin of the World Health Organization, 42:55–67.
47. Davis, S. et al. (2001). Engineered under-dominance allows efficient and economical introgression of traits into pest populations. *J. Theor. Biol.*, 212:83–98
48. Didychuk, A.L., Butcher, S.E., Brow, D.A. (2018). The life of U6 small nuclear RNA, from cradle to grave. *Epub.*, 437–460.
49. Di Matteo, M., Mátrai, J., Belay, E., Firdissa, T., Vandendriessche, T., Chuah, M.K. (2012). PiggyBac toolbox. *Methods Mol Biol.*, 859:241–54.
50. Deredec, A., Burt, A., Godfray, H.C. (2008). The population genetics of using homing endonuclease genes in vector and pest management. *Genetics*, 179:2013–2026.
51. Eckhoff, P.A., Wenger, E.A.H., Godfray, C.H., Burt, A. (2017). Impact of mosquito gene drive on malaria elimination in a computational model with explicit spatial and temporal dynamics. *Proc Natl Acad Sci USA*, 114(2):E255–E264.
52. Facchinelli, L., Valerio, L., Ramsey, J.M., Gould, F., Walsh, R.K., Bond, G., Robert, M.A., Lloyd, A.L., James, A.A., Alphey, L. and Scott, T.W. (2013). Field cage studies and progressive evaluation of genetically-engineered mosquitoes. *PLoS Negl Trop Dis*, 7:e2001. doi:10.1371/journal.pntd.0002001.
53. Foster, G., Whitten, M., Prout, T., Gill, R. (1972). Chromosome rearrangements for the control of insect pests. *Science*, 176(4037):875–880.
54. Fraser, M.J., Ciszczon, T., Elick, T., Bauser, C. (1996). Precise excision of TTAA-specific lepidopteran transposons piggyBac (IFP2) and tagalong (TFP3) from the baculovirus genome in cell lines from two species of Lepidoptera. *Insect Mol Biol.*, 5(2):141–51.
55. Fu, G., Lees, R.S., Nimmo, D., Aw, D., Jin, L., Gray, P., Berendonk, T.U., White-Cooper, H., Scaife, S., Kim Phuc, H., Marinotti, O., Jasinskiene, N., James, A.A., Alphey, L. (2010). Female-specific flightless phenotype for mosquito control. *Proc Natl Acad Sci USA*, 107(10):4550–4.
56. Galizi, R. Hammond, A., Kyrou, K., Taxiarchi, C., Bernardini, F., O'Loughlin, S.M., Papathanos, P.A., Nolan, T., Windbichler, N., Crisanti, A. (2016). A CRISPR-Cas9 sex-ratio distortion system for genetic control. *Sci Rep.*, 6:31139.
57. Galizi, R., Doyle, L.A., Menichelli, M., Bernardini, F., Deredec, A., Burt, A., Stoddard, B.L., Windbichler, N., Crisanti, A. (2014). A synthetic sex

ratio distortion system for the control of the human malaria mosquito. *Nat Commun.,* 5:3977.

58. Gantz, V.M., Bier, E. (2015). Genome editing. The mutagenic chain reaction: A method for converting heterozygous to homozygous mutations. *Science,* 348:442–444.
59. Gantz, V.M., Bier, E. (2016). The dawn of active genetics. *Bioessays,* 38:50–63.
60. Gantz, V.M., Jasinskiene, N., Tatarenkova, O., Fazekas, A., Macias, V.M., Bier, E., James, A.A. (2015). Highly efficient Cas9-mediated gene drive for population modification of the malaria vector mosquito, *Anopheles stephensi. Proc. Natl. Acad. Sci. USA,* 112(49):E6736–43.
61. Guichard, A., Haque, T., Bobik, M., Xu, X.S., Klanseck, C., Kushwah, R.B.S., Berni, M., Kaduskar, B., Gantz V.M., Bier, E. (2019). Efficient allelic-drive in *Drosophila. Nat Commun.,* 10(1):1640.
62. Graumans, W., Jacobs, E., Bousema, T., Sinnis, P. (2020). When is a plasmodium-infected mosquito an infectious mosquito? *Trends Parasitol.,*36(8):705–716.
63. Hammond, A., Galizi, R., Kyrou, K., Simoni, A., Siniscalchi, C., Katsanos, D., Gribble, M., Baker, D., Marois, E., Russell, S., Burt, A., Windbichler, N., Crisanti, A., Nolan, T. (2016). A CRISPR-Cas9 gene drive system targeting female reproduction in the malaria mosquito vector Anopheles gambiae. *Nat Biotechnol.,* 34(1):78–83.
64. Hammond, A.M., Kyrou, K., Bruttini, M., North, A., Galizi, R., Karlsson, X., Kranjc, N., Carpi, F.M., D'Aurizio, R., Crisanti, A., Nolan, T. (2017). The creation and selection of mutations resistant to a gene drive over multiple generations in the malaria mosquito. *PLoS Genet,*13(10):e1007039.
65. Han, Q., Calvo, E., Marinotti, O., Fang, J., Rizzi, M., James, A.A. (2003). Analysis of the wild-type and mutant genes encoding the enzyme kynurenine monooxygenase of the yellow fever mosquito, *Aedes aegypti. Insect Mol Biol.,* 12(5):483–90.
66. Hartl, D.L., Clark, A.G. (1997). Principles of population genetics, 3rd Ed., Sinauer Associates Inc. Sunderland, Massachusetts, ISBN 0-87893-306-9.
67. Heinrich, J., Scott, M. (2000). A repressible female-specific lethal genetic system for making transgenic insect strains suitable for a sterile-release program. *Proc Natl Acad Sci USA,* 97(15):8229–8232.
68. Huang, W., Wang, S., Jacobs-Lorena, M. (2020). Self-limiting paratransgenesis. *PLoS Negl Trop Dis.,* 14(8):e0008542.

69. Isaacs, A.T., Li, F., Jasinskiene, N., Chen, X., Nirmala, X., Marinotti, O., Vinetz, J.M., James, A.A. (2011). Engineered resistance to *Plasmodium falciparum* development in transgenic *Anopheles stephensi*. *PLoS Pathogens*, 7(4):e1002017.
70. Isaacs, A.T., Jasinskiene, N., Tretiakov, M., Thiery, I., Zettor, A., Bourgouin, C., James, A.A. (2012). Transgenic *Anopheles stephensi* co-expressing single-chain antibodies resist *Plasmodium falciparum* development. *Proc. Natl. Acad. Sci. USA*, 109:E1922–E1930. *PNAS PLUS*, 109:11070–11071.
71. James, A.A. (2005). Gene drive systems in mosquitoes: Rules of the road. *Trends in Parasitology*, 21:64–67.
72. James, S., Collins, F.H., Welkhoff, P.A., Emerson, C., Godfray, C. H., Gottlieb, M. J., Greenwood, B., Lindsay, S.W., Mbogo, C.M., Okumu, F.O., Quemada, H., Savadogo, M., Singh, J.A., Tountas, K.H., Touré, Y.T. (2018). Pathway to deployment of gene drive mosquitoes as a potential biocontrol tool for elimination of malaria in sub-Saharan Africa: Recommendations of a scientific working group. *Am J Trop Med Hyg.*, 98(6_Suppl):1–49. doi: 10.4269/ajtmh.18-0083.
73. Jasinskiene, N., Coleman, J., Ashikyan, A., Salampessy, M., Marinotti, O., James, A.A. (2007). Genetic control of malaria parasite transmission: Threshold levels for infection in an avian model system. *Am. J. Trop. Med. Hygiene*, 76:1072–1078.
74. Jinek, M., Chylinski, K., Fonfara, I., Hauer, M., Doudna, J. A., Charpentier, E. (2012). A programmable dual-RNA-guided DNA endonuclease in adaptive bacterial immunity. *Science*, 337:6096, 816–821.
75. Kandul, N., Liu, J., Buchman, A., Gantz, V.M., Bier, E., Akbari, O.S. (2020). Assessment of a split homing based gene drive for efficient knockout of multiple, *Genes G3 (Bethesda)*, 10(2):827–837.
76. Kidwell, M.G., Ribeiro, J.M.C. (1992). Can transposable elements be used to drive disease refractoriness genes into vector populations? *Parasitol. Today*, 8:325–329.
77. Klassen, W., Curtis, C. F. (2005). History of sterile insect technique, in sterile insect technique principles and practice in area-wide integrated pest management. V.A. Dyck, J. Hendrichs, A.S. Robinson (eds.), 3–36, IAEA, Springer, Netherlands.
78. Klasson. L., Kambris, Z., Cook, P.E., Walker, T., Sinkins, S.P. (2009). Horizontal gene transfer between Wolbachia and the mosquito Aedes aegypti. *BMC Genomics*, 10:33. doi: 10.1186/1471-2164-10-33. PMC2647948.

79. Knott, G.J., Doudna, J.A. (2018). CRISPR-Cas guides the future of genetic engineering. *Science*, 361(6405):866–869.

80. Konet, D.D., Anderson, J., Piper, J., Akkina, R., Suchman, E., Carlson, J. (2007). Short-hairpin RNA expressed from polymerase III promoters mediates RNA interference in mosquito cells. *Insect Mol Biol.*, 199–206.

81. Kusano, A., Staber, C., Chan, H.Y.E., Ganetzky, B. (2003). Closing the (Ran)GAP on segregation distortion in Drosophila. *Bioessays*, 108–115.

82. Kyrou, K., Hammond, A. M., Galizi, R., Kranjc, N. Burt, A., Beaghton, A.K., Nolan, T., Crisanti, A. (2018). A CRISPR–Cas9 gene drive targeting doublesex causes complete population suppression in caged *Anopheles gambiae* mosquitoes. *Nature Biotechnology*, 36:1062–1066.

83. Lachance, J. (2008). A fundamental relationship between genotype frequencies and fitnesses. *Genetics*, 180:1087–1093.

84. Li , M., Yang, T., Kandul, N.P., Bui, M., Gamez, S., Raban, R., Bennett, J., Sánchez, H.M., Lanzaro, G.C., Schmidt, H., Lee, Y., Marshall, J.M., Akbari, O.S. (2020). Development of a confinable gene drive system in the human disease vector *Aedes aegypti*. *Elife*, 9:e51701.

85. Lofgren, C. S., Dame, D.A., Breeland, S.G., Weidhaas, D.E., Jeffery, G., R. Kaiser, R., Ford, R., Boston, M.D., Baldwin, K. (1974). Release of chemosterilized males for the control of *Anopheles albimanus* in El Salvador. III. Field methods and population control. *American Journal of Tropical Medicine and Hygiene*, 23:288–297.

86. López Del Amo, V., Bishop, A.L., Sánchez, C.H.M., Bennett, J.B., Feng, X., Marshall, J.M., Bier, E., Gantz, V.M. (2020). A transcomplementing gene drive provides a flexible platform for laboratory investigation and potential field deployment. *Nat Commun*. 11(1): 352.

87. López Del Amo, V., Leger, B.S., Cox, K.J., Gill, S., Bishop, A.L., Scanlon, G.D., Walker, J.A., Gantz, V.M., Choudhary, A. (2020). Small-molecule control of super-mendelian inheritance in gene drives. *Cell Rep.*, 31(13):107841.

88. Lorimer, N., Hallinan, E., Rai, K.S. (1972). Translocation homozygotes in the yellow fever mosquito, *Aedes aegypti*. *J Hered.*, 158–66.

89. Macias, V.M., James, A.A. (2015). Impact of genetic modification of vector populations on the malaria eradication agenda. Adelman, Z. (ed.). In: Genetic control of malaria and dengue. Elsevier Academic Press, Oxford, UK, 423–444.

90. Macias, V.M., Jimenez, A.J., Burini-Kojin, B., Pledger, D., Jasinskiene, N., Phong, C.H., Chu, K., Fazekas, A., Martin, K., Marinotti, O., James, A.A. (2017). *Nanos*-driven expression of *piggyBac* transposase induces

mobilization of a synthetic autonomous transposon in the malaria vector mosquito, *Anopheles stephensi. Insect Biochem. Molec. Biol.*, 87:81–89.

91. Macias, V.M., McKeand, S., Chaverra-Rodrigues, D., Hughes, G.L., Fazekas, A., Jasinskiene, N., James, A.A., Rasgon, J.L. (2020). Cas9-mediated gene-editing in the malaria mosquito *Anopheles stephensi* by ReMOT Control. *G3, Genes, Genomes, Genetics,* 10:1353–1360.
92. Maselko, M., Feltman, N., Upadhyay, A., Hayward, A., Das, S., Myslicki, N., Peterson, A.J., O'Connor, M.B., Smanski, M.J. (2020). Engineering multiple species-like genetic incompatibilities in insects. *Nat Commun.*, 8(1):883.
93. Marinotti, O., Jasinskiene, N., Fazekas, A., Scaife, S., Fu, G., Mattingly, S.T., Chow, K, Brown, D.M., James, A.A. (2013). Development of a population suppression strain of the human malaria vector mosquito, *Anopheles stephensi. Malaria Journal.* 12:142.
94. Marshall, J.M. (2008). A branching process model for the early spread of a transposable element in a diploid population. *Journal of Mathematical Biology,* 57:811–840.
95. Mathur, G., Sanchez-Vargas, I., Alvarez, D., Olson, K.E., Marinotti, O., James, A.A. (2010). Transgene-mediated suppression of dengue viruses in the salivary glands of the yellow fever mosquito, *Aedes aegypti. Insect Mol Biol.*, 19:753–763.
96. Meredith, J.M., Underhill, A., McArthur, C.C., Eggleston, P. (2013). Next-generation site-directed transgenesis in the malaria vector mosquito anopheles gambiae: self-docking strains expressing germline-specific phiC31 integrase. *PLOS ONE,* 8(3):e59264.
97. Moreira, L.A., Iturbe-Ormaetxe, I., Jeffery, J.A., Lu, G., Pyke, A.T., Hedges, L.M., Rocha, B.C., Hall-Mendelin, S., Day, A., Riegler, M., Hugo, L.E., Johnson, K.N., Kay, B.H., McGraw, E.A., van den Hurk, A.F., Ryan, P.A., O'Neill, S.L. (2009). *Cell,* 139(7):1268–78. doi: 10.1016/j.cell.2009.11.042.
98. Oberhofer, G., Tobin, I., Hay, B.A. (2020). Gene drive and resilience through renewal with next generation Cleave and Rescue selfish genetic elements. *Proc Natl Acad Sci USA,* 9013–9021.
99. National Academies of Sciences, Engineering, and Medicine. (2016). Gene drives on the horizon: Advancing science, navigating uncertainty, and aligning research with public values. Washington, DC: The National Academies Press. doi: 10.17226/23405.
100. Nash, A., Urdaneta, G.M., Beaghton, A.K., Hoermann, A., Papathanos, P.A., Christophides, G.K., Windbichler, N. (2019). Integral gene drives for population replacement. *Biol Open,* 8(1):bio037762.

101. Nirmala, X., Marinotti, O., Sandoval, J.M., Phin, S., Gakhar, S., Jasinskiene, N., James, A.A. (2006). Functional characterization of the promoter of the vitellogenin gene, *AsVg1*, of the malaria vector, *Anopheles stephensi. Insect. Biochem. Molec. Biol.,* 36:694–700.

102. Oberhofer, G., Ivy, T., Hay, B.A. (2018). Behavior of homing endonuclease gene drives targeting genes required for viability or female fertility with multiplexed guide RNAs. *Proc Natl Acad Sci USA*, 115(40):E9343–E9352.

103. Oberhofer, G., Ivy, T., Hay, B.A. (2019). Cleave and Rescue, a novel selfish genetic element and general strategy for gene drive. *Proc Natl Acad Sci USA*, 116(13):6250–6259.

104. Osta, M.A., Christophides, G.K., Vlachou, D., Kafatos, F.C. (2004). Innate immunity in the malaria vector Anopheles gambiae: Comparative and functional genomics. *J Exp Biol.*, 207(Pt 15):2551–63.

105. Papathanos, P.A., Windbichler, N., Menichelli, M., Burt, A., Crisanti, A. (2009). The vasa regulatory region mediates germline expression and maternal transmission of proteins in the malaria mosquito Anopheles gambiae: A versatile tool for genetic control strategies. *BMC Mol Biol.,* 10(65).

106. Pham, T.B., Phong, C.H., Bennett, J.B., Hwang, K., Jasinskiene, N., Parker, K., Stillinger, D., Marshall, J.M., Carballar-Lejarazú, R., James, A.A. (2019). Experimental population modification of the malaria vector mosquito, *Anopheles stephensi. PLoS Genetics*, 15(12):e1008440.

107. Price, T.A.R., Windbichler, N., Unckless, R., Sutter, A., Runge, J.N. Ross, P., Pomiankowski, A., Nuckolls, N.L, Montchamp-Moreau, C., Mideo, N., Martin, O.Y., Manser, A., Legros, M., Larracuente, A.M., Holman, L., Godwin, J., Gemmell, N., Courret, C., Buchman, A., Barrett, L.G., Lindholm, A.K. (2020). Resistance to natural and synthetic gene drive systems. *J Evol. Biol.,* 33(10):1345–1360.

108. Rasgon, J., Scott, T.W. (2004). Impact of population age structure on Wolbachia transgene driver efficacy: Ecologically complex factors and release of genetically modified mosquitoes. *Insect Biochem. Mol. Biol.*, 34:707–713.

109. Reeves, R.G., Bryk, J., Altrock, P.M., Denton, J.A., Reed (2014). First steps towards underdominant genetic transformation of insect populations. *PLoS One*, 9(5):e97557

110. Reisen, W.K., Baker, R.H., Sakai, R.K., Mahmood, F., Rathor, H.R., Raana, K., Toqir, G. (1981). *Anopheles culicifacies*: Mating behavior and competitiveness in nature of chemosterilized males carrying a genetic

sexing system. *Annals of the Entomological Society of America,* 74:395–401.

111. Ribeiro, J.M., Kidwell, M.G. (1994). Transposable elements as population drive mechanisms: Specification of critical parameter values. *J Med Entomol.,* 31(1):10–6.
112. Roberts, A., Paes de Andrade, P., Okumu, F., Quemada, H., Savadogo, M., Singh, J.A., James, S. (2017). Results from the workshop "problem formulation for the use of gene drive in mosquitoes", *Am J Trop Med Hyg.,* 96(3):530–533.
113. Robertson, H.M. (1993). The mariner transposable element is widespread in insects. *Nature,* 362:241–245.
114. Sandler, L., Novitski, E. (1957). Meiotic drive as an evolutionary force. *Am. Naturalist,* XCI:105–110.
115. Schmidt, H., Collier, T.C., Hanemaaijer, M.J., Houston, P.D., Lee, Y., Lanzaro, G.C. (2020). Abundance of conserved CRISPR-Cas9 target sites within the highly polymorphic genomes of Anopheles and Aedes mosquitoes. *Nat Commun.,* 11(1):1425.
116. Seawright, J.A., Kaiser, P.E., Dame, D.A., Lofgren, C.S. (1978). Genetic method for the preferential elimination of females of *Anopheles albimanus. Science,* 200:1303–1304.
117. Shropshire, J.D., Bordenstein, S.R. (2019). Two-By-One model of cytoplasmic incompatibility: Syn-thetic recapitulation by transgenic expression of cifA and cifB in Drosophila. *PLoS Genet.* 15(6):e1008221.
118. Simões, M.L., Caragata, E.P., Dimopoulos, G. (2018). Diverse host and restriction factors regulate mosquito-pathogen interactions. *Trends Parasitol.,* 34(7):603–616.
119. Simoni, A., Hammond, A.M., Beaghton, A.K., Galizi, R., Taxiarchi, C., Kyrou, K., Meacci, D., Gribble, M., Morselli, G., Burt, A., Nolan, T., Crisanti, A. (2020). A male-biased sex-distorter gene drive for the human malaria vector *Anopheles gambiae. Nat Biotechnol.,* 38(9):1054–1060.
120. Smith, D.L., Battle, K.E., Hay, S.I., Barker, C.M., Scott, T.W., McKenzie, F.E. (2012). Ross, Macdonald, and a theory for the dynamics and control of mosquito-transmitted pathogens. *PLoS Pathog.,* 8(4):e1002588.
121. Smith, R.C., Barillas-Mury, C. (2016). Plasmodium Oocysts: Overlooked targets of mosquito immunity. *Trends Parasitol.,* 32(12):979–990.
122. Smith, D.L., Perkins, T.A., Reiner, R.C. Jr, Barker, C.M., Niu, T., Chaves, L.F., Ellis, A.M., George, D.B., Le Menach, A., Pulliam, J.R., Bisanzio, D., Buckee, C., Chiyaka, C., Cummings, D.A., Garcia, A.J., Gatton, M.L., Gething, P.W., Hartley, D.M., Johnston, G., Klein, E.Y., Michael, E.,

Lloyd, A.L., Pigott, D.M., Reisen, W.K., Ruktanonchai, N., Singh, B.K., Stoller, J., Tatem, A.J., Kitron, U., Charles, H., Godfray, J., Cohen, J.M., Hay, S.I., Scott, T.W. (2014). Recasting the theory of mosquito-borne pathogen transmission dynamics and control. *Trans R Soc Trop Med Hyg.,* 108(4):185–97.

123. Spielman, A., Pollack, R.J., Kiszewski, A.E., Telford, S.R. (2001). Issues in public health entomology. *Vector Borne Zoonotic Dis.*, 1:3–19.
124. Terenius, O., Juhn, J., James, A.A. (2007). Injection of An. stephensi embryos to generate malaria-resistant mosquitoes. *J Vis Exp*., 5:216.
125. Terradas, G., Buchman, A., Bennett, J.B., Shriner, I., Marshall, J.M., Omar, S. Akbari, O.S., Bier, E. (2021). Inherently confinable split-drive systems in drosophila. *Nat Commun.*, 12:1480.
126. Thomas, D.D., Donnelly, C.A., Wood, R.J., Alphey, L.S. (2000). Insect population control using a dominant, repressible, lethal genetic system. *Science*, 287(5462):2474–6.
127. Ungureanu, E., Killick-Kendrick, R., Garnham, P.C., Branzei, P., Romanescu, C., Shute, P.G. (1977). Pre-patent periods of a tropical strain of *Plasmodium vivax* after inoculations of ten-fold dilutions of sporozoites. *Trans R Soc Trop Med Hyg.,* 70:482–483.
128. Vectorbase (https://vectorbase.org/).
129. Wang, S., Ghosh, A.K., Bongio, N., Stebbings, K.A., Lampe, D.J., and Jacobs-Lorena, M. (2012). Fighting malaria with engineered symbiotic bacteria from vector mosquitoes. *PNAS,* 109(31):12734–12739.
130. Wang, S., Dos-Santos A.L.A., Huang, W., Liu, K.C., Oshaghi, M.A., Wei, G., Agre, P., Jacobs-Lorena, M. (2017). Driving mosquito refractoriness to *Plasmodium falciparum* with engineered symbiotic bacteria. *Science,* 357(6358):1399–1402.
131. Webster, S.H., Vella, M.R., Maxwell, J., Scott, M.J. (2020). Development and testing of a novel killer-rescue self-limiting gene drive system in Drosophila melanogaster. *Proc Biol Sci.*, 287(1925):20192994.
132. Weidhaas, D.E. (1974). Release of chemosterilized males for control of *Anopheles albimanus* in El Salvador. IV. Dynamics of the test population. *American Journal of Tropical Medicine and Hygiene,* 23:298–308.
133. Weidhaas, D.E., Schmidt, C.H., Seabrook, E.L. (1962). Field studies on the release of sterile males for the control of *Anopheles quadrimaculatus. Mosquito News,* 22:283–291.
134. WHO (2014). Procedures for certification of malaria elimination. *Wkly Epidemiol. Rec.*, 89:321–325.

135. Wilson, A.L., Boelaert, M., Kleinschmidt, I., Pinder, M., Scott, T.W., Tusting, L.S., Lindsay S.W. (2015). Evidence-based vector control? Improving the quality of vector control trials. *Trends Parasitol.*, 8:380–90.

136. Windbichler, N., Papathanos, P.A., Catteruccia, F., Ranson, H., Burt, A., Crisanti, A. (2007). Homing endonuclease mediated gene targeting in Anopheles gambiae cells and embryos. *Nucleic Acids Res.*, 35:5922–5933.

137. Windbichler, N., Menichelli, M., Papathanos, P.A., Thyme, S.B., Li, H., Ulge, U.Y., Hovde, B.T., Baker, D., Monnat Jr, R.J., Burt, A., Crisanti, A. (2011). A synthetic homing endonuclease-based gene drive system in the human malaria mosquito. *Nature,* 473:212–215.

138. Wise de Valdez, M.R., Nimmo, D., Betz, J., Gong, H-F., James, A.A., Alphey, L., Black IV, W.C. (2011). Genetic elimination of dengue vector mosquitoes. *Proc. Natl. Acad. Sci. USA*, 108:4772–4775.

139. Wu, B., Luo, L., Gao, X.J. (2016). Cas9-triggered chain ablation of cas9 as a gene drive brake. *Nat Biotechnol.*, (2):137–138.

140. Xu, X.S., Bulger, E.A., Gantz, V.M., Klanseck, C., Heimler, S.R., Auradkar, A., Bennett, J.B., Miller, L.A., Leahy, S., Juste, S.S., Buchman, A., Akbari, O.S., Marshall, J.M., Bier, E. (2020). active genetic neutralizing elements for halting or deleting gene drives. *Mol Cell*, 80(2):246–262.

141. Yamamoto, D.S., Sumitani, M., Hatakeyama, M., Matsuoka, H. (2018). Malaria infectivity of xanthurenic acid-deficient anopheline mosquitoes produced by TALEN-mediated targeted mutagenesis. *Transgenic Res.*, 27:51.

142. Yordanova, I.A., Zakovic, S., Rausch, S., Costa, G., Levashina, E., Hartmann, S. (2018). Micromanaging immunity in the murine host vs. the mosquito vector: Microbiota-dependent immune responses to intestinal parasites. *Front Cell Infect Microbiol*, 8:308.

143. Yuan, Y.W., Wessler, S.R. (2011). The catalytic domain of all eukaryotic cut-and-paste transposase superfamilies. *Proc Natl Acad Sci USA,* 108:7884–7889.

Section III

Gene Drive Mosquito Trials

Chapter 5

Large Cage Trials of Gene Drive Mosquitoes: Does Size Matter?

Mark Q. Benedict
Centers for Disease Control and Prevention (CDC), Division of Parasitic Diseases and Malaria, Entomology Branch, Atlanta, Georgia, USA
mbenedict@cdc.gov

5.1 Introduction

Gene drive offers a tantalizing option of increased power over more traditional genetic control methods available for control of agricultural and public health insects including techniques such as sterile insect technique (SIT) [1], incompatible insect technique (IIT) [2], combinations of [3] transgenic approaches such as late-acting lethal [4], and variations in which only male progeny persist [5]. Gene drive is expected to be more effective by creating mosquitoes that will require much smaller releases to achieve more widespread benefits.

Several teams are developing genetically modified mosquitoes with transgenes that are expected to spread in natural populations

Mosquito Gene Drives and the Malaria Eradication Agenda
Edited by Rebeca Carballar-Lejarazú

ISBN 978-981-4968-33-1 (Hardcover), 978-1-003-30877-5 (eBook)
www.jennystanford.com

via gene drive including population alteration (also called modification or replacement [6]) and population suppression [7]. The original strain development that can be accomplished by a few staff and relatively quickly in the molecular laboratory can quickly be eclipsed by the requirements to satisfy diverse time- and resource-consuming requirements before it is approved for the field: performance testing, production, regulatory, stakeholder engagement, and biosafety. As strains move along this process from the bench to the field, it is essential to determine those strains that have the best chance of satisfying these myriad requirements to avoid spending limited resources on 'products' that will not meet field release demands.

The financial and labor effort that must be committed to a strain increases as it moves from the bench to the field. One method to eliminate poor candidates has been to perform 'large-cage' or 'field-cage' trials. These allow controlled testing in venues that proponents believe are closer to a field deployment than small 'laboratory' cage trials. On the other hand, one can reasonably argue that it is impossible to replicate natural environments. Both arguments seem intuitively plausible. Are there data to support the use of large indoor or field-cage trials? How large must testing cages be to be useful? What lessons have been learned from large-cage trials, particularly in testing genetic control of mosquitoes, that are relevant for gene drive and conventional transgenic strains? Is the proper metric for a cage its volume, physical, and environmental features, or stimulation of behavioral characteristics for a cage that provides the data that is needed?

The value of gene drive will ultimately be proven in the wild, not the laboratory. Not only must the gene drive transgene continue to increase in frequency in the target insect species, but individuals that carry the transgene must exhibit the required natural behaviors, including finding mating sites, selecting mates, finding hosts, and finding resting and oviposition sites. Because the behaviors conferred by the genotype of the modified mosquito must be compatible with that of the target population, matching these can be performed by analytical genetic comparison or introgression of the modified form into the target population prior to release. Species that have well-recognized complex genetic structures such as the *Anopheles gambiae* complex [8, 9] require particularly careful matching. However, inevitable selection for insectary mating behavior during colonization itself can prevent a wholly effective match [10]; this is usually an unavoidable weakness of a genetic control program, although efforts to both measure this effect [11, 12] and ameliorate it [13] have been made.

Gene drive potentially alleviates much of this concern because the transgene copies itself at a high rate into a new, field-derived, target population chromosome of nearly every offspring every generation [14]. Therefore, the transgene is expected to very rapidly 'shed' its original genome with each generation of outcrossing and transgene duplication. While initially, poor mating competitiveness will require larger releases to achieve the same rate of population effect, this should eventually be overcome by the shedding effect, even if alleles that are unfavorable to the field are tightly linked to the transgene. Nevertheless, one should not dismiss the possibility that large effects of, for example, assortative mating might completely prevent the introduction of a gene drive.

The 'shedding' effect will not affect the driving transgene itself, which is necessary as a permanent part of the genotype. The transgene insertion may have enhancer/suppressor effects [15] that occur unexpectedly, due to the insertion of the driving transgene itself. The primary novel phenotype that is intended relative to the native target population is due to the effector itself, such as one conferring disease agent refractoriness or a population suppression effect. Transgenes also often carry a marker, most commonly a fluorescent protein [16]. Besides a marker's obvious utility, it may confer other unidentified effects. The possible effects of markers expressed in the optic lobe and abdominal ganglia such as fluorescent markers under the control of the 3XP3 promoter [16] require special consideration because the organs in which these are expressed might suffer unexpected effects on sense and behavior. Therefore, while genome shedding may reduce concerns about abnormal effects after release, other effects of transgene insertion and its marker cannot easily be dismissed.

Studies of the fitness of transgenes in mosquitoes have typically been confined to measuring life-history traits [17, 18], persistence [19, 20], or both [21] in a laboratory setting. The persistence of a transgene, particularly in over lapping generation populations, allows a greater capacity to detect a more comprehensive suite of subtle effects of a transgene than can be observed by narrow, discrete outcome tests, e.g., for mating competitiveness, fecundity, and longevity.

It is possible that insects containing gene drives will require preliminary testing of their effectiveness in the environment using a non-driving form of the effector released in large numbers ('overwhelming' or 'inundative' release) similar to more conventional genetic control methods to demonstrate safety and effectiveness prior to committing to releases of the gene drive form. While a model may predict an effect on, e.g., malaria transmission due to release of a gene drive mosquito, considerable reassurance would result if a similar population modification was first made without gene drive, and its effects on the population, and perhaps disease transmission, were determined under conditions that would assure containment.

In this chapter, the results of several trials that yielded insight on these questions and will, in the end, assert that the value of large-cage trials may have a less apparent, but demonstrable value, are described. The ideas presented here are an expansion of a blog written for the Malaria World website [22] with the addition of case studies and deeper documentation of observations.

5.2 Large-Cage Trials Are Widely Advised

As part of the phased progression of testing transgenic mosquitoes, the concept of 'large-cage trials' is often an intermediate step that is recommended.

For example:

Phase 1 is anticipated to begin with small-scale laboratory studies. . .followed by testing in larger population cages in a laboratory setting.

Phase 2 initiates confined testing in a more natural setting. . . within a large cage that simulates the disease-endemic setting [23].

Large cage experiments that mimic natural conditions must then be performed and models developed to extrapolate how the system will work following an open release [24].

. . .laboratory studies and confined field tests (or studies that mimic confined field tests such as large cage trials and green-house studies) represent the best approaches to reduce uncertainty in an ecological risk assessment [25].

The assumption has been that large cages provide a 'more realistic' venue for testing than small cages. Underpinning this expectation is perhaps the knowledge that cage trials are usually multigeneration trials that simulate the release and persistence of a transgene in overlapping generation populations that allow more complex interactions than discrete-generation small cage trials.

Recommendations for large-cage trials are based on a widely-held belief that large-cage testing is useful for more rigorous testing after the basic phenotype of the effector of interest has been confirmed in 'small-cage testing' and the strain has demonstrated its potential for further development. Once a candidate strain is chosen to move forward past early phase testing, the costs, risk exposure, and effort can be constrained by large cage testing prior to field trials. The broader programmatic objective for those conducting oversight is to eliminate candidates that show deficiencies (risk, production, or performance) that presage inadequacy upon release into the field, an activity that once committed to, is expensive and lengthy, and exposes the project to risks in public perception, to the scientific community's reputation and to the wider commentariat.

A critical distinction between 'small-cage' and 'large-cage' trials is not simply the cage size. As noted above, a distinction is also the management regime of the mosquitoes that populate the cage, and the duration over which observations are made. Large-cage trials usually consist of ongoing *population studies* conducted over several months that are difficult, but not impossible [26], in small cages. For example, a method for populating cages with 'stable age distribution' populations into which males of interest were released periodically has been developed [27, 28] and has proved useful for further studies to simulate transgene behavior in wild populations [29, 30].

In contrast, preliminary small-cage studies often consist of observations of specific outcomes under controlled, favorable environmental conditions, and uniformity of age, mating status, blood-feeding timing, number, etc. Typically, they are 'discrete generation' studies in which like-age males and females are placed in a cage and eggs are collected, often only once. Therefore, the effects of age on phenotype and life history are minimized. No interactions between adults of different ages or differences among subsequent ovipositions by the same females are possible. Such highly artificial studies provide preliminary proof of the value of further studies, but exclude the possibility of observing more comprehensive and possibly subtle effects of the transgene phenotype.

Large cage population studies necessarily have a much lower level of control, are more demanding for insectary staff and place greater stress on adult mosquitoes. Therefore, they usually provide less certain outcomes, a rather inconvenient fact when statistical analysis is needed.

In the following section, several case studies are highlighted that provide insight into the kinds of effects that can be observed as a result of large-cage studies. These will conclude with a discussion of why their value may have little to do with attempting to create a large cage natural environment.

5.3 Case Studies

The history of genetic control of mosquitoes is full of useful experience related to strain testing, and in many ways, protocols have remained similar up to the present, so much of the experience gained is still relevant. Prior to the 1990s, genetically altered mosquitoes consisted of those carrying a chromosome translocation, displaying cytoplasmic incompatibility (CI), and sterilized by chemosterilization or gamma irradiation (Table S1 in [31]). Among these, only chromosome aberrations were heritable factors that could be monitored over multiple generations; and in that regard, they are most similar to transgenes of temporally limited persistence. For various purposes, a handful of studies attempted to introduce a rare allele or phenotype into a population and to follow it over time [32–34]. The effects of traits conferring sexual sterility were immediately and directly determined by the rate of sterility, of the factor (usually nearly absolute), and the frequency of male mating. Therefore, most of the studies determined mating competitiveness, and these will occupy many of the case studies.

5.3.1 Case Study 1: Success after Only Small Cage Studies: *Culex pipiens* in Burma

Using CI, SIT got off to what seemed like an encouraging start with the near elimination of a population of '*Culex pipiens fatigans*' (*Cx. pipiens*) in Myanmar [35]. While the details of cage studies are scant, only three or four generations were required to eliminate caged populations when males were released in a 1:1 ratio (sterile:fertile) along with virgin females, so it is likely these were discrete generation experiments. These encouraging results led to field releases of the same type of males, which eliminated the target population in about 12 weeks. Although the varied CI mating types that had been observed in *Culex* spp. were not evident in *Aedes* and *Anopheles* at the time, efforts were being made to test other kinds of sterilizing methods that might accomplish similar encouraging results.

5.3.2 Case Study 2: Success after Preliminary Cage Studies: *Culex quinquefasciatus* in Florida

Another notable elimination effort to reduce or eliminate a target population was conducted after outdoor cage testing. *Culex quinquefasciatus* was chemo-sterilized and released on an island in the Gulf of Mexico near Cedar Key, Florida, USA, resulting in the elimination of the target population in one season [36]. Successful preliminary cage studies were noted, but details were unfortunately not provided. Moreover, similar smaller island releases predated this program, so that the effectiveness of the mosquito production, sterilization, and release had already been demonstrated [37]. In this case, equal mating competitiveness of chemosterilized and wild males had been observed in outdoor cages and this allowed an assumption to be made based on numbers released and the amount of sterility observed in the field, which was used to estimate the actual release ratios of sterile to wild males that were achieved.

5.3.3 Case Study 3: Failure without Cage Trials: *Aedes aegypti* in Florida

A prominent effort, but ultimately an ineffective program, was the release of 4.6 million radio-sterilized *Aedes aegypti* over 43 weeks in Pensacola, Florida, USA [38]. This was a complex release program using mosquitoes that were sterilized as pupae in Savannah, GA, USA, and transported as pupae by air to Pensacola for release. Laboratory studies in Savannah had demonstrated that a 20:1 sterile to fertile male ratio resulted in 98.5% sterility in 1 gallon (3.8 l) paper cartons [39]. According to the competitiveness calculation method of Fried [40] in which values <1 are less competitive than the comparator, this would be average competitiveness of 1.38 (a 95% confidence interval of 0.47 of the mean in trials of eight different sterile to untreated male ratios, my calculation), a super-competitiveness value that, while arguably implausible, certainly would not be worrisome.

These encouraging demonstrations had been performed with males irradiated with a dose of an unusually high variance that resulted from the physical configuration of the irradiator. The doses ranged 10.5–17.0 roentgens (equivalent to 92–150 Gy), which resulted in complete sexual sterility, however, the minimum sterilizing dose was never determined. Rather, a dose that did not kill an unacceptable number of pupae or reduce eclosion was chosen. While the males had been tested in 1 gallon paper buckets in the laboratory where they were sterilized, there was no record that similar tests were performed on the mosquitoes at the destination after transport.

Upon release into the field, no effect on the target populations was observed. The failure was attributed to male age, low dispersal, immigration, and movement of larval source containers into the release area. The irradiation itself caused a mortality rate of 10%, which is quite high. In tests of the shipping method, 10% pupal mortality was observed so the combination of irradiation and shipping undoubtedly stressed the mosquitoes that would be released.

A key issue may have been that, at least in part, the minimum radiation doses used for sterilization were twice that required to achieve full sterility [41, 42]. This likely greatly reduced mating competitiveness, a fact that was not evident in small laboratory cages but which might have been evident from large cage studies in which flight performance is more stringently tested. This experience should be compared with observations of the transgenic *Anopheles gambiae* Ag(DSM)2 and *Ae. aegypti* OX3604C strains discussed below.

5.3.4 Case Study 4: Failure without Large Cage Trials: *Anopheles gambiae* in Burkina Faso

To a modern reader who has become familiar with the *An. gambiae* species complex and observations of assortative swarming behavior [43], it may seem unthinkable that one would attempt the following experiments. However, in the late 1960s, practitioners of genetic control were certainly aware of the complex CI mating incompatibilities that were common in *Culex* spp. and that these had been used to conduct a highly successful genetic control trial [35]. Possibly encouraged by this, researchers used the observation that by crossing two members of the *An. gambiae* complex, mainly males were produced and these were sexually sterile when crossed to females of another type; this phenomenon was the basis to attempt suppression of a wild population of yet a third member of the complex [44]. No effect on the wild population was observed in spite of successful laboratory cage trials [45], and details on the sizes of the 'laboratory cages' were not provided. The authors concluded that assortative mating was strong and that few matings with the target females had occurred, and ethological effects that were apparent in the wild had not been detected in laboratory cage trials. It is likely that only in cages that stimulated and allowed swarming and possibly assortative mating would the effect have been observed.

5.3.5 Case Study 5: Indoor Cage Trials Were Encouraging but Outdoor Cage Studies in Mexico Ended Development

Oxitec Ltd. developed and tested a 'flightless female' strain of *Ae. aegypti* that was evaluated in large indoor and outdoor 'field cages.' The OX3604C strain carried a tetracycline-suppressible flightless phenotype specific to females [46]. The transgene could be passed on by transgenic male progeny which were able to fly, thus potentially extending a multi-generational suppressive effect. In large indoor cages, target populations were eliminated in 10–20 weeks, representing a promising success [27]. While there were no calculations of mating competitiveness, the appearance of transgenic progeny demonstrated mating at an unreported rate.

The OX3604C males were then tested in 6 m × 6 m × 2 m (72 m^3) outdoor cages in Mexico [47], where they were not as competitive in mating as wild-type males. While even the field-cage trial was an elaborate undertaking, its findings forewarned against large-scale field trials that would have entailed even greater commitment and likely would not have succeeded. This progressive strategy of testing saved the project from likely failure at greater cost and effort.

5.3.6 Case Study 6: *Culex tarsalis* in California, USA

A prolific team of researchers in California performed a series of field releases of sterile *Culex tarsalis* in California during the 1980s. These studies compose an interesting series of experiments during which the research team systematically attempted to suppress semi-isolated field populations, making adjustments based on previous experience and exploring colonization and larval diet issues that might reduce effectiveness.

Ainsley et al. [48] initially determined an effective dose of irradiation to sterilize males and tested the competitiveness of irradiated *carmine* eye color males against wild colonized males in a 1:1 ratio in 60 cm × 45 cm × 60 cm laboratory cages. Matings were assessed by hatch and the *carmine* phenotype of progeny. At the dose finally chosen, 46% of the matings were by irradiated males, only slightly less than being equally competitive. In the same series of experiments, competitions at a 1:1 ratio were also performed in larger outdoor field cages (5.5 m × 3 m × 6 m). Two trials with different numbers of males were performed. In both, competitiveness was excellent (calculated at ~1.1) until after the first 4 or 5 days when competitiveness decreased to 0.63 and 0.38. Allowing the additional ovipositions revealed an effect of age that had not been detected in small cages. The authors speculated that the decline could be due to the number of ovipositions by females that were mated to either type of male, differential male longevity, or a possible diapause interaction.

These experiments were the basis for field trials to introduce sterility into wild populations, or hopefully suppression if favorable results were obtained. Prior experience had indicated that colonized males were rapidly selected for laboratory mating behavior [49], so to avoid this, males were collected from a prolific larval site in the area, and irradiated and released. The first experiment achieved measurable sterility in the field populations [50] and this effort was scaled up the following year, again using wild-caught males. These were found to mate competitively, but reliance on collections of wild males was admittedly not a tenable approach to a large-scale suppression program. No effect of population suppression was observed, due possibly to simply releasing too few males.

In these experiments, the combination of field and laboratory experience allowed the research team to understand why males were performing poorly and to adjust accordingly. The conclusion that the colonization process itself selected for males that competed poorly in the field led to subsequent studies to attempt to identify ways to mitigate this effect. In the end, one set of measures used to create more natural environments in the laboratory had a little positive effect on maintaining genetic heterogeneity [13].

The initial observation that males competed well in the laboratory was based on selection for strains that mated well in the laboratory! This serves as a caution for those who are colonizing mosquitoes that mate assortatively in swarms in the wild but are forced to adapt to mating in small cages in the laboratory where they may not.

5.3.7 Case Study 7: *Anopheles gambiae* Ag(DSM)2

The case of cage size affecting mating competitiveness has been demonstrated [51] for a transgenic strain in which males are rendered sterile by a homing endonuclease that cuts the X chromosome in gametes of both sexes. The initial evaluations of competitiveness demonstrated equal rates in small cages [52]. However, further evaluation in larger cages demonstrated that mating competitiveness was decreased with increasing cage size, but this was not affected by whether swarming was observed [51].

According to Reisen et al. [49], a reasonable hypothesis is that laboratory strains will preferably be colonized under conditions that allow swarming in order to make mating rate projections that are realistic from large cage trials.

5.3.8 Case Study 8: *Anopheles albimanus* in El Salvador

A technique that has gone out of fashion but which was common for many of the previous century's field trials was to release marked virgin females along with the males to be tested. This provides a large number of identifiable females from which competitive mating data can be obtained; the method was used to test chemosterilized *An. albimanus* males in El Salvador [53]. The team recognized the conundrum of comparing laboratory results using laboratory-adapted strains to results that would be obtained in the field. Their methods and results were unusually illuminating.

Sterile males were released into the field at a ratio of 5:1:1 (sterile male:wild male:virgin female) and captured for sterility analysis. The 'wild males' were from a strain called CAMPO, which had been specially reared to maintain field characteristics by adding field material daily; in fact, it was reportedly not well adapted to the laboratory. Competitiveness of the sterile males in the field was calculated to be 0.785.

Cage studies were conducted similarly but with a 1:1:1 ratio. In this case, the sterile, laboratory adapted males outperformed the wild-type CAMPO males and were more competitive than the wild males, and had competitiveness of 1.36. This outcome is consistent with the unsurprising assertion that mosquitoes will perform well in environments to which they have been adapted.

This finding echoes the *Cx. tarsalis* findings and illustrates the fact that laboratories located in areas where there is not ready access to vector populations will be forced to conduct cage trials with mosquitoes that probably do not represent the behavior of the target populations. When evaluating strains in the insectary, field-collected mosquitoes of many species – the most realistic comparator – would fail every quality test conducted in the laboratory. Field-collected mosquitoes will seldom mate, and often will not blood-feed [54] or oviposit. These issues are likely more important for twilight-swarming, highly nocturnal species such as *Culex* and *Anopheles* than *Aedes* spp [55].

5.4 Colonization Considerations

Genetic evidence that colonization diminishes genetic variation [11, 12] exists as does evidence that this effect can sometimes be mitigated. In trials of Oxitec OX3604A (a tetracycline repressible flightless female strain) and dengue-resistant transgenic strains in which a 'Genetically Diverse Laboratory Strain' was created by intercrossing multiple independent strains [27, 56]. Although such an approach may increase heterogeneity and is useful to some extent, all of the donor strains are still laboratory-adapted strains that may have undergone similar selection processes and are similarly diverged in important behavioral and physiological traits from field populations.

Perhaps a good rule of thumb is that the more easily a species can be colonized, the more likely laboratory colonies are to meaningfully predict field performance. That does not mean that cage studies with other species are not meaningful, but rather that for species that are difficult to colonize, the effect of the transgene or other effector is being compared in a very narrow sense with a comparator that is closely matched. The fact that both easily colonized and those that are not will have behaviors, genotypes, and physiology that are laboratory-adapted must be considered.

It follows that the most meaningful predictive cage trials will be conducted in settings that are conducive to all of the natural behaviors and have ready access to wild mosquitoes to refresh the 'wild type' colony used for the cage trials. A good practice is to transition testing from laboratories that do not have these capabilities to those that do early in the testing and development process.

Efforts to elicit and retain natural behaviors with existing laboratory stocks can also be made. It has previously been observed that even a very old laboratory stock of *An. gambiae* has not lost the ability to swarm [51], though routine stock-keeping methods do not permit swarming and the frequency of this behavioral capacity in the laboratory stock is unknown. Various efforts have been made to equip large cages with resting shelters [28, 30, 57], swarming stimuli [51, 58], and environmental complexity [59] in order to create more natural environments, but these scarcely simulate nature. The conditions should preferably elicit assortative mating. However, even if swarming does not account for differences related to cage size, cage size itself is a stressor that promotes more sensitive discrimination of deleterious effects. I am aware of only one stock in Africa that is maintained in large outdoor cages that would allow natural mating behavior related to both cage size, lighting, and ground features to be maintained. Therefore, stock development itself seriously hobbles meaningful testing of field mating performance but is still useful for demonstrating the integrity of the phenotype of most interest, e.g., pathogen refractoriness or sex ratio distortion.

The limitations of laboratory research lead to what might be called the 'colonization conundrum.' Research teams typically create strains predominated by unnatural behavior and evaluate their quality by measuring, e.g., mating, egg hatching rate, blood feeding, and oviposition, at which none of the target population would be successful due to the laboratory environment. Yet, we hope to make meaningful predictions about transgenes in the field using these artificial strains.

There also admittedly exists a logical flaw in the progression of testing from small to large indoor and outdoor cages and finally to the field. Failure in small cages disqualifies strains from further testing in larger cages, and we do not know in fact that failure at the large cage stage precludes success in the field since that failure stops further development. However, as described in some of the cases above, cage trials can often identify specific deficiencies that cannot be overcome by releasing mosquitoes in more natural environments. For example, females that consistently have low egg-hatching rates due to a transgene will not be likely to perform better in the field. When cage studies are conducted, it is worth considering which physiological, genetic, and behavioral qualities are likely to differ between the laboratory and field and to what extent these will affect the success of releases.

5.5 Defining 'Large'

Consistency of phenotypes such as *Plasmodium* refractoriness, sexual sterility, or sex bias in different cage testing venues seems likely because these are determined by relatively simple interactions between one or a few effector molecules and specific organisms or genes. To address the effect of environmental conditions such as temperature that might affect expression, experiments can be conducted in incubators to determine to what extent a single well-defined parameter might affect the phenotype. Nevertheless, the consistency of those traits has no bearing on the fate of a transgene in mixed populations. Less easily observed, but highly relevant biological characteristics affecting the persistence or spread of a transgene might be affected by cage size. These are reasonable concerns because, as described above, usually small-cage testing does not involve overlapping 'stable' populations in which the age distribution and physiological state of adults are mixed, and outcomes are often constrained by the experimental design. The objective of most large-cage trials is primarily not to assess the phenotype of the effector or transference rate of a transgene to their progeny; it is to assess the effectiveness (phenotype, frequency) of the modified mosquito in a population context. These are very different. The latter necessitates all of the natural behaviors that influence the fitness of the transgenic mosquito. The time variable is usually weeks or months rather than generations, which are unknown.

In the following section, arguments for why 'large-cage' testing makes sense will be presented. These arguments are consistent with biology, experience, or reason. Readers can consider whether they are relevant concerns for their target species based on their experience.

But, what in fact is 'large'? The range of sizes called 'large' that are used for mosquito trials is informative. Large outdoor cages were 6 m × 6 m × 2 m (72 m^3) for outdoor trials of the OX3604C *Ae. aegypti* strain in Mexico [60]. Indoor large cage for evaluating an *An. gambiae* transgenic strain was 5.0 m × 1.2 m × 2.6 m (15.6 m^3) [47]. Smaller 1.75 m^2 footprint cages in a greenhouse (probably about 3.5 m^3) were used to examine the competitiveness of irradiated *Anopheles coluzzii* males [61]. In spite of the range of sizes, all were described by the authors as 'large'. Perhaps the best way to frame the question for choosing cages and conditions is, "Which elicits a wider range of activities and stresses that could not have been obtained in typical small laboratory cages?"

In order to answer this question, it is enlightening to consider the natural environment where the target species occurs. Even the most favorable natural environments present numerous challenging conditions that are not usually present in the laboratory: temperature and humidity variation; resting sites that are home to predators and spider webs; blood meal sources that are mobile and at least somewhat defensive and protected; mating sites that must be found and suitable mates selected; oviposition sites that need to be located, etc. One cannot hope to create all of these challenges in any but the largest outdoor cages, likely located where the species is endemic [62]. Nonetheless, to the extent that large cages elicit a wider range of behaviors than occur in small cages and create conditions that are more similar to natural environments, they can reveal deficiencies that might raise concerns about mosquito behavior and life history in nature.

If we wished to define behaviors, conditions, and interactions not needed for success in small cages and therefore to complement those studies with useful information, what must be elicited to make the experiments relevant? Following are some characteristics that we and others have observed differ between 'large' and 'small' cages when testing *An. gambiae* or which can be manipulated, though many of the same issues are likely relevant to *Aedes*.

Locating sugar: Sugar water is easy to locate in small cages because during normal flight activity, it is encountered by chance; smelling and searching are not necessary. However, the most commonly used sugars for feeding in the laboratory, sucrose and glucose, are difficult for adults to find in large cages. Why? They have no fragrance, the cue that adults use to locate it. We have found that in large cages in order to promote sugar feeding, one must add an attractant such as honey [58], or verbenone to the sugar meal [63], or the adults die within a few days. Plants that provide sugar (*Tithonia diversifolia* and *Senna didymobotrya*) have also been used and are a more natural option [57, 63].

Swarm mating: Mating is the initial and vital interface between released mosquitoes and the wild population; it is an essential component of any genetic control method. Provisioning for swarm mating is particularly important for *Anopheles* and *Culex* mosquitoes that employ this behavior. Even for *Ae. aegypti* that mate around a host, the presence of a real or simulated host might elicit a more natural behavior. For *An. gambiae* and *Anopheles funestus,* swarming behavior has been elicited in both relatively small [64, 65] and large cages [51, 57] with special lighting provisions.

Mating competition: We have observed that cage size has a strong effect on mating competitiveness [51]. While initial testing in small cages showed no effect of a transgene on competitiveness [52], testing in large cages later demonstrated strong inferiority.

Assortative mating: There exists a report of assortative mating in 5-l cages [66], but we have observed that in small laboratory cages, members of the *An. gambiae* complex readily mate with one another, an event that rarely occurs in nature [43]. Some cage sizes and designs could be sufficient to allow for assortative mating because differences in swarm markers have been observed, and further efforts to understand what is necessary to elicit this behavior would be useful.

Locating oviposition sites: We observed in large cage experiments in Italy that females could not find the same oviposition dishes that we used in small cages; they required special illumination or eggs were laid on the reflective aluminum floor. While special illumination is not 'natural', this observation indicates that the ease of finding oviposition sites in small cages does not translate to larger cages and requires visual and water-sensing capabilities that are less important in small cages. The use of 'aged water' has also been implemented in large cages to attract females to oviposition sites that were small relative to the entire space [57].

Resting site seeking and use: Small cages almost never require an adult to locate a suitable site for resting and adults are forced to rest on the surfaces of the cage. Usually, there is no consideration for providing any preferred structure although finding suitable resting sites is essential for wild mosquitoes. It requires them to sense humidity, light, and temperature, and to move there at the correct time. In the absence of that, they are subject to predation and desiccation. Wet clay pots [58] and brick and clay tile shelters [28] effectively attract adults and when supplied with water by soaking or providing a wet sponge to create a humid and dark shelter that is quite attractive to *An. gambiae s.l.* When these can be viewed from the outside, they also provide a useful tool to estimate the population size since the vast majority of adults reside in these shelters during daylight [28, 67].

Variable temperature and humidity: Even indoors, few studies attempt to vary the temperature and humidity even though this is easily done with most modern programmable control systems. The stresses of flying and the need for sugar water and resting sites are strongly affected by temperature and humidity. Variations in temperature and humidity have been provided in some experimental systems [28, 57].

Locating blood sources: Host seeking is stimulated by a combination of heat, odors, and CO_2 [68], and the use of these different cues differs in relation to distance from the host [69]. A female's ability to locate blood, using even only one of these, is usually adequate in a small cage. Large cages require longer-distance sensing than small ones and the addition of scent such as a dirty sock [70]. Human skin scent on a membrane and CO_2 in the vicinity of the blood source is helpful to attain high blood-feeding rates in large cages.

Spatial complexity: An interior with no interrupting structure is typical for insectary cages. Spatial complexity – objects, plants, and partitions – increases the need for adults to navigate around obstacles, follow less direct plumes to locate blood, sugar, and oviposition sites, provides alternative resting places and, in certain cases, may stimulate swarms. Complexity presents a more difficult environment that necessitates a greater sensory range and more complex behaviors. It also requires more knowledge on the part of the research team regarding how to provide sensory cues that will stimulate natural behavior.

Flight distance: In spite of the direct relationship of 'large' to this parameter, it is perhaps the most difficult to objectively parameterize. The typical energy reserve replenishment of wild mosquitoes may not be tested in cages that are too small. However, it was observed that OX3604C transgenic males had inferior flight performances compared to their wild-type counterparts [71] and failed to eradicate populations of *Ae. aegypti* in large cages in the field [47], due to their poor mating capabilities. This was possibly due to reduced flight capacity that resulted in poor mating performance in large cages. However, it is unclear what dimension challenges mosquito flight. Even relatively small 'large cages' (9.1 m^3) seem to have different demands on energetics, and the researchers question the relevance of life history observations in small cages [63].

This is a shortlist of attributes of large-cage testing that are far less important in small cages; similar issues have been articulated previously [58]. There are certainly more subtle ones, but even this list strongly supports the utility of varied large-cage studies.

5.6 Conclusion

While one can debate whether large-cage testing provides anything 'more natural', it provides something which should be standard in transgenic mosquito development: independent trials of transgene performance under radically different conditions that elicit a wide range of behaviors. If so-called 'independent' trials are performed under similar controlled conditions, for example, in different laboratories using the same experimental design, lighting, temperature, and humidity, their value is questionable. Disparate trials that confirm the expected effects build our confidence that the salient behaviors will be robustly reproduced under an even wider variety of natural conditions. Performing trials in many different conditions that evoke a wider range of behaviors that will be exercised and conditions that will be experienced in natural settings approaches providing the answer that is desired; whether the mosquitoes will satisfy expectations upon release.

References

1. Lees RS, Gilles JR, Hendrichs J, Vreysen MJ, Bourtzis K. Back to the future: The sterile insect technique against mosquito disease vectors. *Curr Opin Insect Sci* [Internet]. 2015; 10:156–62. Available from: http://dx.doi.org/10.1016/j.cois.2015.05.011
2. Bourtzis K, Dobson SL, Xi Z, Rasgon JL, Calvitti M, Moreira LA, et al. Harnessing mosquito-*Wolbachia* symbiosis for vector and disease control. *Acta tropica*. 2014; 132:S150–163.
3. Dimopoulos G. Combining sterile and incompatible insect techniques for *Aedes albopictus* suppression. *Trends Parasitol*. 2019; 35:671–3.
4. Phuc HK, Andreasen MH, Burton RS, Vass C, Epton MJ, Pape G, et al. Late-acting dominant lethal genetic systems and mosquito control. *BMC Biol* [Internet]. 2007; 5:11–1. Available from: http://bmcbiol.biomedcentral.com/articles/10.1186/1741-7007-5-11
5. Oxitec's 2nd generation technology will be used across mosquito and agriculture pest applications globally. www.oxitec.com [Internet]. 2018. Available from: https://www.oxitec.com/en/news/oxitec-transitioning-friendly-self-limiting-mosquitoes-to-2nd-generation-technology-platform-paving-way-to-new-scalability-performance-and-cost-breakthroughs
6. Pham TB, Phong CH, Bennett JB, Hwang K, Jasinskiene N, Parker K, et al. Experimental population modification of the malaria vector mosquito, Anopheles stephensi. *PLoS Genet*. 2019; 15:e1008440.

7. Kyrou K, Hammond AM, Galizi R, Kranjc N, Burt A, Beaghton AK, et al. A CRISPR–Cas9 gene drive targeting *doublesex* causes complete population suppression in caged *Anopheles gambiae* mosquitoes. *Nature* [Internet]. 2018; 36:1062–6. Available from: http://www.nature.com/doifinder/10.1038/nbt.4245
8. Coetzee M, Hunt RH, Wilkerson R, Torre della A, Coulibaly MB, Besansky NJ. Anopheles coluzzii and *Anopheles amharicus*, new members of the *Anopheles gambiae* complex. *Zootaxa*. 2013; 3619:246–74.
9. Davidson G. *Anopheles gambiae*, a complex of species. Bull WHO. 1964; 31:625–34.
10. Reisen WK, Milby MM, Asman SM, Bock ME, Bock F, Meyer RP, et al. Attempted suppression of a semi-isolated *Culex tarsalis* population by the release of irradiated males: A second experiment using males from a recently colonized strain. *Mosquito News*. 1982; 42:565–75.
11. Norris DE, Shurtleff AC, Touré YT, Lanzaro GC. Microsatellite DNA polymorphism and heterozygosity among field and laboratory populations of *Anopheles gambiae s.s.* (Diptera: Culicidae). *J Med Entomol*. 2001; 38:336–40.
12. Azrag RS, Ibrahim K, Malcolm C, Rayah EE, Sayed el B. Laboratory rearing of *Anopheles arabiensis*: Impact on genetic variability and implications for sterile insect technique (SIT) based mosquito control in northern Sudan. *Malaria J*. 2016; 15:432–8.
13. Asman SM, Knop NF, Blomquist RE. Preliminary studies to identify selection factors in the laboratory colonization of Culex tarsalis. *J Florida Anti-Mosq Assoc*. 1983; 54:16–21.
14. Sinkins SP, Gould F. Gene drive systems for insect disease vectors. *Nat Rev Genet*. 2006; 7:427–35.
15. O'Brochta DA, Pilitt KL, Harrell RA, Aluvihare C, Alford RT. Gal4-based enhancer-trapping in the malaria mosquito *Anopheles stephensi*. *G3 (Bethesda)*. 2012; 2:1305–15.
16. Berghammer AJ, Klingler M, Wimmer EA. A universal marker for transgenic insects. *Nature*. 1999; 402:370–1.
17. Hun L, Luckhart S, Riehle MA. Increased *Akt* signaling in the fat body of *Anopheles stephensi* extends lifespan and increases lifetime fecundity through modulation of insulin-like peptides. *J Insect Physiol*. 2019; 103932.
18. Irvin N, Hoddle MS, O'Brochta DA, Carey B, Atkinson PW. Assessing fitness costs for transgenic *Aedes aegypti* expressing the GFP marker and transposase genes. *Proc Nat Acad Sci USA*. 2004; 101:891–6.

19. Catteruccia F, Godfray HCJ, Crisanti A. Impact of genetic manipulation on the fitness of *Anopheles stephensi* mosquitoes. *Science*. 2003; 299:1225–7.
20. Paton D, Underhill A, Meredith J, Eggleston P, Tripet F. Contrasted fitness costs of docking and antibacterial constructs in the EE and EVida3 strains validates two-phase *Anopheles gambiae* genetic transformation system. 2013; 8:e67364–10. Available from: https://dx.plos.org/10.1371/journal.pone.0067364
21. Li C, Marrelli MT, Yan G, Jacobs-Lorena M. Fitness of transgenic *Anopheles stephensi* mosquitoes expressing the SM1 peptide under the control of a vitellogenin promoter. *J Heredity*. 2008; 99:275–82.
22. Benedict MQ. Size matters: Do "large cages" for mosquito trials have any objectively meaningful dimensions? [Internet]. 2019 [cited 2020 Dec 20]. Available from: https://malariaworld.org/blog/size-matters-do-"large-cages-"mosquito-trials-have-any-objectively-meaningful-dimensions
23. WHO-TDR, FNIH. Guidance framework for testing of genetically modified mosquitoes. Geneva: WHO. 2014.
24. James AA. Gene drive systems in mosquitoes: Rules of the road. *Trends Parasitol* [Internet]. 2005; 21:64–7. Available from: http://linkinghub.elsevier.com/retrieve/pii/S1471492204002909
25. Committee on Gene Drive Research in Non-Human Organisms: Recommendations for Responsible Conduct, Board on Life Sciences, Division on Earth and Life Studies, National Academies of Sciences, Engineering, and Medicine. Gene Drives on the Horizon: Advancing Science, Navigating Uncertainty, and Aligning Research with Public Values. 2016.
26. Valerio L, North A, Collins CM, Mumford JD, Facchinelli L, Spaccapelo R, et al. Comparison of model predictions and laboratory observations of transgene frequencies in continuously-breeding mosquito populations. *Insects*. 2016; 7:47.
27. Wise de Valdez MR, Nimmo D, Betz J, Gong H-F, James AA, Alphey L, et al. Genetic elimination of dengue vector mosquitoes. *Proc. Natl. Acad. Sci. U.S.A.* 2011; 108:4772–5. Available from: http://www.pnas.org/cgi/doi/10.1073/pnas.1019295108
28. Klein TA, Windbichler N, Deredec A, Burt A, Benedict MQ. Infertility resulting from transgenic I-PpoI male *Anopheles gambiae* in large cage trials. *Pathog Glob Health*. 2012; 106:20–31.
29. Facchinelli L, North AR, Collins CM, Menichelli M, Persampieri T, Bucci A, et al. Large-cage assessment of a transgenic sex-ratio distortion strain

on populations of an African malaria vector. *Parasites & Vectors*. 2019; 12:137.

30. Pollegioni P, North AR, Persampieri T, Bucci A, Minuz RL, Groneberg DA, et al. Detecting the population dynamics of an autosomal sex ratio distorter transgene in malaria vector mosquitoes. *J Appl Ecol*. 2020; 57:2086–2096.
31. Benedict MQ. Sterile insect technique: Lessons from the past. *J Med Entomol*. 2021;58(5):1974–1979.
32. Reisen WK, Bock ME, Milby MM, Reeves WC. Attempted insertion of a recessive autosomal gene into a semi-isolated population of Culex tarsalis (Diptera: Culicidae). *J. Med. Entomol*. 1985; 22(3): 250–260.
33. Hanson SM, Mutebi JP, Craig GB, Novak RJ. Reducing the overwintering ability of *Aedes albopictus* by male release. *J Am Mosquito Control Assoc*. 1993; 9:78–83.
34. Lorimer N. Long-term survival of introduced genes in a natural population of *Aedes aegypti* (L.) (Diptera: Culicidae). *Bull Entomol Res*. 1981; 71:129–32.
35. Laven H. Eradication of *Culex pipiens fatigans* through cytoplasmic incompatibility. *Nature*. 1967; 216:383–4.
36. Patterson RS, Weidhaas DE, Ford HR, Lofgren CS. Suppression and elimination of an island population of *Culex pipiens quinquefasciatus* with sterile males. *Science*. 1970; 168:1368–70.
37. Patterson RS, Ford HR, Lofgren CS, Weidhaas DE. Sterile males: Their effect on an isolated population of mosquitoes. *Mosquito News*. 1970; 30:23–7.
38. Morlan HB, McCray EM Jr, Kilpatrick JW. Field tests with sexually sterile males for control of *Aedes aegypti*. *Mosquito News* [Internet]. 1962; 22:295–300. Available from: http://www.osti.gov/scitech/ biblio/4053382
39. McCray EM Jr, Jensen JA, Schoof HF. Cobalt-60 sterilization studies with *Aedes aegypti* (L.). *Mosquito News*. 1961; 110–115.
40. Fried M. Determination of sterile-insect competitiveness. *Journal of Economic Entomology*. 1971; 64:869–872.
41. Hallinan E, Rai KS. Radiation sterilization of *Aedes aegypti* in nitrogen and implications for sterile male technique. *Nature*. 1973; 244:368–369.
42. Yamada H, Maïga H, Bimbilé Somda NS, Carvalho DO, Mamai W, Kraupa C, et al. The role of oxygen depletion and subsequent radioprotective effects during irradiation of mosquito pupae in water. *Parasites & Vectors*. 2020; 13:198–10.

43. Sawadogo PS, Namountougou M, Toé KH, Rouamba J, Maïga H, Ouédraogo KR, et al. Article In Press. *Acta tropica*. 2013; 1–11.
44. Davidson G, Odetoyinbo JA, Colussa B, Coz J. A field attempt to assess the mating competitiveness of sterile males produced by crossing 2 members of the *Anopheles gambiae* complex. Bull WHO. 1970; 42:55–67.
45. Davidson G. Genetical control of Anopheles gambiae. 1969; 7: 151–4. Available from: https://www.cabdirect.org/cabdirect/abstract/19721000392
46. Fu G, Lees RS, Nimmo D, Aw D, Jin L, Gray P, et al. Female-specific flightless phenotype for mosquito control. *Proc Nat Acad Sci USA*. 2010; 107:4550–4554.
47. Facchinelli L, Valerio L, Ramsey JM, Gould F, Walsh RK, Bond G, et al. Field cage studies and progressive evaluation of genetically-engineered mosquitoes. 2013; 7:e2001. Available from: http://dx.plos.org/10.1371/journal.pntd.0002001.s011
48. Ainsley RW, Asman SM, Meyer RP. The optimal radiation dose for competitive males of *Culex tarsalis* (Diptera: Culicidae)1. *J Med Entomol*. 1980; 17:122–125.
49. Reisen WK, Sakai RK, Baker RH, Rathor HR, Raana K, Azra K, et al. Field competitiveness of *Culex tritaeniorhynchus* giles males carrying a complex chromosomal aberration: A second experiment. *Env Entomol*. 1980; 73:479–484.
50. Asman SM, Zalom FG, Meyer RP: A field release of irradiated male *Culex tarsalis* in California. Proceedings and papers of the Forty-eighth Annual Conference of the California Mosquito and Vector Control Association, Inc. 1980; 1–2.
51. Facchinelli L, Valerio L, Lees RS, Oliva CF, Persampieri T, Collins CM, et al. Stimulating *Anopheles gambiae* swarms in the laboratory: Application for behavioral and fitness studies. *Malaria J*. 2015; 14:271.
52. Windbichler N, Papathanos PA, Crisanti A. Targeting the X chromosome during spermatogenesis induces Y chromosome transmission ratio distortion and early dominant embryo lethality in *Anopheles gambiae*. *PLoS Genet*. 2008; 4:e1000291.
53. Kaiser PE, Bailey DL, Lowe RE, Seawright JA, Dame DA. Mating competitiveness of chemosterilized males of a genetic sexing strain of Anopheles albimanus in laboratory and field tests. Free Download, Borrow, and Streaming: Internet Archive. 1979; 39:768–75. Available from: https://archive.org/details/cbarchive_117430_matingcompetitivenessofchemost1979/page/n1/mode/2up

54. Armstrong JA, Bransby-Williams WR. The maintenance of a colony of Anopheles gambiae, with observations on the effects of changes in temperature. Bull WHO. 1961; 24:427–35.

55. Ng'habi KR, John B, Nkwengulila G, Knols BGJ, Killeen GF, Ferguson HM. Effect of larval crowding on mating competitiveness of Anopheles gambiae mosquitoes. *Malaria J.* [Internet]. 2005; 4:49. Available from: http://www.malariajournal.com/content/4/1/49

56. Franz AWE, Sanchez-Vargas I, Raban RR, Black WC, James AA, Olson KE. Fitness impact and stability of a transgene conferring resistance to dengue-2 virus following introgression into a genetically diverse *Aedes aegypti* strain. *PLoS Negl Trop Dis.* 2014; 8:e2833.

57. Jackson BT, Stone CM, Ebrahimi B, Briet OJT, Foster WA. A low-cost mesocosm for the study of behavior and reproductive potential in Afrotropical mosquito (Diptera: Culicidae) vectors of malaria. *Med Vet Entomol* [Internet]. 2014. Available from: http://doi.wiley.com/10.1111/mve.12085

58. Stone CM, Taylor RM, Foster WA. An effective indoor mesocosm for studying populations of *Anopheles gambiae* in temperate climates. *J Am Mosquito Control Assoc.* 2009; 25:514–6.

59. Knop NF, Asman SM, Reisen WK, Milby MM. Changes in the biology of *Culex tarsalis* (Diptera: Culicidae) associated with colonization under contrasting regimes. *Env Entomol.* 1987; 16:405–14.

60. Facchinelli L, Valerio L, Bond JG, Wise de Valdez MR, Harrington LC, Ramsey JM, et al. Development of a semi-field system for contained field trials with *Aedes aegypti* in Southern Mexico. *Am J Trop Med Hyg.* 2011; 85:248–56.

61. Maïga H, Damiens DD, Niang A, Sawadogo SP, Fathcrhaman O, Lees RS, et al. Mating competitiveness of sterile male *Anopheles coluzzii* in large cages. *Malaria J.* 2014; 13:460.

62. Knols BG, Njiru BN, Mathenge EM, Mukabana WR, Beier JC, Killeen GF. MalariaSphere: A greenhouse-enclosed simulation of a natural *Anopheles gambiae* (Diptera: Culicidae) ecosystem in western Kenya. *Malaria J.* 2002; 1:1–13.

63. Stone CM, Hamilton IM, Foster WA. A survival and reproduction trade-off is resolved in accordance with resource availability by virgin female mosquitoes. *Animal Behavior.* 2011; 81:765–74.

64. Marchand RP. A new cage for observing mating behavior of wild *Anopheles gambiae* in the laboratory. *J Am Mosquito Control Assoc.* 1985; 1:1–3.

65. Peloquin JJ, Asman SM. Use of a modified Marchand cage to study mating and swarming behavior in *Culex tarsalis*, with reference to colonization. *J Am Mosquito Control Assoc.* [Internet]. 2002; 4:516–9. Available from: http://eutils.ncbi.nlm.nih.gov/entrez/eutils/elink.fcgi?dbfrom=pubmed&id=3225570&retmode=ref&cmd=prlinks

66. Aboagye-Antwi F, Alhafez N, Weedall GD, Brothwood J, Kandola S, Paton D, et al. Experimental swap of Anopheles gambiae's assortative mating preferences demonstrates key role of x-chromosome divergence island in incipient sympatric speciation. *PLoS Genet.* 2015; 11:e1005141–19.

67. Stone CM, Taylor RM, Roitberg BD, Foster WA. Sugar deprivation reduces insemination of Anopheles gambiae (Diptera: Culicidae), despite daily recruitment of adults, and predicts decline in model populations. *J Med Entomol.* 2009; 46:1327–37.

68. Takken W, Verhulst NO. Host preferences of blood-feeding mosquitoes. *Annu. Rev. Entomol.* 2013; 58:433–53.

69. Takken W. The role of olfaction in host-seeking of mosquitoes: A review. *Int J Trop Insect Sci.* 1991; 12:287–95.

70. Facchinelli L, North AR, Collins CM, Menichelli M, Persampieri T, Bucci A, et al. Large-cage assessment of a transgenic sex-ratio distortion strain on populations of an African malaria vector. *Parasites & Vectors.* 2019; 12:1–14.

71. Bargielowski I, Kaufmann C, Alphey L, Reiter P, Koella J. Flight performance and teneral energy reserves of two genetically-modified and one wild-type strain of the yellow fever mosquito *Aedes aegypti*. *Vector Borne Zoonotic Dis.* 2012; 12:1053–8.

Chapter 6

Field Trial Site Selection for Mosquitoes with Gene Drive: Geographic, Ecological, and Population Genetic Considerations

Gregory C. Lanzaro,[a] Melina Campos,[a] Marc Crepeau,[a] Anthony Cornel,[a] Abram Estrada,[a] Hans Gripkey,[a] Ziad Haddad,[a,b] Ana Kormos,[a] and Steven Palomares[a]

[a] *Vector Genetics Laboratory, Department of Pathology, Microbiology and Immunology, School of Veterinary Medicine, University of California, Davis, California, USA*
[b] *California Institute of Technology, Jet Propulsion Laboratory, Pasadena, California, USA*
gclanzaro@ucdavis.edu

6.1 Introduction

The purpose of this chapter is to outline the framework used by the University of California Malaria Initiative for the selection of sites for conducting field trials of genetically engineered mosquitoes (GEMs) designed for the control of malaria in sub-Saharan Africa. The framework is articulated within the nine sub-sections of this chapter, each informing the next and together, offering a comprehensive and intentional approach to a complex issue.

Mosquito Gene Drives and the Malaria Eradication Agenda
Edited by Rebeca Carballar-Lejarazú

ISBN 978-981-4968-33-1 (Hardcover), 978-1-003-30877-5 (eBook)
www.jennystanford.com

What is evaluated in a GEM field trial?

The GEM approach aims to develop a malaria control strategy that is safe, cost-effective, sustainable, and equitable. To achieve this goal, a GEM includes effector genes capable of reducing malaria infections below a threshold required to sustain pathogen transmission. This may be achieved by either reducing mosquito numbers (population suppression) or by engineering mosquitoes that are unable to transmit the parasite without eliminating the mosquito (population modification).

The second component of a GEM is an efficient gene drive. The gene drive serves two critical purposes: to successfully establish the effector genes at high frequency in the target population into which it is initially introduced, and to facilitate its spread into neighboring populations via normal mosquito dispersal and gene flow. The second aspect requires a so-called low-threshold gene drive, meaning one with a maximum capability for spreading across the environment (invasiveness). Invasiveness is an essential feature toward achieving sustainability, effectiveness, and equity where the intent is malaria control on a large spatial scale and at low cost. If these are the minimal characteristics required for an effective final product, then it follows that the GEM to be evaluated in a field trial should meet them.

One difficulty in applying highly invasive low-threshold drives is their ability to spread beyond the boundaries designated for their use, posing post-release surveillance and regulatory challenges. It has been suggested that field trials can progress in a stepwise fashion with early trials using high-threshold drives, such as split-drive systems with limited invasiveness [26, 85]. Threshold-dependent drives have their place in controlling vectors on a small spatial scale, such as in urban settings [72] or as laboratory tools to assist final GEM product development [74]. However, deploying a high threshold drive to achieve disease control in the real world would require repeated releases for even small-scale regional control. This would drive up the cost and reduce sustainability; therefore, they are simply not a viable option for large-scale malaria control in sub-Saharan Africa [58]. In addition, ideal field sites for GEM trials are a limited resource, and characterizing such sites is a time-consuming and costly endeavor, as we outline in this chapter. We, therefore, conclude that they should be used to evaluate GEMs that possess the characteristics of the final product. Furthermore, the issue of GEM invasion into undesired locations outside the field trial target site can be managed by selecting the appropriate site. Henceforth, when we refer to GEM we mean a GEM with a low threshold, highly invasive gene drive.

Initial considerations: First things first

Our initial focus is on defining biological and physical characteristics that would make a site ideal, or as near to ideal, as possible. Recent discussions concerning field site selection have focused on regulatory and engagement issues [66, 88, 91]. Ethical, social, and legal issues are critically important, and no field test can be undertaken before these are addressed. However, we follow earlier published recommendations that the first consideration toward identifying sites should be based on biological, geographic, and geologic attributes [5, 65, 99]. These factors are critical to identifying the best sites. There is little point in spending time and resources addressing the ethical, social, and legal aspects of conducting a field trial if the site itself cannot be justified on well-reasoned scientific grounds. Here we describe a set of criteria that may be applied through thoughtful consideration and assessment of potential field trial sites. When completed, this framework should provide a cogent justification for why a particular site was selected.

6.2 Defining the Goal

We define the primary goal of field trial site selection as the identification of a site(s) that maximizes the prospects for success, minimizes risk, and serves as a fair, valid, and convincing test of the efficacy and impacts of the GEM being evaluated.

The purpose of a field trial is to describe the behavior of GEM when introduced into a natural population of the target species, *Anopheles gambiae* and/or its sister species *A. coluzzii*. These two species are identified as targets because they are among the most important vectors of human malaria in Africa, have been the focus of GEM production, and their development is close to complete.

A protocol for the development and evaluation of GEMs for disease control has been outlined by the World Health Organization (WHO) [116]. This protocol has been widely endorsed [2, 86] and serves as the foundation for the field site selection framework described here. The protocol proceeds through PHASES, usually depicted as a linear, non-overlapping series of activities moving from laboratory studies to deployment and post-implementation surveillance. We prefer thinking of these PHASES as overlapping, interactive, and iterative with information garnered in one phase used to improve or modify the preceding one (Fig. 6.1). In addition to constructing the GEM, PHASE 1 of the WHO pathway includes an initial evaluation of its efficacy. This evaluation includes assessing the phenotype generated by the transgenes, transgene inheritance (especially as it relates to the efficiency of the gene drive component), stability of the construct over time, and a rudimentary assessment of overall fitness. These assessments are conducted by the laboratories in which GEMs are designed and constructed and they involve the introduction of GEMs into small cages at various ratios of transgenic to wild type [24, 42].

PHASE 2 involves further testing, under natural conditions, of those GEMs that show promise based on PHASE 1 results. Early guidelines recommended that initial tests be conducted in large, highly contained, greenhouse-like cages designed to simulate natural conditions [5, 14, 34, 99]. Artificial, physically contained environments (cages) have limited value because they do not allow the analysis of community- and ecosystem-level interactions in any meaningful sense, they cannot replicate food web structure, and they do not permit examination of ecological phenomenon (e.g., dispersal) across spatial scales [25, 119]. Critically, experiments conducted in such environments often yield highly replicable, but spurious results, which can be counterproductive if applied to situations as they occur in nature [96]. The limitations of using large field cages as part of PHASE 2 GEM trials were recognized in later guidelines and their use is now suggested as optional, unless required by regulatory authorities [58, 59]. From our perspective, conducting trials in large cages does not satisfy our goal that

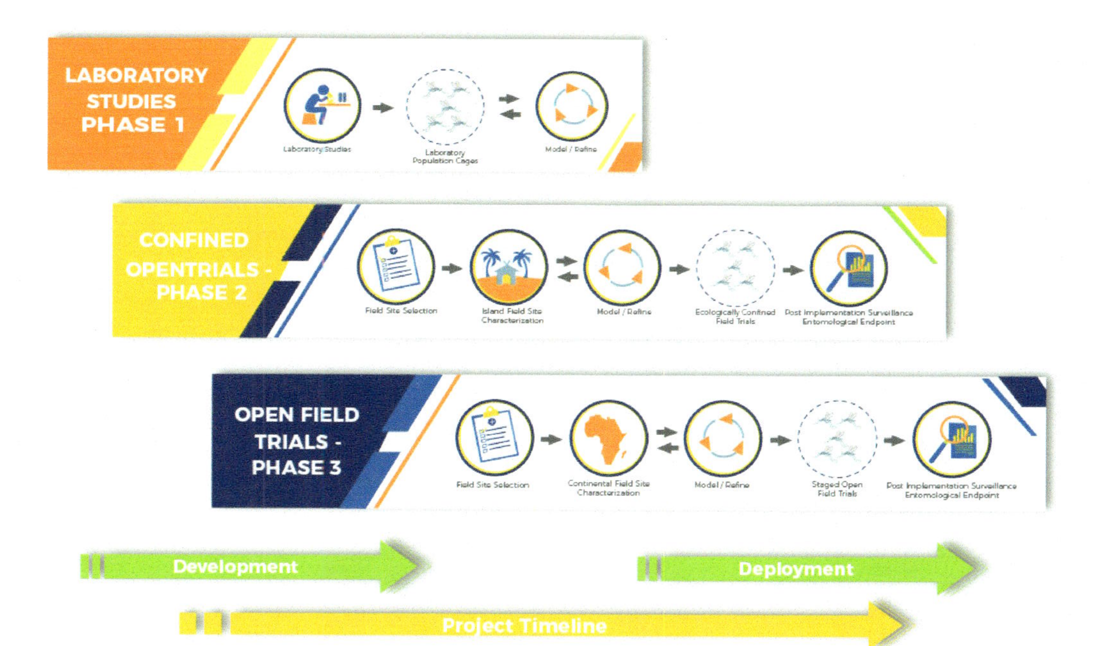

Figure 6.1 Pathway to GEM development and deployment (modified from ref. [116]).

field tests be valid and convincing; therefore, we suggest that GEM evaluation should move from PHASE 1 directly to ecologically confined PHASE 2 field trials.

The efficacy of a GEM can be measured using entomological and/or epidemiological endpoints. Whereas an epidemiological endpoint demonstrates a reduction in the incidence of infection, an entomological endpoint measures a reduction in the risk of malaria transmission. It is generally agreed that entomological endpoints should be the goal for PHASE 2 trials [116]. The entomological endpoint may be measured by a reduction in the entomological inoculation rate inferred from a reduction in vector population size (population suppression strategies) or pathogen replication within the vector (population modification strategies) as well as transgene frequency and GEM fitness in nature.

6.3 Framework for Field Site Selection

An ecologically confined site offers geographical, environmental, and/or biological confinement [116]. Aside from the frequently articulated ethical and regulatory issues [70, 91], there are important biological attributes that favor an ecologically confined area. Physical islands have been recognized as ideal for this purpose [86, 99]. We argue that ecologically confined, island field sites are best suited for a PHASE 2 GEM field trial because of several factors outlined here, including biogeography and climate, lower species complexity, lower genetic complexity, and high containment. Therefore, our strategy for identifying field sites for the evaluation of GEM begins with the identification of potential island sites. We then proceed by defining and justifying a set of criteria on which our evaluations are based. Characterization and evaluation of each potential site using this set of criteria are used to identify candidate sites, which are then recommended for further characterization to be conducted during site visits, culminating in the identification of sites that we deem as suitable for field trials involving GEM release.

Site selection was initiated with the identification of all potential island sites, which we define broadly as any island associated

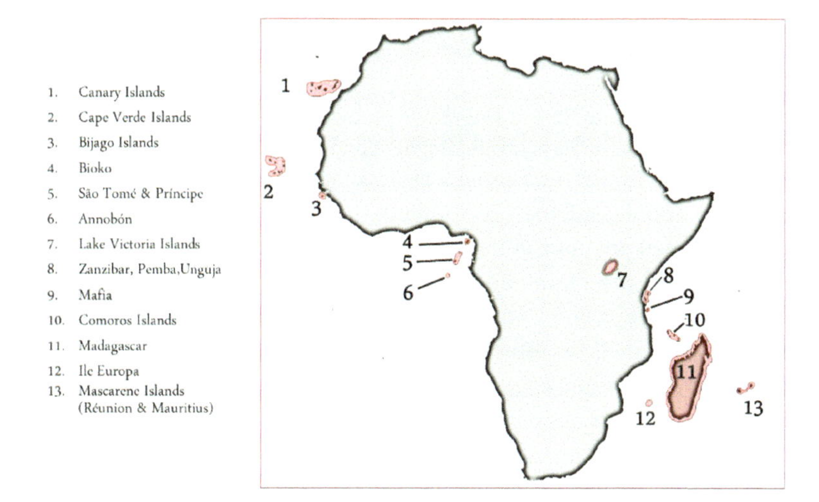

Figure 6.2 African islands and island groups considered potential field sites for GEMs for malaria eradication.

with the continent of Africa. We evaluated 23 potential field sites, including 5 individual islands, islands within 7 archipelagos, and 4 islands within Lake Victoria (Fig. 6.2).

Islands have served as model ecosystems for many plant and animal species. Collectively, these studies have given rise to the field of island biogeography, which provides valuable insights in considering islands for GEM field trials. The key geographical features of islands are that they have a relatively small size, distinct boundaries, simplified biotas, relative geological youth, and are geographically distant and isolated from the continental mainland. In population genetics, gene flow reflects the direct exchange of heritable information through migration between related populations. Distance and geographic isolation are barriers to migration and limit inter-population gene flow resulting in genetic isolation. These features make up the foundation of the "Dynamic Equilibrium Theory of Island Biogeography" [36, 75, 79, 95]. This theory served as our guide to the evaluation of sites.

6.3.1 Physical Features

Geological history

Geographic and geologic parameters for all potential island field sites, including those related to geological history, are summarized in Table 6.1. The islands considered are of three types: continental, oceanic, and lacustrine. Continental or land bridge islands are unsubmerged portions of the continental shelf and were at one time connected to the mainland. Oceanic islands arise from the ocean floor and were never connected to the mainland. Oceanic islands may be classified as either volcanic islands or coral islands and atolls. All the oceanic islands under consideration here are volcanic. Lacustrine islands are islands within lakes and are typically formed by deposits of sedimentary rock, as the Lake Victoria Islands are considered here. The islands in Lake Victoria were included in this study because the Lake Victoria is the largest tropical lake in the world. The Lake Victoria region is large enough to generate its own weather system and supports some of the most densely settled areas of human occupation in sub-Saharan Africa. In addition, islands within the lake have been previously considered as potential field sites for GEM trials [15, 64, 78].

To a large extent, the genetics and demographics of *Anopheles* populations on islands reflect the geological history of those islands [98]. Glacial Maximum Mainland Connection (GMMC) is a proxy variable for island geological history, which indicates whether an island was connected to the mainland during the Last Glacial Maximum (LGM), around 21,500 years ago. GMMC is expressed as 1 if true for the island being described or 0 if false; these are provided for each island in Table 6.1. The continental and lacustrine islands considered here were connected to the continent during the LGM, which allowed biotic interchange and homogenization between the landmasses. Islands with a GMMC score of 0 are isolated or oceanic islands. Since oceanic islands have never been connected to the mainland, they typically have preserved unique species lineages, including adaptive radiations on volcanic archipelagos. This could be advantageous to a GEM field trial program because it suggests limited dispersal of mosquitoes between islands within an archipelago and therefore the potential for multiple treatment

Table 6.1 Bioclimatic and isolation index values used for the evaluation of potential island field sites

Island	Country	Archipelago	Island type	Latitude (DD)	Longitude (DD)	Population	Area (km^2)	Distance to mainland (km)	UNEP Isolation Index	SLMP	GMMC	Elevation max (m)	Temp. (°C)	Precip. (mm)
Canary Islands	Spain	Canary Islands	Oceanic	28.29°N	16.63°W	2,153,389	7509.66	116.63	30.4	0.812	0	3705	20.21	320.36
Cape Verde	Cape Verde	Cape Verde Islands	Oceanic	16.54°N	23.04°W	560,000	4088.52	586.53	55	0.466	0	2813	24.45	269.43
Bijagos	Guinea-Bissau	Bijagos Islands	Continental	11.29°N	15.97°W	32,424	1944.72	0.83	10.8	1.123	1	59	26.42	2268.14
Bioko	Equatorial Guinea	Cameroon Line	Continental	3.62°N	8.75°E	334,463	1950.46	73.03	17	1.148	1	3011	25.5	2677
Annobón	Equatorial Guinea	Cameroon Line	Oceanic	1.42°S	5.62°E	5,232	15.7	350	45	-	0	587	26.8	1000–1500
São Tomé	São Tomé & Príncipe	Cameroon Line	Oceanic	0.19°N	6.61°E	185,660	854.8	283.63	39	0.753	0	1977	25.5	3693
Príncipe	São Tomé & Príncipe	Cameroon Line	Oceanic	1.61°N	7.41°E	8,052	143.16	221.72	39	0.86	0	934	24.8	1992
Bugala	Uganda	Lake Victoria	Lacustrine	0.42°S	32.24°E	17,355	296	3.7	5.487	-	1	160	18	1125–2250
Koome	Uganda	Lake Victoria	Lacustrine	0.09°S	32.75°E	16,000	100	14.3	10.688	-	1	180	25.5	2112
Mfangano	Kenya	Lake Victoria	Lacustrine	0.46°S	34.01°E	18,600	66	7.4	10.414	-	1	551	25	700–1200
Ukara	Tanzania	Lake Victoria	Lacustrine	1.85°S	33.06°E	34,181	80	22.8	15.623	-	1	162	27.2	1524

(*Continued*)

Table 6.1 (*Continued*)

Island	Country	Archipelago	Island type	Latitude (DD)	Longitude (DD)	Population	Area (km^2)	Distance to mainland (km)	UNEP Isolation Index	SLMP	GMMC	Elevation max (m)	Temp. (°C)	Precip. (mm)
Pemba	Tanzania	Zanzibar	Continental	5.03°S	39.78°E	406,848	987.08	68.6	31.231	1.178	0	149	26.3	1940
Unguja	Tanzania	Zanzibar	Continental	6.14°S	39.36°E	303,043	1591.5	50.9	17	1.337	1	133	27.8	1811
Mafia	Tanzania	Mafia	Continental	7.87°S	39.75°E	46,483	443.24	36.06	29.432	1.194	1	66	26.9	1719
Grand Comore	Comoros	Comoros	Oceanic	11.65°S	43.33°E	433,437	1021.61	307.45	49	0.736	0	2368	25.4	2410
Moheli	Comoros	Comoros	Oceanic	12.34°S	43.73°E	56,932	212.09	340.61	49	0.754	0	793	25.3	1998
Anjouan	Comoros	Comoros	Oceanic	12.21°S	44.44°E	360,409	432.08	417.78	49	0.706	0	1591	25.6	1876
Mayotte	France	Comoros	Oceanic	12.83°S	45.17°E	279,471	371.42	490.16	47	0.669	0	636	26.1	1394
Madagascar	Madagascar	Madagascar	Oceanic	18.77°S	46.87°E	25,680,342	590547.4	780.51	58	0.46	0	2876	27.6	3373
Ile Europa	France	French Territory	Oceanic	22.37°S	40.35°E	0	32.64	492.95	67.941	0.736	0	20	24.5	530
Réunion	France	Mascarene Islands	Oceanic	21.11°S	55.54°E	859,959	2512.65	1699.32	73	0.467	0	3066	23.7	2109
Mauritius	Mauritius	Mascarene Islands	Oceanic	20.35°S	57.55°E	1,221,921	1868.44	1874.49	87	0.399	0	816	24	2099

Note: DD, decimal degrees; UNEP, United Nations Environment Program; SLMP, Surrounding Landmass Proportion; GMMC, Glacial Maximum Mainland Connection a proxy variable for island geological history, which indicates whether an island was connected to the mainland during the LGM (1 = true and 0 = false); refers to missing or incomplete data. Additional data sources: Bugala [64, 83, 104, 120]; Mfango [45]; Ukara [77, 82, 101]; and Koome [12, 39, 57, 84, 108].

*In the Gulf of Guinea, the Cameroon line consists of six offshore volcanic swells that have formed the islands of Annobón, São Tomé, Príncipe, and Bioko. On the mainland, the line starts with Mount Cameroon.

or control sites. The GMMC value for the Lake Victoria Islands was determined from Tryon et al., 2016 [107]. The GMMC value for Annobón was determined by Jones, 1994 [62]. We have found that GMMC values correlate well with *A. gambiae* and *A. coluzzii* genetics and with *Anopheles* species richness on the islands studied here and therefore this value is a good metric to evaluate the suitability of an island as a GEM field trial site.

Describing geographic isolation

We characterized our potential island sites using three measures of geographic isolation. The first is simply the geographic distance to the nearest mainland (referred to Table 6.1 as "Distance to Mainland"). This metric is generally considered sufficient as an isolation index at a global scale. Its significance is that the nearest mainland is assumed to be the richest gene pool and the source of species on the islands [54, 111]. Colonization of islands by mainland species is crucial for island biology and poor dispersers are unlikely to reach distant islands [31, 54, 111]. Distances for individual islands reported in Table 6.1 were calculated as the shortest great circular distance between an island's mass centroid and the mainland coast. For archipelagos, distances from the nearest island to the mainland were used [111]. Distance to the mainland for each Lake Victoria island and Annobón was determined using Google Earth's distance and area measuring tool. The two closest points on the mainland and island shores were used as measuring points. As a rule of thumb, we excluded from consideration any island that had a distance to a mainland less than 10 km.

A second metric used to describe the degree of isolation for an island is the United Nations Environment Program (UNEP) Isolation Index. It is calculated as "the sum of the square roots of the distances to the nearest equivalent or larger island, the nearest group or archipelago, and the nearest continent." Where one of these categories does not exist, the next higher distance is repeated [29]. The higher the value, the more geographically isolated the island is. For the Cape Verde Islands, the UNEP Isolation Index is based on 9 of the 11 islands. Brava and Santa Laiza islands had no isolation index in the UNEP island directory. An isolation index was calculated for the Lake Victoria Islands, Pemba, Mafia, and Île Europa, according to the UNEP's calculation criteria outlined above using independently acquired values.

Surrounding land mass proportion (SLMP) is another isolation index where the isolation of the focal island is proportional to the area of the surrounding landmass [111]. SLMP is a preferred index for the analysis of species variation on a focal island. The equilibrium theory of island biogeography supports this index as individual islands may act as stepping-stones for species dispersal and establishment, which this index accounts for by shortening the distance between an island and potential source populations [79]. SLMP is calculated as the sum of the proportions of landmass within buffer distances of 100, 1000, and 10,000 km around the island perimeter. Additionally, SLMP accounts for the coastline shape of large landmasses by considering only regions that extend into the measured buffers. SLMP values for the Canary Islands, Cape Verde Islands, and Bijagós Islands were represented as the average of all islands in their respective archipelagos [111].

A larger SLMP value indicates that an island is surrounded by more landmass within the buffer regions. For this study, we are focusing on islands with a lower SLMP value since these islands will have less surrounding landmass, which could facilitate mosquito dispersal into or out of the target island.

Geography and Climate

Geographic and climatic data for all potential island field sites are summarized in Table 6.1. Information for bioclimatic variables was derived from the WorldClim (worldclime.org) database [35]. The temperature values in Table 6.1 were averaged for the Canary Islands, Cape Verde, and the Bijagós Islands. The mean temperature and precipitation for the Lake Victoria and Annobón were obtained from various sources [12, 45, 83]. Population data for islands was collected from country census data, the United Nations, and/or other sources in the literature [1, 46–52, 87, 105].

The maximum values for annual mean temperature (Temp.) and annual mean precipitation (Precip.) are key factors in shaping mosquito bionomics and mosquito-stages of *Plasmodium* development. All the island sites under consideration lie in the tropics under bioclimatic (temperature and precipitation) conditions acceptable for genetically engineered *A. gambiae* and/or *A. coluzzii* field trials.

6.3.2 Biological Features

Species complexity/Anopheline species richness

Table 6.2 provides a comparison of *Anopheles* species diversity among specific islands and a sample of mainland African countries. It consists of a list of primary and secondary malaria vectors, as well as species for which vector status is uncertain. Also included are notes on malaria case incidence for each of the countries and islands to illustrate the status of malaria control efforts, and to evaluate the potential for measuring epidemiological endpoints. Species composition data are summarized on the map in Fig. 6.3. Mainland countries where *Anopheles* species diversity was compared to that of islands included those countries neighboring the islands under consideration or countries where GEM field trials are already under consideration. The islands selected represent the potential field sites listed in Fig. 6.2.

It is generally agreed that potential field sites with the fewest number of non-target *Anopheles* species are desirable [19, 59]. If multiple sister species are present, there exists the possibility that the transgene will move between species via natural hybridization [71, 112], which could add an additional level of complexity to post-release assessments. Although the movement of transgene elements between malaria vector species may be considered desirable, it raises the specter of horizontal transfer, which is generally identified as a risk to this technology. Assessment of entomological endpoints following a GEM release requires repeated mosquito collections to quantify changes in the ratio of GEM to wild-type mosquitoes. This necessitates sorting large numbers of individual field-collected mosquito samples to separate target from non-target species. For members of sibling species complexes, which are morphologically indistinguishable, this requires the application of PCR-based diagnostics to each individual specimen. If collections include larvae, time-consuming microscopic examination to identify species is required even for those species distinguishable morphologically in the adult stage. Logistically, these procedures

Table 6.2 Comparison of mainland versus island *Anopheles* species diversity and malaria incidence

MAINLAND					
Country	**Primary vectors**	**Secondary vectors**	**Other *Anopheles*(non-vector or status unclear)**	**References**	**Malaria cases (per year)**
Burkina Faso	*An. arabiensis, coluzzii, funestus, gambiae, nili*	*An. brunnipes, coustani, cydippis, hancocki, leesoni, maculipalpis, paludis, pharoensis, rivulorum, rufipes, pretoriensis, sergentii, squamosus, theileri, ziemanni*	*An. argenteolobatus, brohieri, brumpti, domicolus, dureni, flavicosta, freetownensis, implexus, longipalpis, murphyi, natalensis, obscurus, rhodesiensis, somalicus, wellcomei*	1,2	6,840,864 (2000) 8,602,187 (2010) 7,245,827 (2015) 7,859,000 (2019)
Cameroon	*An. arabiensis, coluzzii, funestus, gambiae, moucheti, nili*	An. *bervoetsi, brunnipes, carnevalei, coustani, cydippis, demeilloni, hancocki, leesoni, maculipalpis, marshallii, melas, obscurus, ovengensis, paludis, pharoensis, pretoriensis, rivulorum, rivulorum*-like, *rufipes, sergentii, squamosus, ziemanni*	*An. brohieri, buxtoni, christyi, cinctus, concolor, deemingi, domicolus, dualaensis, eouzani, flavicosta, freetownensis, hargreavesi, implexus, jebudensis, longipalpis, mousinhoi, multicinctus, namibiensis, natalensis, okuensis, rageaui, rhodesiensis, smithii, somalicus, tenebrosus, wellcomei*		6,291,500 (2000) 5,909,335 (2010) 5,777,768 (2015) 6,291,256 (2019)

Mali	*An. arabiensis, coluzzii, funestus, gambiae, nili*	*An. brunnipes, coustani, dthali, hancocki, leesoni, maculipalpis, paludis, pharoensis, pretoriensis, rivulorum, rufipes, sergentii, squamosus, ziemanni*	*An. brohieri, domicolus, flavicosta, obscurus, rhodesiensis, somalicus, wellcomei*	4,446,769 (2000) 5,772,983 (2010) 6,833,022 (2015) 6,560,000 (2019)
Tanzania	*An. arabiensis, funestus, gambiae, moucheti, nili*	*An. aruni, brunnipes, cinereus, coustani, cydippis, demeilloni, gibbinsi, leesoni, maculipalpis, marshallii, merus, paludis, parensis, pharoensis, pretoriensis, quadriannulatus, rivulorum, rufipes, squamosus, theileri, ziemanni*	*An. ardensis, argenteolobatus, christyi, confusus, distinctus, erepens, garnhami, implexus, keniensis, kingi, letabensis, longipalpis, lovettae, machardyi, namibiensis, natalensis, njombiensis, rhodesiensis, schwetzi, seydeli, swahilicus, tenebrosus, walravensi, wellcomei, wilsoni*	11,514,222 (2000) 5,917,848 (2010) 7,298,719 (2015) 6,453,096 (2019)
Uganda	*An. arabiensis, funestus, gambiae, moucheti, nili*	*An. bwambae, cinereus, coustani, cydippis, demeilloni, gibbinsi, hancocki, leesoni, maculipalpis, marshallii, paludis, parensis, pharoensis, pretoriensis, quadriannulatus, rivulorum, rufipes, squamosus, symesi, theileri, ziemanni*	*An. ardensis, brohieri, christyi, domicolus, garnhami, hargreavesi, harperi, implexus, keniensis, kingi, longipalpis, natalensis, obscurus, rhodesiensis, tenebrosus, vinckei, wellcomei*	11,522,961 (2000) 13,277,279 (2010) 9,690,714 (2015) 11,629,246 (2019)

(*Continued*)

Table 6.2 (*Continued*)

ISLANDS					
Island	**Primary vectors**	**Secondary vectors**	**Other *Anopheles* (non-vector or status unclear)**		**Malaria cases (year)**
Bijagos Islands (Guinea-Bissau)	*An. arabiensis, coluzzii, gambiae*	*An. melas, cinereus, coustani, maculipalpis, pharoensis, rufipes, squamosus, ziemanni*	*An. dancailicus, hargreavesi, smithii*	3, 4	
Bioko (Equatorial Guinea)	*An. coluzzii, funestus, gambiae, moucheti*	*An. brunnipes, carnevalei, leesoni, melas, ovengensis*	*An. cinctus, lloreti, obscurus, smithii*	5, 6	
Canary Islands (Spain)		*An. multicolor, sergentii*	*An. hispaniola* (historically a vector in Europe)	7	
Cape Verde	*An. arabiensis*	*An. pretoriensis*		8	144 (2000) 47 (2010) 7 (2015) 0 (2019)
Anjouan (Comoros)	*An. funestus, gambiae*	*An. coustani, mascarensis, pretoriensis*		9, 10, 11	35,309 (2000) 36,538 (2010) 1,300 (2015) 17,599 (2019)
Grand Comore (Comoros)	*An. gambiae*	*An. pretoriensis*			

Moheli (Comoros)	*An. funestus, gambiae*	*An. coustani, maculipalpis, mascarensis, pretoriensis*			
Lake Victoria islands	*An. arabiensis, funestus, gambiae*	*An. coustani, pharoensis, symesi, ziemanni*		12, 13, 14	
Mayotte (France)	*An. funestus, gambiae*	*An. coustani, maculipalpis, mascarensis, pretoriensis*	*An. comorensis*	1	>2,000 (2000) 433 (2010) 14 (2015) 22 (2016)
Île Europa (France)	*An. gambiae*			15	
Madagascar	*An. arabiensis, funestus, gambiae, coustani*	*An. brunnipes, cydippis, maculipalpis, mascarensis, merus, pharoensis, pretoriensis, rufipes, squamosus*	*An. flavicosta, fuscicolor, grassei, grenieri, griveaudi, lacani, milloti, notleyi, pauliani, radama, ranci, roubaudi, tenebrosus*	1	901,335 (2000) 893,540 (2010) 1,897,533 (2015) 2,052,071 (2019)
Pemba (Tanzania)	*An. arabiensis, gambiae*	*An. merus*		17	
Príncipe	*An. coluzzii*			18, 19	31,975 (2000) 2,740 (2010) 2,058 (2015) 2,446 (2019)

(*Continued*)

Table 6.2 (*Continued*)

São Tomé	*An. coluzzii, funestus, gambiae*	*An. coustani, melas, paludis, pharoensis*			
Annobón	*An. coluzzii*			20	
Mauritius	*An. arabiensis*	*An. coustani, maculipalpis, merus*		21	All imported: 52 (2010) 33 (2012)
Réunion (France)	*An. arabiensis*	*An. coustani*		22	Eliminated (1979)
Zanzibar (Tanzania)	*An. arabiensis, funestus, gambiae*	*An. aruni, coustani, leesoni, maculipalpis, marshallii, merus, paludis, parensis, pretoriensis, rivulorum, squamosus, ziemanni*	*An. longipalpis, obscurus, quadriannulatus, swahilicus, tenebrosus, wellcomei*	2	3,528 (2005) 2,572 (2012) 3,814 (2015) 3,025 (2016)

1. Irish, S. R., Kyalo, D., Snow, R. W., & Coetzee, M. (2020). Updated list of Anopheles species (Diptera: Culicidae) by country in the Afrotropical Region and associated islands. Zootaxa, 4747(3), zootaxa 4747 4743 4741. Retrieved from https://www.ncbi.nlm.nih.gov/pubmed/32230095. doi:10.11646/zootaxa.4747.3.1
2. Kyalo, D., Amratia, P., Mundia, C. W., Mbogo, C. M., Coetzee, M., & Snow, R. W. (2017). A geo-coded inventory of anophelines in the Afrotropical Region south of the Sahara: 1898-2016. Wellcome Open Res, 2, 57. Retrieved from https://www.ncbi.nlm.nih.gov/pubmed/28884158. doi:10.12688/wellcomeopenres.12187.1
3. Ant, T., Foley, E., Tytheridge, S., Johnston, C., Goncalves, A., Ceesay, S., et al. (2020). A survey of Anopheles species composition and insecticide resistance on the island of Bubaque, Bijagos Archipelago, Guinea-Bissau. *Malar J, 19*(1), 27. Retrieved from https://www.ncbi.nlm.nih.gov/pubmed/31941507. doi:10.1186/s12936-020-3115-1
4. Sanford, M. R., Cornel, A. J., Nieman, C. C., Dinis, J., Marsden, C. D., Weakley, A. M., et al. (2014). Plasmodium falciparum infection rates for some Anopheles spp. from Guinea-Bissau, West Africa. *F1000Res, 3*, 243. Retrieved from https://www.ncbi.nlm.nih.gov/pubmed/25383188. doi:10.12688/f1000research.5485.2
5. Berzosa, P. J., Cano, J., Roche, J., Rubio, J. M., Garcia, L., Moyano, E., et al. (2002). Malaria vectors in Bioko Island (Equatorial Guinea): PCR determination of the members of Anopheles gambiae Giles complex (Diptera: Culicidae) and pyrethroid knockdown resistance (kdr) in An. gambiae sensu stricto. *J Vector Ecol, 27*(1), 102-106. Retrieved from https://www.ncbi.nlm.nih.gov/pubmed/12125862.
6. Guerra, C., Fuseini, G., Donfack, O., Smith, J., Mifumu, T., Akadiri, G., et al. (2020). Malaria outbreak in Riaba district, Bioko Island: lessons learned. *Malaria Journal, 19*. doi:10.1186/s12936-020-03347-w

7. Baez, M., & Fernandez, J. M. (1980). Notes on the mosquito fauna of the Canary Islands (Diptera: Culicidae) *Mosquito Systematics, 12*(3), 349-355.
8. Alves, J., Gomes, B., Rodrigues, R., Silva, J., Arez, A. P., Pinto, J., et al. (2010). Mosquito fauna on the Cape Verde Islands (West Africa): an update on species distribution and a new finding. *J Vector Ecol, 35*(2), 307-312. Retrieved from https://www.ncbi.nlm.nih.gov/pubmed/21175936. doi:10.1111/j.1948-7134.2010.00087.x
9. Brunhes, J. (1977). Les moustiques de l'archipel des Comores. *Cahiers ORSTOM. Série Entomologie Médicale et Parasitologie, 25*(2), 131-152.
10. Brunhes, J., Le Goff, G., & Geoffroy, B. (1997). Anophèles afro-tropicaux : 1. Descriptions d'espèces nouvelles et changements de statuts taxonomiques (Diptera : Culicidae). *Annales de la Societe Entomologique de France, 33.*
11. Coetzee, M., Hunt, R. H., Wilkerson, R., Della Torre, A., Coulibaly, M. B., & Besansky, N. J. (2013). Anopheles coluzzii and Anopheles amharicus, new members of the Anopheles gambiae complex. *Zootaxa, 3619*, 246-274. Retrieved from https://www.ncbi.nlm.nih.gov/pubmed/26131476.
12. Ajamma, Y. U., Villinger, J., Omondi, D., Salifu, D., Onchuru, T. O., Njoroge, L., et al. (2016). Composition and Genetic Diversity of Mosquitoes (Diptera: Culicidae) on Islands and Mainland Shores of Kenya's Lakes Victoria and Baringo. *Journal of Medical Entomology, 53*(6), 1348-1363. Retrieved from https://www.ncbi.nlm.nih.gov/pubmed/27402888. doi:10.1093/jme/tjw102
13. Lukindu, M., Bergey, C. M., Wiltshire, R. M., Small, S. T., Bourke, B. P., Kayondo, J. K., et al. (2018). Spatio-temporal genetic structure of Anopheles gambiae in the Northwestern Lake Victoria Basin, Uganda: implications for genetic control trials in malaria endemic regions. *Parasit Vectors, 11*(1), 246. Retrieved from https://www.ncbi.nlm.nih.gov/pubmed/29661226. doi:10.1186/s13071-018-2826-4
14. Ogola, E., Villinger, J., Mabuka, D., Omondi, D., Orindi, B., Mutunga, J., et al. (2017). Composition of Anopheles mosquitoes, their blood-meal hosts, and Plasmodium falciparum infection rates in three islands with disparate bed net coverage in Lake Victoria, Kenya. *Malar J, 16*(1), 360. Retrieved from https://www.ncbi.nlm.nih.gov/pubmed/28886724. doi:10.1186/s12936-017-2015-5
15. Boussès, P., Dehecq, J. S., Brengues, C., & Fontenille, D. (2013). Inventaire actualisé des moustiques (Diptera : Culicidae) de l'île de La Réunion, océan Indien. *Bulletin de la Société de pathologie exotique, 106*(2), 113-125. Retrieved from https://doi.org/10.1007/s13149-013-0288-7. doi:10.1007/s13149-013-0288-7
16. World Health Organization. (2020). *World malaria report 2020: 20 years of global progress and challenges*. Retrieved from Geneva:
17. Haji, K. A., Khatib, B. O., Smith, S., Ali, A. S., Devine, G. J., Coetzee, M., et al. (2013). Challenges for malaria elimination in Zanzibar: pyrethroid resistance in malaria vectors and poor performance of long-lasting insecticide nets. *Parasit Vectors, 6*, 82. Retrieved from https://www.ncbi.nlm.nih.gov/pubmed/23537463. doi:10.1186/1756-3305-6-82
18. Campos, M., Hanemaaijer, M., Gripkey, H., Collier, T., Lee, Y., Cornel, A., et al. (2021). *The origin of island populations of the African malaria mosquito, Anopheles coluzzii*: Communications Biology, in press.
19. Loiseau, C., Melo, M., Lee, Y., Pereira, H., Hanemaaijer, M. J., Lanzaro, G. C., et al. (2019). High endemism of mosquitoes on São Tomé and Príncipe Islands: evaluating the general dynamic model in a worldwide island comparison. *Insect Conservation and Diversity, 12*(1), 69-79. Retrieved from https://onlinelibrary.wiley.com/doi/abs/10.1111/icad.12308. doi:https://doi.org/10.1111/icad.12308
20. Salgueiro, P., Moreno, M., Simard, F., O'Brochta, D., & Pinto, J. (2013). New insights into the population structure of Anopheles gambiae s.s. in the Gulf of Guinea Islands revealed by Herves transposable elements. *PLoS One, 8*(4), e62964. Retrieved from https://www.ncbi.nlm.nih.gov/pubmed/23638171. doi:10.1371/journal.pone.0062964
21. Iyaloo, D. P., Elahee, K. B., Bheecarry, A., & Lees, R. S. (2014). Guidelines to site selection for population surveillance and mosquito control trials: a case study from Mauritius. *Acta Tropica, 132 Suppl*, S140-149. Retrieved from https://www.ncbi.nlm.nih.gov/pubmed/24280144. doi:10.1016/j.actatropica.2013.11.011
22. WHO Expert Committee on Malaria & World Health Organization. (1979). *WHO Expert Committee on Malaria : seventeenth report* (9241206403). Retrieved from Geneva: https://apps.who.int/iris/handle/10665/41359

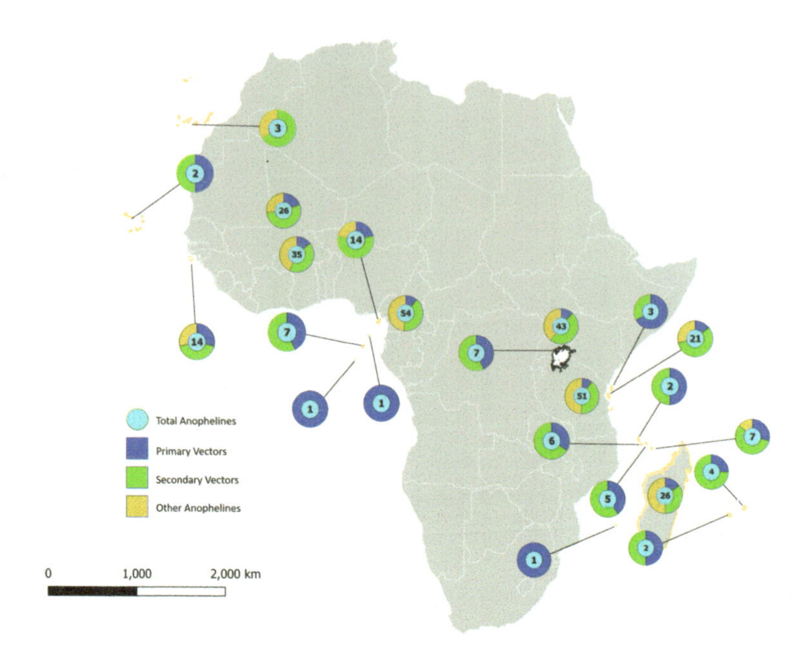

Figure 6.3 *Anopheles* species complexity in Africa including island and select mainland sites. Map locations and summary of data are presented in Table 6.2. Cyan circle = total number of *Anopheles* spp.; blue = proportion of primary vector species; green = proportion of secondary vectors; and yellow = proportion of species identified as non-vector or for which vector status unknown (taken from Lanzaro G.C., et al. (2021). *Evolutionary Applications*, 14(9), 2147–2161).

are greatly simplified where fewer non-target *Anopheles* species are present, positively impacting the time and resources required for successful assessment.

Although entomological endpoints are the main consideration in evaluating the outcome of a PHASE 2 trial, epidemiological impacts should be considered where feasible. If epidemiological endpoints are to be assessed, the presence of multiple primary and secondary vectors is problematic as they can lengthen the season of malaria transmission [9]. Therefore, their presence can mask the effects that GEMs might have on transmission at a field site by maintaining the rate of transmission, even if the parasite is not present in the target mosquito species. If a site is selected in which very few malaria vector species occur, it becomes more likely that the GEM release will have a measurable impact on the level of malaria transmission.

Published compilations of Afrotropical *Anopheles* species distribution were used to assemble the information for mainland countries presented in Table 6.2 [53, 67]. Data for island sites was collected from various sources and there are several important considerations to note when reviewing the table that we outline here. First, it is likely that additional *Anopheles* species from those listed are present in the Lake Victoria Islands because they are present in adjacent mainland sites. In the Bijagós Islands, *A. coluzzii* is the most common member of the *A. gambiae* complex, followed by *A. melas* and *A. gambiae. Anopheles arabiensis* is rarely found on the Bijagós Islands. It is not known if other *Anopheles* species that occur in mainland Guinea-Bissau have ever been collected in the Bijagós, although they are likely present. In the Comoros Islands, *A. comorensis* was described and named by Brunhes et al. [21] as a member of the *A. gambiae* complex, but its status is in question pending further analysis [27]. Recent collections on the islands of São Tomé and Príncipe confirmed the presence of *A. gambiae* on São Tomé but not on Príncipe [23]. *Anopheles funestus, A. melas, A. pharoensis,* and *A. paludis* have been recorded in São Tomé and Príncipe but likely no longer occur on the islands because they have not been collected there for many years, despite regular mosquito surveillance. *Anopheles gambiae* and *A. funestus* while previously present on the Bioko Island, are now considered to have been eliminated [40].

Second, species listed in Table 6.2 are designated as primary or secondary vectors or as "other" if they are non-vectors or their status as vectors is not clear. Species that commonly had sporozoite infection rates above 1%, as determined by salivary gland dissections, CSP ELISA, or PCR of head and thorax were listed as primary vectors. Species with infection rates of <1% were listed as secondary vectors. Our knowledge of the population structure and biology of almost all the secondary vectors is very limited and their role as secondary vectors throughout their distributions is variable. Sporozoite positivity in a secondary vector species in one location is not necessarily indicative of its vector status in another.

Mainland African countries have considerably larger numbers of primary and secondary *Anopheles* vector species than any of the islands, which is consistent with island biogeography theory regarding species richness [79]. Mainland countries with multiple ecological zones, especially those with tropical deciduous forests, tropical mountain systems, and tropical dry forests, such as Cameroon and Uganda [110], have the highest *Anopheles* species richness. The forest-dwelling *A. moucheti* and *A. nili* are considered major vectors in forests because they are believed to be responsible for >95% of the total malaria transmission in these habitats in Central and West Africa [9]. *Anopheles melas,* a West African coastal, brackish water-dwelling mosquito, and member of the *A. gambiae* complex, is designated a primary, secondary, or non-vector in the literature [4, 9, 38, 80]. This is evidence that *A. melas* is part of a group of cryptic species, which likely contributes to the confusion concerning its vector status [30]. The East African species *A. merus,* is often considered a minor vector [38]; however, sporozoite rates higher than 3% have been recorded in this species in some locations in Mozambique [28] and Kenya [81]. *Anopheles bwambae,* another member of the *A. gambiae* complex, restricted to the Semliki Forest in Uganda and considered herein a secondary vector, sometimes blood feeds on humans and has been recorded to have a sporozoite infection rate of 0.7% [113]. The Islands of Annobón, Île Europa, and Príncipe are the only islands with a single malaria vector species. Grande Comore has one confirmed vector species, *A. gambiae,* but the role of *A. pretoriensis* as a secondary vector on this island is currently unknown. *Anopheles gambiae* is rare but still considered a major vector on São Tomé [23] and the role of *A. coustani* as a malaria vector on São Tomé is currently unknown.

Finally, the numbers of malaria cases reported in Table 6.2 include all case counts of imported and locally acquired infections. Reductions in malaria morbidity have generally plateaued since 2015, with no significant changes in the number of malaria cases between 2015 and 2019 [118].

Predictably, mainland populations have a dramatically higher anopheline species richness compared with islands. Table 6.2 illustrates the important difference in *Anopheles* species diversity between African island and mainland sites. The lower numbers of non-target species provide strong evidence to support the selection of an island field trial site for optimal evaluation and assessment of a GEM product.

Genetic complexity

The capacity of gene drive constructs to spread through wild-type populations in small cage experiments has yielded promising results [24, 68]. However, these studies utilize populations of mosquitoes derived from long-standing laboratory colonies that do not replicate populations as they occur in nature [10, 17]. Notably, these colonies lack the genetic diversity present as standing genetic variation (SGV) in natural populations [69, 89, 92]. In general, GEMs introduced into natural populations will need to interact with the wide range of highly polymorphic genomes present as SGV in natural populations. A portion of this SGV occurs within potential CRISPR-Cas9 target sites [97] and may represent drive resistance alleles, whose presence can have serious impacts on gene drive performance [32, 109]. In addition, SGV may include mutations in genomic sites other than the target site resulting in unintended DNA cleavage. Such off-target effects may include unexpected and unwanted phenotypes [59]. The abundance of drive resistance alleles and off-target sequences provide two examples of undesirable polymorphisms. In summary, field trial sites containing populations of the target species that have the lowest possible levels of SGV are desirable.

We conducted an analysis of genomic diversity to compare mainland, lacustrine, continental, and oceanic island populations of the two target species. The locations and sample sizes per site are provided in Fig. 6.4. Methods for DNA extraction, sequencing, and data analysis are as described by Schmidt et al, 2020 [98]. Genetic diversity was measured using the nucleotide diversity statistic (π), defined as the average number of pairwise nucleotide differences per nucleotide site. Our estimates of π were made from 10 kb non-overlapping windows on the euchromatic regions of chromosome 3 for both *A. coluzzii* and *A. gambiae*. Chromosome 3 was used to avoid potential confounding effects of paracentric inversions common on chromosome 2.

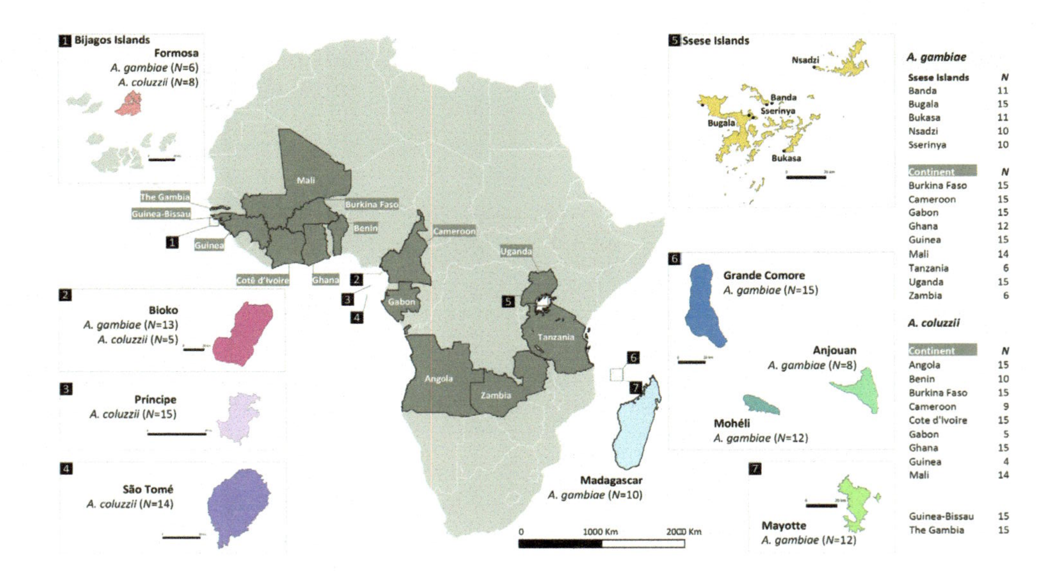

Figure 6.4 Study sampling locations. Samples of *A. gambiae* and *A. coluzzii* were from 12 countries (dark shade) in continental Africa: Angola, Benin, Burkina Faso, Cameroon, Cotê d'Ivoire, Gabon, Ghana, Guinea, Tanzania, Uganda, and Zambia. Populations from Guinea-Bissau and The Gambia (dark shade) were included with no species assignment. A table on the right displays the number of samples for each of the mainland populations. The insert maps show African islands sampled in this study: (1) Formosa islands within Bijagos archipelago; (2) Bioko, (3) Príncipe and São Tomé, (4) islands in the Gulf of Guinea, (5) five islands in Ssese islands in Lake Victoria in Uganda, (6) Comoros (Anjouan, Mohéli and Grande Comore), and (7) Mayotte. Madagascar island is shown on the main map. The number of samples included for each island is shown in parenthesis. Insert maps contain a scale of 20 km in length (taken from Lanzaro G.C., et al. (2021). *Evolutionary Applications*, 14(9), 2147–2161).

Mean nucleotide diversities (π) on oceanic islands (*A. gambiae* – 0.80%, *A. coluzzii* – 0.88%) were significantly lower ($p < 0.0001$) than mainland population means (*A. gambiae* – 1.17%, *A. coluzzii* – 1.11%) for both species (Fig. 6.5A,B). Comparisons among island types yielded results that were consistent with island biogeography theory. Nucleotide diversity in continental island populations did not differ from mainland populations, and lacustrine islands had only slightly, but statistically significant, lower values for π. These observations are expected given the geologic history and proximity of continental and lacustrine islands to the coast. Anjouan island populations presented the lowest (0.73%) nucleotide diversity (π) for *A. gambiae* and Príncipe Island for *A. coluzzii* (0.66%), likely due to their small size and a high degree of isolation.

In general, the lower biocomplexity on isolated islands includes reduced genetic variation [36]. Our results are concordant with this observation (Fig. 6.5). Selecting field sites with populations containing the lowest levels of variation should decrease the potential for transgene/genome interactions that negatively impact GEM performance.

Containment: Anthropogenic dispersal

Although some mosquito species are known to disperse on prevailing winds over long distances [44, 100], there are, to our knowledge, no reliable reports of open-ocean wind dispersal of malaria vector species over the distances (hundreds of kilometers) separating the oceanic islands under consideration here. For these species, the level of genetic isolation of mosquito populations on isolated oceanic islands is generally high [98] suggesting that dispersal of the islands is low. Nonetheless, a dispersal that may occur between the island and mainland populations is most likely to rely on anthropogenic conveyance [13, 100]. From the standpoint of those planning PHASE 2 trials of GEMs, the possibility of inadvertent anthropogenic conveyance of mosquitoes out of the trial area should be considered.

Efforts to enhance ecological confinement by restricting anthropogenic conveyance of mosquitoes should be scaled to the magnitude of the risk of such conveyance. Attempts have been made to estimate the level of accidental transport of mosquito

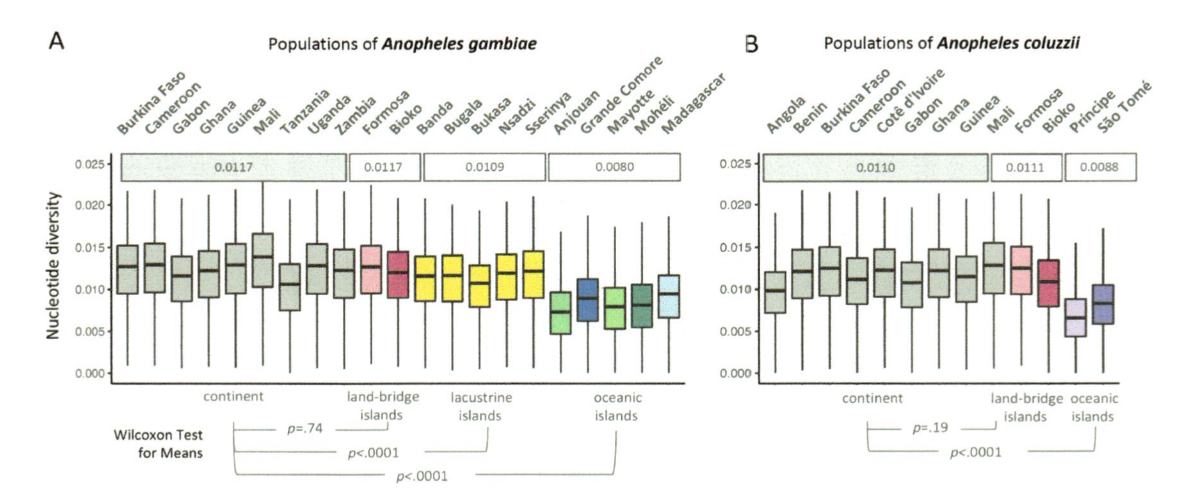

Figure 6.5 Population diversity. Metric is grouped by sampling locations of (**A**) *A. gambiae* and (**B**) *A. coluzzii* populations from island and mainland (gray boxplots). Boxplot of nucleotide diversity (π) performed in 10 kb windows of euchromatic regions of chromosome 3. The midline in all boxplots represents the median, with upper (75th percentile) and lower (25th percentile) limits, whiskers show maximum and minimum values, and outliers are not shown. Mean nucleotide diversity for a set of populations are shown above the boxplots; *A. gambiae* populations were divided into four groups: mainland continental (gray), land-bridge (pink), lacustrine (yellow), and oceanic (green/blue) islands; *A. coluzzii* in three: mainland continental (gray), land-bridge (pink), and oceanic (blue) islands. P-values for testing of means between islands and mainland are shown below. The geographic location for each site and the number of genomes analyzed per site are provided in Fig. 6.4 (taken from Lanzaro G.C., et al. (2021). *Evolutionary Applications*, 14(9), 2147–2161).

species by human activity, relying on both empirical observations and mathematical modeling. Many instances of interception of such species on aircraft and ships have been documented [43, 102, 103, 106]. Species bionomics is likely to be a key factor affecting the extent to which a particular species is able to take advantage of anthropogenic dispersal.

For example, *Anopheles* and other night-flying species are less likely to obtain transport on aircraft arriving and departing strictly during daylight hours [102]. Container breeding species such as *Aedes aegypti* are frequently transported unintentionally in reservoirs of water present in seagoing shipments of used tires [76, 117]. The rise of international air travel widened the risk of dispersal to a broader variety of vector species by allowing for the adventitious rapid transport of adults to distant habitats. For *Anopheles* species, the risk posed by air transport may thus eclipse that posed by sea transport.

Most cases of live adult *Anopheles* found in cabins or cargo holds of aircraft, report very few insects. However, for the purposes of enhancing ecological containment of a low threshold gene drive, even a single insect carrying the drive may pose a risk. Quantitative assessment of the risk is feasible, although accurate estimation of modeling parameters, particularly with respect to the number of mosquitoes per ship or plane, may present challenges. Any risk assessment model for determining the necessity and estimating efficacy of ecological containment, will require an analogous estimate of the transgene frequency, either as a function of time throughout the experiment, or as an overall expected maximum.

Frequency of air and sea departure to interim and final destinations with receptive environments and conspecific mosquito populations can be estimated based on records available from national civil aviation and port authorities or private vendors of flight and maritime transport data. Islands, due to their smaller human populations and geographic areas (and commensurately reduced levels of trade and travel) will generally originate less transboundary air and sea traffic, giving them inherently lower risk levels for these modes of anthropogenic dispersal.

Perhaps more important than air and sea transport is passive dispersal by rail and road. This poses a significant risk for mainland sites, where extensive in-country and transboundary connections exist [22, 33, 37]. Risk by this mode of mosquito dispersal is reduced to zero for oceanic island test sites. However, some of the continental (Bijagós Islands, Zanzibar) and lacustrine islands (Sese Islands in Lake Victoria) are serviced by auto ferries connecting them to the mainland. Obviously, these forms of conveyance may be of significant consequence in these locations.

Containment: Genetic isolation

Single nucleotide polymorphism (SNP) data was analyzed to reveal relationships among populations and results were visualized using principal components analysis (PCA). The position of individuals in the space defined by the principal components can be interpreted as revealing levels of genetic similarity/dissimilarity among the populations from which those individuals were sampled. Populations occupying the same space are presumed to be very similar genetically and those widely separated, very different.

Results of the PCA for *A. gambiae* populations are illustrated in Fig. 6.6A. This analysis reveals a high degree of genetic similarity among individuals from both lacustrine and continental islands relative to bordering mainland populations. Conversely, oceanic islands (Comoros archipelago and Madagascar) form discrete individual clusters indicating that they are genetically distinct both from the mainland and each other. Differences between eastern and western mainland populations of *A. gambiae* can be clearly observed with very high degrees of similarity and tight clustering among individuals from various locations in the west (Burkina Faso, Cameroon, Gabon, Ghana, Guinea, and Mali) and greater differences among individuals from different locations in the east (Tanzania, Uganda, and Zambia), with each eastern population forming a discrete cluster (Fig. 6.6A). Differences between eastern and western populations of *A. gambiae* have been noted by others [7] and are consistent with the hypothesis of a western origin of *A. gambiae* with eastward expansion via a series of population bottlenecks [98].

Results of the PCA for *A. coluzzii* confirm that populations on continental islands form tight clusters which include mainland populations, but oceanic islands form discrete clusters indicating

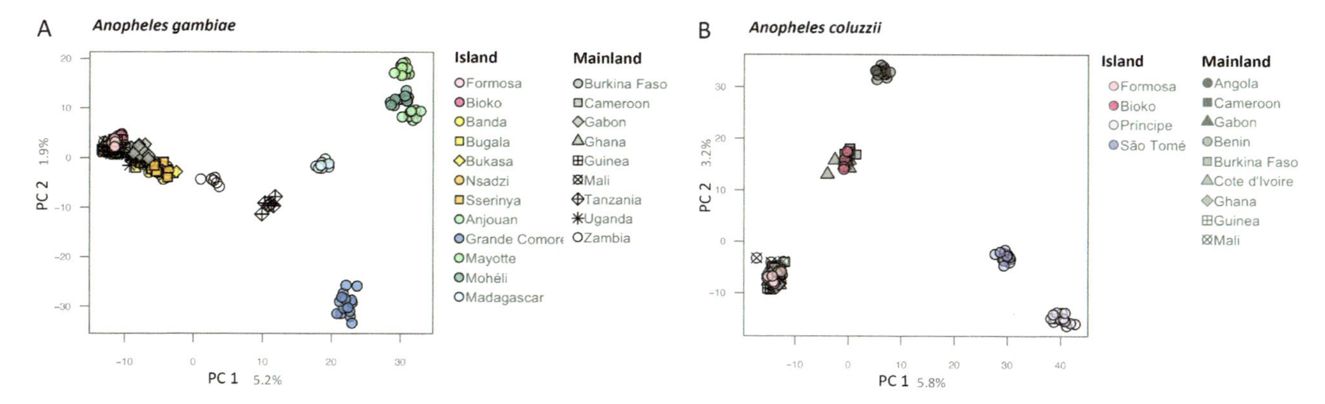

Figure 6.6 Population structure analyses by PCA. 2D-plot of *A. gambiae* (**A**) and *A. coluzzii* (**B**) from islands and mainland populations across Africa. Analyzes were based on 50,000 biallelic SNPs from euchromatic regions on chromosome 3. Each marker represents one individual mosquito. The geographic location for each site and the number of genomes analyzed per site are provided in Fig. 6.4 (taken from Lanzaro G.C., et al. (2021). *Evolutionary Applications*, 14(9), 2147–2161).

genetic divergence from mainland populations and from each other. Individuals from *A. coluzzii* populations on the mainland were divided into three geographically related population clusters: (i) Benin, Burkina Faso, Cotê d'Ivoire, Ghana, Guinea, and Mali forming a western group, (ii) Cameroon and Gabon forming a central group, and (iii) Angola representing a southern group (Fig. 6.6B). These results indicate high levels of genetic isolation for oceanic island populations of both *A. coluzzii* and *A. gambiae*.

The extent to which individuals move (migrate) between two populations can be approximated by measuring the level of genetic divergence between those populations. Migration (m) can be thought of as including the genotypes of the individuals doing the moving and, in this context, migration results in gene flow. Genetic divergence can be described using the statistic F_{ST}, which is the genetic variance in a subpopulation (S) relative to the total variance (T). F_{ST} values range between 0 and 1 and are higher when populations have considerably diverged. The relationship between F_{ST} and m is complex, but excluding the effects of drift and selection, the more gene flow between two populations, the lower the F_{ST} value. So, in general, gene flow has a homogenizing effect on the genetic composition of subpopulations. All pairwise F_{ST} values for the populations of *A. gambiae* and *A. coluzzii* analyzed in this study (Fig. 6.4) are presented in Fig. 6.7. Overall, the results are consistent with the population genetic structure described in the PCA (Fig. 6.6).

Pairwise F_{ST} values for *A. gambiae* (Fig. 6.7A) between west-central African mainland populations (Burkina Faso, Ghana, Guinea, Mali, and Cameroon) are very low (0.000–0.009). The divergence between the continental island of Formosa in the Bijagó archipelago and nearby mainland sites in Guinea-Bissau is likewise very low ($F_{ST} = 0.001$). However, the F_{ST} value for the continental island of Bioko and the nearest mainland sites in Cameroon is substantially higher ($F_{ST} = 0.036$).

Divergence among east African mainland populations of *A. gambiae* (Tanzania, Uganda, and Zambia) is substantially higher (F_{ST} = 0.033–0.090) than among west-central populations (Fig. 6.7A). Values for pairwise F_{ST} between the five lacustrine Ssese Islands in the Lake Victoria (Banda, Bugala, Bukasa, Nsadzi, and Sserinya) are lower (0.010–0.029), but are about the same or lower than

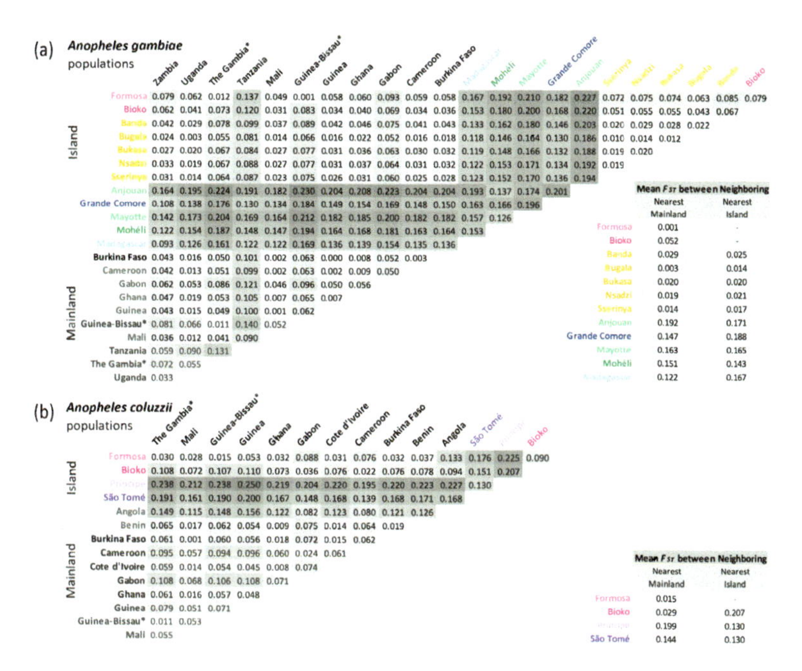

(a) ***Anopheles gambiae* populations**

		Zambia	Uganda	The Gambia*	Tanzania	Mali	Guinea-Bissau*	Guinea	Ghana	Gabon	Cameroon	Burkina Faso	Madagascar	Mohéli	Mayotte	Grande Comore	Anjouan	Sserinya	Nsadzi	Bukasa	Bugala	Banda	Bioko
Island	Formosa	0.079	0.062	0.012	0.137	0.049	0.001	0.058	0.060	0.093	0.059	0.058	0.167	0.192	0.210	0.182	0.227	0.072	0.075	0.074	0.063	0.085	0.079
	Bioko	0.062	0.041	0.073	0.120	0.031	0.083	0.034	0.040	0.069	0.034	0.036	0.153	0.180	0.200	0.168	0.220	0.051	0.055	0.055	0.043	0.067	
	Banda	0.042	0.029	0.078	0.099	0.037	0.089	0.042	0.046	0.075	0.041	0.043	0.133	0.162	0.180	0.146	0.203	0.020	0.029	0.028	0.022		
	Bugala	0.024	0.003	0.055	0.081	0.014	0.066	0.016	0.022	0.052	0.016	0.018	0.118	0.146	0.164	0.130	0.186	0.010	0.014	0.012			
	Bukasa	0.027	0.020	0.067	0.084	0.027	0.077	0.031	0.036	0.063	0.030	0.032	0.119	0.148	0.166	0.132	0.188	0.019	0.020				
	Nsadzi	0.033	0.019	0.067	0.088	0.027	0.077	0.031	0.037	0.064	0.031	0.032	0.122	0.153	0.171	0.134	0.192	0.019					
	Sserinya	0.031	0.014	0.064	0.087	0.023	0.075	0.026	0.031	0.060	0.025	0.028	0.123	0.152	0.170	0.136	0.194						
	Anjouan	0.164	0.195	0.224	0.191	0.182	0.230	0.204	0.208	0.223	0.204	0.204	0.193	0.137	0.174	0.201							
	Grande Comore	0.108	0.138	0.176	0.130	0.134	0.184	0.149	0.154	0.169	0.148	0.150	0.163	0.166	0.196								
	Mayotte	0.142	0.173	0.204	0.169	0.164	0.212	0.182	0.185	0.200	0.182	0.182	0.157	0.126									
	Mohéli	0.122	0.154	0.187	0.148	0.147	0.194	0.164	0.168	0.181	0.163	0.164	0.153										
	Madagascar	0.093	0.126	0.161	0.122	0.122	0.169	0.136	0.139	0.154	0.135	0.136											
Mainland	Burkina Faso	0.043	0.016	0.050	0.101	0.002	0.063	0.000	0.008	0.052	0.003												
	Cameroon	0.042	0.013	0.051	0.099	0.002	0.063	0.002	0.009	0.050													
	Gabon	0.062	0.053	0.086	0.121	0.046	0.096	0.050	0.056														
	Ghana	0.047	0.019	0.053	0.105	0.007	0.065	0.007															
	Guinea	0.043	0.015	0.049	0.100	0.001	0.062																
	Guinea-Bissau*	0.081	0.066	0.011	0.140	0.052																	
	Mali	0.036	0.012	0.041	0.090																		
	Tanzania	0.059	0.090	0.131																			
	The Gambia*	0.072	0.055																				
	Uganda	0.033																					

Mean F_{ST} between Neighboring	Nearest Mainland	Nearest Island
Formosa	0.001	-
Bioko	0.052	-
Banda	0.029	0.025
Bugala	0.003	0.014
Bukasa	0.020	0.020
Nsadzi	0.019	0.021
Sserinya	0.014	0.017
Anjouan	0.192	0.171
Grande Comore	0.147	0.188
Mayotte	0.163	0.165
Mohéli	0.151	0.143
Madagascar	0.122	0.167

(b) ***Anopheles coluzzii* populations**

		The Gambia*	Mali	Guinea-Bissau*	Guinea	Ghana	Gabon	Cote d'Ivoire	Cameroon	Burkina Faso	Benin	Angola	São Tomé	Príncipe	Bioko
Island	Formosa	0.030	0.028	0.015	0.053	0.032	0.088	0.031	0.076	0.032	0.037	0.133	0.176	0.225	0.090
	Bioko	0.108	0.072	0.107	0.110	0.073	0.036	0.076	0.022	0.076	0.078	0.094	0.151	0.207	
	Príncipe	0.238	0.212	0.238	0.250	0.219	0.204	0.220	0.195	0.220	0.223	0.227	0.130		
	São Tomé	0.191	0.161	0.190	0.200	0.167	0.148	0.168	0.139	0.168	0.171	0.168			
Mainland	Angola	0.149	0.115	0.148	0.156	0.122	0.082	0.123	0.080	0.121	0.126				
	Benin	0.065	0.017	0.062	0.054	0.009	0.075	0.014	0.064	0.019					
	Burkina Faso	0.061	0.001	0.060	0.056	0.018	0.072	0.015	0.062						
	Cameroon	0.095	0.057	0.094	0.096	0.060	0.024	0.061							
	Cote d'Ivoire	0.059	0.014	0.054	0.045	0.008	0.074								
	Gabon	0.108	0.068	0.106	0.108	0.071									
	Ghana	0.061	0.016	0.057	0.048										
	Guinea	0.079	0.051	0.071											
	Guinea-Bissau*	0.011	0.053												
	Mali	0.055													

Mean F_{ST} between Neighboring	Nearest Mainland	Nearest Island
Formosa	0.015	-
Bioko	0.029	0.207
Príncipe	0.199	0.130
São Tomé	0.144	0.130

Figure 6.7 Population structure analysis by F_{ST} analyses. Genetic differentiation between island and mainland African populations of *A. gambiae* (**A**) and *A. coluzzii* (**B**). Analyzes were based on biallelic SNPs on euchromatic regions on chromosome 3. Gray shades highlight low (light) to high (dark) values. Insert tables highlights the mean F_{ST} between each island population and its geographically proximal mainland and island populations. The geographic location for each site and the number of genomes analyzed per site are provided in Fig. 6.4. *Population with *A. coluzzii*/*A. gambiae* hybrid individuals (taken from Lanzaro G.C., et al. (2021). *Evolutionary Applications*, 14(9), 2147–2161).

between the islands and nearby mainland sites in Uganda (0.003–0.029). The divergence between the east African oceanic islands in the Comoros archipelago (Anjouan, Grande Comore, Moheli, and Mayotte) and nearby mainland sites in Tanzania is considerably higher (F_{ST} = 0.130–0.169). The divergence between these islands and Madagascar is likewise high (F_{ST} = 0.153–0.193) as is divergence among island pairs within the archipelago (F_{ST} = 0.126–0.196).

The F_{ST} values for populations of *A. coluzzii* are likewise consistent with the PCA (Fig. 6.7B) suggesting that mainland populations form three groups: a northwestern group (Benin, Burkina Faso, Côte d'Ivoire, Ghana, Guinea, and Mali), a central group (Cameroon and Gabon) and Angola in the south. The F_{ST} values were higher between groups than within (Fig. 6.7B). The divergence between the continental island of Formosa and the nearest mainland populations in Guinea-Bissau and between the island of Bioko and the nearest sites in Cameroon are low ($F_{ST} = 0.015$ and 0.022, respectively). Populations of *A. coluzzii* on the oceanic islands of São Tomé and Príncipe were by far the most isolated from mainland populations ($F_{ST} = 0.144$ and 0.199, respectively). In addition, the two islands were highly diverged from each other ($F_{ST} = 0.130$).

Taken together, the data summarized in Figs. 6.6 and 6.7 reveal a high degree of genetic isolation among oceanic islands compared with either continental or lacustrine islands. These results are consistent with expectations based on island biogeography theory as described above and reinforce the benefits of selecting a contained island site for conducting GEM field trials.

6.4 Evaluation of Potential Island Field Sites

An initial evaluation of all identified potential field trial sites was based on published information describing various characteristics of the islands under consideration to determine if they met the set of predetermined site selection criteria listed in Box 6.1. These characteristics include descriptions of entomological, geographic, and geophysical data.

Site selection criteria (Box 6.1) are organized by priority and availability of data in three tiers. Tier one criteria must be satisfied before moving forward to evaluate a site under tiers two and three. Tier two criteria are important but not crucial for the success of the project. They should, as much as possible, be satisfied before travel to the location is considered. It is unlikely that tier three criteria can be evaluated without a site visit.

Within the context of the geologic, geographic, and biologic features of islands, as described in the preceding narrative, we

Box 6.1 Criteria for the selection of field sites

Tier 1

Presence of *Anopheles gambiae*/*A. coluzzii*

1. Geographic isolation
2. Genetic isolation
3. Island size
4. Topography

Tier 2

1. Other Anopheline species
2. Insecticide susceptibility/resistance
3. Travel feasibility
4. Security
5. *Plasmodium* prevalence
6. Presence of endangered species

Tier 3

1. Mosquito colonizability
2. Willingness to collaborate
3. Human and material resources

here describe our approach to the evaluation of each of the tier 1 considerations for all potential island field sites.

6.4.1 Evaluation of Tier 1 Criteria

Presence of target species

The anopheline mosquito fauna present at selected mainland continental and island sites is presented in Table 6.2. The first criterion for field site selection is the presence of the target species, which are, in our case, *Anopheles gambiae sensu stricto* and/or its sister species *Anopheles coluzzii*. These species are absent from the Canary Islands, Cape Verde, Mauritius, and Réunion; therefore, we exclude them from further consideration.

Geographic isolation

Isolation of a field site from non-target sites is generally considered a key criterion in GEM field site selection. Emigration of GEMs out of the field trial site into neighboring non-target sites on the mainland poses a problem, especially as it relates to risk and regulatory concerns. Equally important is the immigration of wild-type individuals from neighboring sites into the trial site.

Immigration in this case will confound efforts to measure GEM invasiveness and could potentially render the gene drive inefficient or even ineffective. The dynamic equilibrium theory predicts that choosing a remote island as an initial field test site greatly reduces the potential for gene flow between vector populations both into and out of the island site. This is further supported by the results of our population genomics assessment, as discussed below. We excluded any island with a UNEP Isolation Index of less than 15. We used a Surrounding Landmass Proportion (SLMP) value of 1 as a cutoff, so islands with an SLMP value >1 were considered unacceptable. Sites considered unacceptable based on these criteria were the Bijago Islands, Bugala, Koome, Mfangano, Pemba, Unguja, and Mafia.

Genetic isolation

We present original results of a population genomics-based assessment of genetic isolation for the potential island sites and a sample of mainland populations for comparison. These analyzes were based on 50,000 SNPs on chromosome 3 in the *A. coluzzii* and *A. gambiae* genomes. A total of 144 *A. coluzzii* and 246 *A. gambiae* genomes plus an additional 30 hybrid genomes from Guinea-Bissau and The Gambia were included (Fig. 6.4). These genome sequences were obtained from multiple sources: 196 were taken from the Ag1000G database, 167 were from the UC Davis Vector Genetics Laboratory, and 57 were from Bergey et al. 2019 [15]. Results of these analyzes are presented in Figs. 6.6 and 6.7. Both PCA and pairwise F_{ST} analyzes revealed that populations on oceanic islands were highly diverged from neighboring mainland populations. Analysis of populations on both continental and lacustrine islands did not exhibit an equivalent level of genetic isolation. From these results, we conclude that the oceanic islands included in this analysis exhibit a high degree of genetic isolation and that gene flow between oceanic islands and the mainland is minimal.

Island size

There are no well-defined criteria to guide decisions with respect to an appropriately sized area for a GEM field trial. One important consideration is the mosquito flight range. To evaluate the dispersal capacity of a GEM, the site should exceed the flight range of the target species. For our considerations, we assumed a maximal daily flight range of 10 km for *A. gambiae* [63]. Generally, we aimed to identify sites small enough to be manageable, but large enough to be convincing, keeping the following considerations as a guide.

Area (km^2) is an important parameter influencing the biology of populations residing on an island. Large island areas typically include more habitat types and can support larger populations. This characteristic can increase the rate of speciation and lower extinction rates over time [95]. Island size information presented in Table 6.1 is taken from the publication by Weigelt et al. [111]. The methods they employed are briefly described here, but readers interested in detail should review their publication [111]. In describing island size, the database of global administrative area (GADM) was used to obtain high-resolution island polygons. The area was calculated for each GADM polygon in a cylindrical equal-area projection. Areas for archipelagos (Canary Islands, Bijagos, and Cape Verde) were reported here as the sum of all islands in each archipelago [111]. The area for Annobón was obtained from the United Nations Environmental Program [29]. The areas for the Lake Victoria Islands (excluding Koome) were taken from the literature [45, 77, 120]. The area for Koome Island was approximated using Google Earth's distance and area measuring tool. Using island size as a criterion, we exclude the Islands of Annobón and Île Europa as being too small and consider Madagascar as too large.

Topography

As the frequency of a transgene and drive construct increases toward fixation, the offspring of any wild type individuals migrating into the target field site will rapidly (over a short number of generations) acquire the construct through mating with a GEM. It is important that the GEMs should disperse into neighboring populations outside the initial release site and rapidly affect them, thereby achieving regional control. The spread of GEMs with gene drive is an essential component of their utility, increasing sustainability, and decreasing cost of deployment. A convincing trial should measure the dispersal capability of GEMs over terrain that poses a challenge to dispersal, as would occur at most sites throughout mainland Africa. Therefore, an ideal island site should include topographical features that may pose a challenge to mosquito dispersal and gene flow among sites within the island.

Elevation relates to the number of available habitats because of differences between windward and leeward sites, temperature decrease with altitude, and high precipitation regimes at certain altitudes. As a measure of topographic complexity and as a proxy for environmental heterogeneity, the difference between the elevation maximum and minimum of each island measured from sea level is reported in the "Elevation" column in Table 6.1. These were obtained from the AW3D30 of the Japan Aerospace Exploration Agency [60]. GeoTIFF files were downloaded and the highest elevation of each island/archipelago was identified. Island topography was further described using 90 m resolution elevation data from the SRTM 90 m DEM Digital Elevation database [61]. In this case, the altitude and magnitude of the steepest gradient measurements were used as descriptors of topography. A representative sample of these analyzes for the Islands of Grande Comore and São Tomé are presented in Fig. 6.8A,B to illustrate acceptable topography and for the Islands of Zanzibar and Mafia in Fig. 6.8C,D to illustrate a lack of topographic complexity that is unacceptable. Based on topography, we exclude the Bijago Islands, all of the Lake Victoria Islands, Zanzibar, Pemba, Unguja, and Mafia.

6.4.2 Other Considerations – Tier 2 Criteria

Other Anopheline species

An important criterion is a number of non-target Anopheline species present (species richness) at the site, with fewer being the most desirable. In their seminal publication, Macarthur and Wilson [79] applied insights from population biology (birth and death processes) to island biogeography through immigration and extinction processes. Island biogeographic features determine species richness via the island area effect and island distance effect (distance is a proxy variable for isolation). Overall, the theory predicts that species richness generally increases with island area and decreases with distance from the mainland [54, 79, 95]. Anopheline mosquito species richness on islands off the coast of

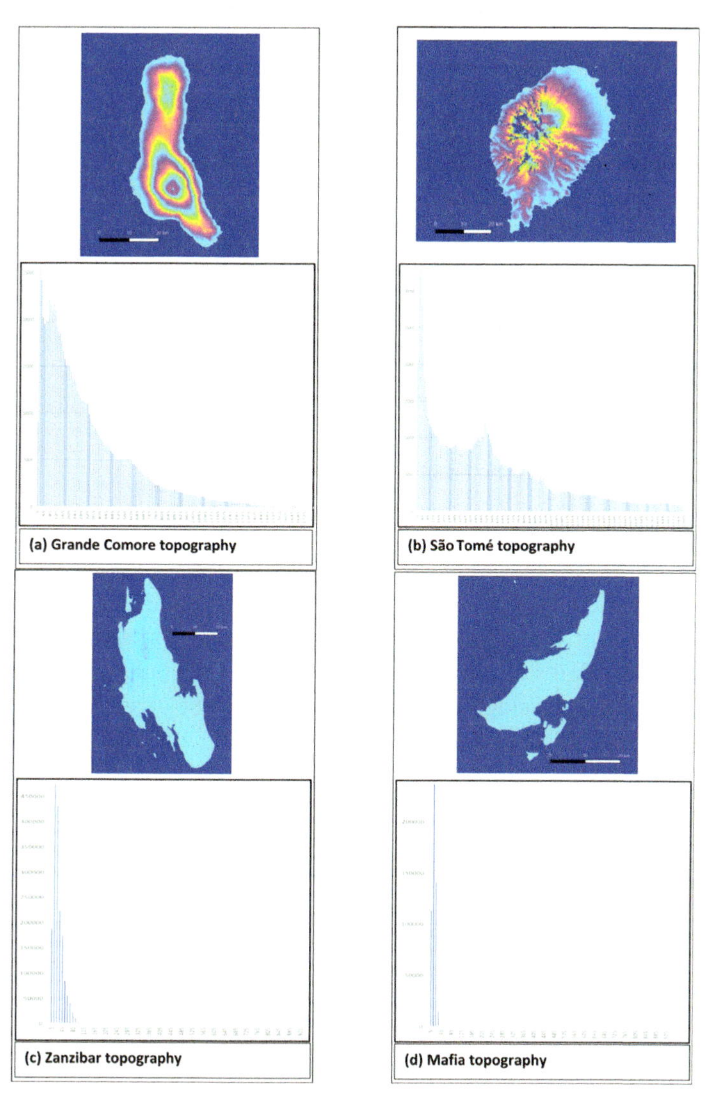

Figure 6.8 Island topography. Satellite images and elevation histograms of Grande Comore (**A**), São Tomé, (**B**), Zanzibar (**C**), and Mafia (**D**). The satellite images show the relief, or variations in topographical complexity, with warmer colors signifying high elevation points and cooler colors signifying low elevation. Histograms show "magnitude of steepest gradient" that display the amount of vertical change per unit of horizontal distance in the direction of maximum change (taken from Lanzaro G.C., et al. (2021). *Evolutionary Applications*, 14(9), 2147–2161).

Africa is consistent with this prediction. As illustrated in Fig. 6.3, oceanic islands have the lowest Anopheline species richness and on this basis, these islands are favored for GEM field trials.

Insecticide resistance

One desirable feature for an island release site is that the local mosquito populations be susceptible to the insecticides routinely used for mosquito control. This would facilitate the application of insecticides to remove GEMs if desired. Susceptibility to insecticides can be assessed by reviewing the literature for the frequency of genes in local mosquito populations known to be related to resistance (e.g., *kdr*) or published results of bioassays. However, resistance is dynamic and a regular schedule of resistance bioassay experiments should be established at candidate sites.

Travel feasibility

Assessment of the practicality of travel to a candidate island includes factors such as availability and regularity of flights, complexity of flight connections, visa requirements, and presence of U.S. consulates or embassies. Frequent travel would be required between GEM development institutions and the selected site as well as shipment of equipment and supplies to the site and personnel and samples from the site. Regular travel and shipping channels are necessary to maintain reasonable study costs.

Security

Researchers and other individuals involved in the project should be able to conduct their work in a safe environment free of hostility. Regular monitoring of the recommendations from the U.S. Department of State regarding travel to island sites is advised as well as checking in with resident collaborators.

Plasmodium prevalence

The endpoints for field evaluations may be entomological endpoints that represent a reduction of disease transmission risk as measured by the proportion of the mosquito population carrying anti-parasite transgenes.

Entomological endpoints are the goal of PHASE 2 studies. Epidemiological endpoints measure an actual reduction in malaria transmission and are most significant in PHASE 3. Nonetheless, it would be of value in a Phase 2 field trial to estimate *P. falciparum* infection rates in mosquitoes pre- and post-release. Having said this, we do not consider high rates of malaria transmission among indigenous human populations to be an essential criterion for our PHASE 2 trials.

Endangered species

Protection of endangered species and understanding of existing regulations and governance surrounding endangered species are also important considerations in the final site selection process. Particularly, in the case of GEM intended for population suppression, the role of the target species within the overall local ecology may require special assessment, as it relates to endangered species, where the risk of negative impacts is more urgently consequential. Species listed as endangered or critically endangered by the International Union for Conservation of Nature (IUCN) Red List should be referenced when determining if an island is suitable for experimental mosquito release [55].

6.4.3 Other Considerations – Tier 3 Criteria

Mosquito colonizability

One strategy to improve GEM fitness is based on moving the transgene construct onto a genetic background as similar as possible to the wild type present at the target release site. This may be achieved by repeated backcrossing of the transgenic strain to a wild-type strain developed from the target site population followed by selection for transgenic individuals prior to release. An alternative method is to establish a wild-type laboratory strain directly from the target field site and transform these *de novo* after maintaining them in the lab for the shortest number of generations feasible. Either of these approaches involves the establishment of laboratory colonies from mosquitoes collected from the island site. In our experience, the ease with which laboratory colonies can be established from field-collected material varies among sites of origin. Some field populations can be readily colonized; whereas, others are extremely difficult to establish and/or maintain. This trait can, at this time, only be assessed empirically. Work toward establishing colonies from indigenous mosquito populations may take time and should be initiated as soon as candidate sites are selected and can be visited.

Willingness to collaborate

The success of a field trial is dependent on the willingness of the local government and resident non-governmental organizations engaged in public health activities to assist with the logistics of field operations and to lead the effort in communicating with resident communities. Stakeholders, local organizations, and communities should also be open to engaging in dialog regarding the potential use of genetically modified organisms.

Human resources

There is a general paucity of well-qualified medical entomologists across sub-Saharan Africa [115]. This is especially true for many of the island sites under investigation here. Nonetheless, there are personnel at institutions of higher learning and malaria control programs who have extensive field experience and can serve as critical resources to the fieldwork involved in conducting PHASE 2 field trials. When candidate sites are identified, identification of these stakeholders and early engagement with them are important in the establishment of relationships and assessing human resource capacity at a site.

Material resources

It will be necessary to rear GEMs on location prior to release. Therefore, assessment of existing insectary facilities available on target islands is important. If these facilities are deemed inadequate upon inspection or are absent, assessment of the availability of space and adequate infrastructure for their construction should be prioritized. Basic Biosafety Level 2 (BSL-2) laboratory space will be required for the pre- and post-evaluation of GEMs at our field site. Holding GEMs in insectary facilities prior to release will require containment meeting local regulations for containment. A thorough review of available space and evaluation of other materials needed for all proposed onsite work should be conducted.

6.5 Conclusion

Each potential site was evaluated based on the criteria listed in Box 6.1. Evaluations were based on information available from the literature or calculated by us as summarized in the narrative above. In addition, new population genomic analyzes conducted by the Vector Genetics Laboratory at UC Davis and presented above were used in our evaluations. Sites that fail to meet all tier 1 criteria were eliminated from further consideration. Those sites that met all tier 1 criteria were raised from potential status to candidate status. Some criteria require further analysis or site visits before a final evaluation can be completed. Sites visits are recommended for candidate sites only. Evaluation of insecticide resistance should be conducted during site visits. Security at candidate sites is dynamic and should be evaluated regularly before and during trials. Evaluation of potential impacts on endangered species requires knowledge about the extent to which these overlap ecologically with *A. coluzzii* and/or *A. gambiae* which can only be thoroughly evaluated by mosquito collections made during early site visits. Assessment of tier 3 criteria can only be evaluated after initial site visits are made.

Overall evaluations are presented in Box 6.2. Evaluation of all 23 potential field sites indicates that Bioko, São Tomé & Príncipe, and the Comoros Islands (Anjouan, Grand Comore, Mayotte, and Moheli) can be elevated from "potential" to "candidate" GEM field trial sites. The Mascarene and Cape Verde Islands fit many criteria, but *Anopheles gambiae* does not occur in these islands. Annobón scores well based on a number of our criteria, but travel was determined to be infeasible and the island was deemed too small to represent a trial, which would provide compelling outcomes.

Therefore, we propose the following as the lead candidate sites for a PHASE 2 GEM field trial: the Comoros Islands, São Tomé and Príncipe, and Bioko.

The framework described here has been applied by the University of California Malaria Initiative (UCMI) as it enters PHASE 2 of GEM research. It is our belief that this comprehensive framework provided identification of site(s) that will maximize the prospect for success, minimize risk, and will serve as a fair, valid, and convincing

Box 6.2 Overall summary of the evaluation of potential island sites. Boxes marked + and shaded green = site meets criterion; boxes marked - and shaded red = site fails to meet criterion; boxes marked ? and shaded orange = data not available; boxes marked * and shaded gray = criterion to be determined when a site is visited

	TIER 1					TIER 2						TIER 3		
ISLAND	A. coluzzii/A. gambiae present	Geographic isolation	Genetic isolation	Size (area)	Topography	Other Anopheline species	Insecticide susceptibility/resistance	Travel feasibility	Security	Plasmodium prevalence	Endangered species	Mosquito colonizability	Willingness to collaborate	Human/material resources
Canary Islands	-	+	?	+	+	+	*	+	*	-	*	*	*	*
Cape Verde	-	+	?	+	+	+	*	+	*	-	*	*	*	*
Bijagos	+	-	+	+	-	-	*	+	*	+	*	*	*	*
Bioko	+	+	+	+	+	-	*	+	*	+	*	*	*	*
Annobón	+	+	+	-	+	+	*	-	*	+	*	*	*	*
São Tomé	+	+	+	+	+	+	*	+	*	+	*	*	*	*
Príncipe	+	+	+	+	+	+	*	+	*	+	*	*	*	*
Bugala	+	-	-	+	-	+	*	+	*	+	*	*	*	*
Koome	+	-	-	+	-	+	*	+	*	+	*	*	*	*
Mfangano	+	-	-	+	-	+	*	+	*	+	*	*	*	*
Ukara	+	-	-	+	-	+	*	+	*	+	*	*	*	*
Zanzibar		-	?	+	-	+	*	+	*	+	*	*	*	*
Pemba	+	-	?	+	-	+	*	+	*	+	*	*	*	*
Unguja	+	-	?	+	-	+	*	+	*	+	*	*	*	*
Mafia	+	-	?	+	-	-	*	+	*	+	*	*	*	*
Grand Comore	+	+	+	+	+	+	*	+	*	+	*	*	*	*
Moheli	+	+	+	+	+	+	*	+	*	+	*	*	*	*
Anjouan	+	+	+	+	+	+	*	+	*	+	*	*	*	*
Mayotte	+	+	+	+	+	+	*	+	*	+	*	*	*	*
Ile Europa	?	+	?	-	-	?	*	-	*	?	*	*	*	*
Madagascar	+	+	+	-	+	-	*	+	*	+	*	*	*	*
Mauritius	-	+	?	+	+	+	*		*		*	*	*	*

test of the efficacy and impacts of the UCMI GEM product, meeting the goal of a PHASE 2 field trial. Furthermore, this process provides a well-reasoned, science-based justification for selecting these sites for GEM field trials.

References

1. African Development Bank. (2018). *Comoros Data Portal.* Retrieved from: https://comoros.opendataforafrica.org/
2. African Union and NEPAD. (2018). *Gene Drives for Malaria Control and Elimination in Africa.* Retrieved from https://www.nepad.org/publication/gene-drives-malaria-control-and-elimination-africa
3. Ajamma, Y. U., Villinger, J., Omondi, D., Salifu, D., Onchuru, T. O., Njoroge, L., et al. (2016). Composition and genetic diversity of mosquitoes (diptera: culicidae) on islands and mainland shores of Kenya's Lakes Victoria and Baringo. *Journal of Medical Entomology, 53*(6), 1348–1363. Retrieved from https://www.ncbi.nlm.nih.gov/pubmed/27402888. doi:10.1093/jme/tjw102
4. Akogbéto, M. (2000). Le paludisme côtier lagunaire à Cotonou: données entomologiques. *Cahiers d'études et de recherches francophones / Santé, 10*(4), 267–275.
5. Alphey, L., Beard, C. B., Billingsley, P., Coetzee, M., Crisanti, A., Curtis, C., et al. (2002). Malaria control with genetically manipulated insect vectors. *Science, 298*(5591), 119–121. Retrieved from https://www.ncbi.nlm.nih.gov/pubmed/12364786. doi:10.1126/ science.1078278
6. Alves, J., Gomes, B., Rodrigues, R., Silva, J., Arez, A. P., Pinto, J., et al. (2010). Mosquito fauna on the Cape Verde Islands (West Africa): An update on species distribution and a new finding. *J Vector Ecol., 35*(2), 307–312. Retrieved from https://www.ncbi.nlm.nih.gov/pubmed/21175936. doi:10.1111/j.1948-7134.2010.00087.x
7. Anopheles Gambiae 1000 Genomes Consortium. (2020). Genome variation and population structure among 1142 mosquitoes of the African malaria vector species Anopheles gambiae and Anopheles coluzzii. *Genome Research, 30,* 1–14. doi:10.1101/gr.262790.120
8. Ant, T., Foley, E., Tytheridge, S., Johnston, C., Goncalves, A., Ceesay, S., et al. (2020). A survey of Anopheles species composition and insecticide resistance on the island of Bubaque, Bijagos Archipelago,

Guinea-Bissau. *Malar J, 19*(1), 27. Retrieved from https://www.ncbi.nlm.nih.gov/pubmed/31941507. doi:10.1186/s12936-020-3115-1

9. Antonio-Nkondjio, C., Kerah, C. H., Simard, F., Awono-Ambene, P., Chouaibou, M., Tchuinkam, T., et al. (2006). Complexity of the malaria vectorial system in Cameroon: Contribution of secondary vectors to malaria transmission. *Journal of Medical Entomology, 43*(6), 1215–1221. Retrieved from https://www.ncbi.nlm.nih.gov/pubmed/17162956. doi:10.1603/0022-2585(2006)43[1215:cotmvs]2.0.co;2
10. Baeshen, R., Ekechukwu, N. E., Toure, M., Paton, D., Coulibaly, M., Traore, S. F., et al. (2014). Differential effects of inbreeding and selection on male reproductive phenotype associated with the colonization and laboratory maintenance of Anopheles gambiae. *Malar J, 13*, 19. Retrieved from https://www.ncbi.nlm.nih.gov/pubmed/24418094. doi:10.1186/1475-2875-13-19
11. Baez, M., & Fernandez, J. M. (1980). Notes on the mosquito fauna of the Canary Islands (Diptera: Culicidae). *Mosquito Systematics, 12*(3), 349–355.
12. BakamaNume, B. B. (2010). A Contemporary Geography of Uganda. Mkuki Na Nyota.
13. Belkin, J. N. (1962). The Mosquitoes of the South Pacific: (Diptera, Culicidae). University of California Press.
14. Benedict, M., D'Abbs, P., Dobson, S., Gottlieb, M., Harrington, L., Higgs, S., et al. (2008). Guidance for contained field trials of vector mosquitoes engineered to contain a gene drive system: Recommendations of a scientific working group. *Vector Borne and Zoonotic Diseases, 8*(2), 127–166. Retrieved from https://www.ncbi.nlm.nih.gov/pubmed/18452399. doi:10.1089/vbz.2007.0273
15. Bergey, C. M., Lukindu, M., Wiltshire, R. M., Fontaine, M. C., Kayondo, J. K., & Besansky, N. J. (2020). Assessing connectivity despite high diversity in island populations of a malaria mosquito. *Evolutionary Applications, 13*(2), 417–431. Retrieved from https://onlinelibrary.wiley.com/doi/abs/10.1111/eva.12878. doi:https://doi.org/10.1111/eva.12878
16. Berzosa, P. J., Cano, J., Roche, J., Rubio, J. M., Garcia, L., Moyano, E., et al. (2002). Malaria vectors in Bioko Island (Equatorial Guinea): PCR determination of the members of Anopheles gambiae Giles complex (Diptera: Culicidae) and pyrethroid knockdown resistance (kdr) in An. gambiae sensu stricto. *J Vector Ecol., 27*(1), 102–106. Retrieved from https://www.ncbi.nlm.nih.gov/pubmed/12125862.

17. Boete, C. (2009). Anopheles mosquitoes: Not just flying malaria vectors... especially in the field. *Trends in Parasitology, 25*(2), 53–55. Retrieved from https://www.ncbi.nlm.nih.gov/pubmed/19095498. doi:10.1016/j.pt.2008.10.005
18. Boussès, P., Dehecq, J. S., Brengues, C., & Fontenille, D. (2013). Inventaire actualisé des moustiques (Diptera : Culicidae) de l'île de La Réunion, océan Indien. *Bulletin de la Société de pathologie exotique, 106*(2), 113–125. Retrieved from https://doi.org/10.1007/s13149-013-0288-7. doi:10.1007/s13149-013-0288-7
19. Brown, D. M., Alphey, L. S., McKemey, A., Beech, C., & James, A. A. (2014). Criteria for identifying and evaluating candidate sites for open-field trials of genetically engineered mosquitoes. *Vector Borne and Zoonotic Diseases, 14*(4), 291–299. Retrieved from https://www.ncbi.nlm.nih.gov/pubmed/24689963. doi:10.1089/vbz.2013.1364
20. Brunhes, J. (1977). Les moustiques de l'archipel des Comores. *Cahiers ORSTOM. Série Entomologie Médicale et Parasitologie, 25*(2), 131–152.
21. Brunhes, J., Le Goff, G., & Geoffroy, B. (1997). Anophèles afro-tropicaux : 1. Descriptions d'espèces nouvelles et changements de statuts taxonomiques (Diptera : Culicidae). *Annales de la Societe Entomologique de France, 33.*
22. Campos, E. G., Trevino, H. A., & Strom, L. G. (1961). The dispersal of mosquitoes by railroad trains involved in international traffic. *Mosquito News, 21*(3), 190–192.
23. Campos, M., Hanemaaijer, M., Gripkey, H., Collier, T., Lee, Y., Cornel, A., et al. (2021). The origin of island populations of the African malaria mosquito, Anopheles coluzzii. Submitted to Communications Biology.
24. Carballar-Lejarazú, R., Ogaugwu, C., Tushar, T., Kelsey, A., Pham, T. B., Murphy, J., et al. (2020). Next-generation gene drive for population modification of the malaria vector mosquito, <em> Anopheles gambiae </em>. *Proceedings of the National Academy of Sciences, 117*(37), 22805–22814. Retrieved from https://www.pnas.org/ content/pnas/117/37/22805.full.pdf. doi:10.1073/pnas.2010214117
25. Carpenter, S. R. (1996). Microcosm experiments have limited relevance for community and ecosystem ecology. *Ecology, 77*(3), 677–680. Retrieved from http://www.jstor.org/stable/ 2265490. doi:10.2307/2265490
26. Cisnetto, V., Barlow, J. (2020). The development of complex and controversial innovations: Genetically modified mosquitoes for

malaria eradication. *Research Policy, 49*(3), 103917. Retrieved from https://www.sciencedirect.com/science/article/pii/S0048733319302355. doi:https://doi.org/10.1016/j.respol.2019.103917

27. Coetzee, M., Hunt, R. H., Wilkerson, R., Della Torre, A., Coulibaly, M. B., Besansky, N. J. (2013). Anopheles coluzzii and Anopheles amharicus, new members of the Anopheles gambiae complex. *Zootaxa, 3619*, 246–274. Retrieved from https://www.ncbi.nlm.nih.gov/pubmed/26131476.
28. Cuamba, N., Mendis, C. (2009). The role of Anopheles merus in malaria transmission in an area of southern Mozambique. *J Vector Borne Dis., 46*(2), 157–159. Retrieved from https://www.ncbi.nlm.nih.gov/pubmed/19502697.
29. Dahl, A. L. (1991). Island directory (Prelim. ed.). Nairobi, Kenya: UNEP.
30. Deitz, K. C., Athrey, G. A., Jawara, M., Overgaard, H. J., Matias, A., & Slotman, M. A. (2016). Genome-wide divergence in the West-African malaria vector Anopheles melas. *G3 (Bethesda), 6*(9), 2867–2879. Retrieved from https://www.ncbi.nlm.nih.gov/pubmed/27466271. doi:10.1534/g3.116.031906
31. Diver, K. C. (2008). ORIGINAL ARTICLE: Not as the crow flies: assessing effective isolation for island biogeographical analysis. *Journal of Biogeography, 35*(6), 1040–1048. Retrieved from https://onlinelibrary.wiley.com/doi/abs/10.1111/j.1365-2699.2007.01835.x. doi:https://doi.org/10.1111/j.1365-2699.2007.01835.x
32. Drury, D. W., Dapper, A. L., Siniard, D. J., Zentner, G. E., & Wade, M. J. (2017). CRISPR/Cas9 gene drives in genetically variable and nonrandomly mating wild populations. *Sci Adv., 3*(5), e1601910. Retrieved from https://www.ncbi.nlm.nih.gov/pubmed/28560324. doi:10.1126/sciadv.1601910
33. Eritja, R., Palmer, J., Roiz, D., Sanpera-Calbet, I., Bartumeus, F. (2017). Direct evidence of adult aedes albopictus dispersal by car. *Scientific Reports, 7*, 14399. doi:10.1038/s41598-017-12652-5
34. Facchinelli, L., Valerio, L., Ramsey, J. M., Gould, F., Walsh, R. K., Bond, G., et al. (2013). Field cage studies and progressive evaluation of genetically-engineered mosquitoes. *PLoS Negl Trop Dis., 7*(1), e2001. Retrieved from https://www.ncbi.nlm.nih.gov/pubmed/23350003. doi:10.1371/journal.pntd.0002001
35. Fick, S. E., Hijmans, R. J. (2017). WorldClim 2: new 1-km spatial resolution climate surfaces for global land areas. *International Journal of Climatology, 37*(12), 4302–4315. Retrieved from

https://rmets.onlinelibrary.wiley.com/doi/abs/10.1002/joc.5086. doi:https://doi.org/10.1002/joc.5086

36. Frankham, R. (1997). Do island populations have less genetic variation than mainland populations? *Heredity (Edinb), 78* (3), 311–327. Retrieved from https://www.ncbi.nlm.nih.gov/pubmed/9119706. doi:10.1038/hdy.1997.46
37. Frean, J., Brooke, B., Thomas, J., Blumberg, L. (2014). Odyssean malaria outbreaks in Gauteng Province, South Africa, 2007–2013. *S Afr Med J, 104*(5), 335–338. Retrieved from https://www.ncbi.nlm.nih.gov/pubmed/25212198. doi:10.7196/samj.7684
38. Gillies, M. T., De Meillon, B. (1968). *The Anophelinae of Africa South of the Sahara: Ethiopian Zoogeographical Region*. South African Institute for Medical Research.
39. Google Earth (Cartographer). Map showing location of Kome Island. Retrieved from https://earth.google.com/web/search/Koome+Island,+Uganda/@-0.08724243,32.74244361,1211.47630539a,48380.8541521d,35y,16.3547808h,11.41826421t,0r/data=CigiJgokCQxDZ9yeljVAEQtDZ9yeljXAGRf30zaOXUJAIW6hBeDnvlDA
40. Guerra, C., Fuseini, G., Donfack, O., Smith, J., Mifumu, T., Akadiri, G., et al. (2020). Malaria outbreak in Riaba district, Bioko Island: lessons learned. *Malaria Journal, 19*. doi:10.1186/s12936-020-03347-w
41. Haji, K. A., Khatib, B. O., Smith, S., Ali, A. S., Devine, G. J., Coetzee, M., et al. (2013). Challenges for malaria elimination in Zanzibar: Pyrethroid resistance in malaria vectors and poor performance of long-lasting insecticide nets. *Parasit Vectors, 6*, 82. Retrieved from https://www.ncbi.nlm.nih.gov/pubmed/23537463. doi:10.1186/1756-3305-6-82
42. Hammond, A., Galizi, R., Kyrou, K., Simoni, A., Siniscalchi, C., Katsanos, D., et al. (2016). A CRISPR-Cas9 gene drive system targeting female reproduction in the malaria mosquito vector Anopheles gambiae. *Nature Biotechnology, 34*(1), 78–83. Retrieved from https://www.ncbi.nlm.nih.gov/pubmed/26641531. doi:10.1038/nbt.3439
43. Highton, R. B., van Someren, E. C. (1970). The transportation of mosquitos between international airports. *Bull World Health Organ, 42*(2), 334–335. Retrieved from https://www.ncbi.nlm.nih.gov/pubmed/5310146.
44. Huestis, D., Dao, A., Diallo, M., Sanogo, Z. L., Samake, D., Yaro, A., et al. (2019). Windborne long-distance migration of malaria mosquitoes in the Sahel. *Nature*.

45. Idris, Z. M., Chan, C. W., Kongere, J., Gitaka, J., Logedi, J., Omar, A., et al. (2016). High and heterogeneous prevalence of asymptomatic and sub-microscopic malaria infections on islands in Lake Victoria, Kenya. *Sci Rep., 6*, 36958. Retrieved from https://www.ncbi.nlm.nih.gov/pubmed/27841361. doi:10.1038/srep36958
46. Instituo Nacional de Estatística. (2018). *Dados Demograficos 1970 á 2016*. Retrieved from: https://www.ine.st/index.php/component/phocadownload/category/51-demograficas
47. Institut National de la Statistique de Madagascar. (2020). Madagascar en chiffre. In (pp. Online table).
48. Institut National de la Statistique et des Études Économiques et Démographiques. (2020). *Estimation de la population au 1er janvier 2020* [Population information]. Retrieved from: https://www.insee.fr/fr/statistiques/1893198
49. Instituto Nacional de Estadística. (2010). *RGPH 2010 - Cabo Verde em Numeros*. Retrieved from: http://ine.cv/en/quadros/rgph-2010-cabo-verde-em-numeros/
50. Instituto Nacional de Estadística. (2020). *Población residente por fecha, sexo y edad* [Population information]. Retrieved from: https://www.ine.es/jaxiT3/Tabla.htm?t=31304
51. Instituto Nacional de Estadística de Guinea Ecuatorial. (2015). *CENSO 2015* [Population census]. Retrieved from: https://inege.gq/index.php/estadisticas/#46-demografia-y-poblacion
52. Instituto Nacional de Estatiística e Censos (Guinea-Bissau). (2009). *Recenseamento geral da população e habitação Guiné-Bissau : III RGPH/2009*. Retrieved from: http://arks.princeton.edu/ark:/88435/dsp01w6634600z
53. Irish, S. R., Kyalo, D., Snow, R. W., Coetzee, M. (2020). Updated list of Anopheles species (Diptera: Culicidae) by country in the Afrotropical Region and associated islands. *Zootaxa, 4747*(3), zootaxa 4747 4743 4741. Retrieved from https://www.ncbi.nlm.nih.gov/pubmed/32230095. doi:10.11646/zootaxa.4747.3.1
54. Itescu, Y., Foufopoulos, J., Pafilis, P., & Meiri, S. (2020). The diverse nature of island isolation and its effect on land bridge insular faunas. *Global Ecology and Biogeography, 29*(2), 262–280. Retrieved from https://onlinelibrary.wiley.com/doi/abs/10.1111/geb.13024. doi: https://doi.org/10.1111/geb.13024
55. IUCN. (2020). *The IUCN Red List of Threatened Species*. Retrieved from: https://www.iucnredlist.org

56. Iyaloo, D. P., Elahee, K. B., Bheecarry, A., Lees, R. S. (2014). Guidelines to site selection for population surveillance and mosquito control trials: A case study from Mauritius. *Acta Tropica, 132 Suppl*, S140–149. Retrieved from https://www.ncbi.nlm.nih.gov/pubmed/24280144. doi:10.1016/j.actatropica.2013.11.011
57. Jackson, G., Gartlan, J. S. (1965). The flora and fauna of lolui island, Lake Victoria: A study of vegetation, men and monkeys. *Journal of Ecology, 53*(3), 573–597. Retrieved from http://www.jstor.org/stable/2257622. doi:10.2307/2257622
58. James, S., Collins, F. H., Welkhoff, P. A., Emerson, C., Godfray, H. C. J., Gottlieb, M., et al. (2018). Pathway to deployment of gene drive mosquitoes as a potential biocontrol tool for elimination of malaria in Sub-Saharan Africa: Recommendations of a scientific working group (dagger). *American Journal of Tropical Medicine and Hygiene, 98*(6), 1–49. Retrieved from https://www.ncbi.nlm.nih.gov/pubmed/29882508. doi:10.4269/ajtmh.18-0083
59. James, S. L., Marshall, J. M., Christophides, G. K., Okumu, F. O., Nolan, T. (2020). Toward the definition of efficacy and safety criteria for advancing gene drive-modified mosquitoes to field testing. *Vector Borne and Zoonotic Diseases, 20*(4), 237–251. Retrieved from https://www.ncbi.nlm.nih.gov/pubmed/32155390. doi:10.1089/vbz.2019.2606
60. Japan Aerospace Exploration Agency. (2020). *ALOS Global Digital Surface Model (DSM) ALOS World 3D-30m (AW3D30) Version 3.1.* Retrieved from: https://www.eorc.jaxa.jp/ALOS/en/aw3d30/aw3d30v31_product_e_a.pdf
61. Jarvis A., H. I. R., A. Nelson, E. Guevara. (2008). *Hole-filled seamless SRTM data V4.* Retrieved from: https://srtm.csi.cgiar.org/
62. Jones, P. J. (1994). Biodiversity in the Gulf of Guinea: An overview. *Biodiversity & Conservation, 3*(9), 772–784. Retrieved from https://doi.org/10.1007/BF00129657. doi:10.1007/BF00129657
63. Kaufmann, C., Briegel, H. (2004). Flight performance of the malaria vectors Anopheles gambiae and Anopheles atroparvus. *J Vector Ecol., 29*(1), 140–153. Retrieved from https://www.ncbi.nlm.nih.gov/pubmed/15266751.
64. Kayondo, J. K., Mukwaya, L. G., Stump, A., Michel, A. P., Coulibaly, M. B., Besansky, N. J., et al. (2005). Genetic structure of Anopheles gambiae populations on islands in northwestern Lake Victoria, Uganda. *Malar J, 4*, 59. Retrieved from https://www.ncbi.nlm.nih.gov/pubmed/16336684. doi:10.1186/1475-2875-4-59

65. Knols, B., Bossin, H. (2006). Identification and characterization of field sites for genetic control of disease vectors. Dordrecht, The Netherlands: Springer, 2006:203–209.

66. Kolopack, P. A., Lavery, J. V. (2017). Informed consent in field trials of gene-drive mosquitoes. *Gates Open Res, 1*, 14. Retrieved from https://www.ncbi.nlm.nih.gov/pubmed/29355214. doi:10.12688/gatesopenres.12771.1

67. Kyalo, D., Amratia, P., Mundia, C. W., Mbogo, C. M., Coetzee, M., Snow, R. W. (2017). A geo-coded inventory of anophelines in the Afrotropical Region south of the Sahara: 1898–2016. *Wellcome Open Res, 2*, 57. Retrieved from https://www.ncbi.nlm.nih.gov/pubmed/28884158. doi:10.12688/wellcomeopenres.12187.1

68. Kyrou, K., Hammond, A., Galizi, R., Kranjc, N., Burt, A., Beaghton, A., et al. (2018). A CRISPR-Cas9 gene drive targeting doublesex causes complete population suppression in caged Anopheles gambiae mosquitoes. *Nature Biotechnology, 36*. doi:10.1038/nbt.4245

69. Lainhart, W., Bickersmith, S. A., Moreno, M., Rios, C. T., Vinetz, J. M., Conn, J. E. (2015). Changes in genetic diversity from field to laboratory during colonization of anopheles darlingi root (diptera: culicidae). *American Journal of Tropical Medicine and Hygiene, 93*(5), 998–1001. Retrieved from https://www.ncbi.nlm.nih.gov/pubmed/26283742. doi:10.4269/ajtmh.15-0336

70. Lavery, J. V., Harrington, L. C., Scott, T. W. (2008). Ethical, social, and cultural considerations for site selection for research with genetically modified mosquitoes. *American Journal of Tropical Medicine and Hygiene, 79*(3), 312–318. Retrieved from https://www.ncbi.nlm.nih.gov/pubmed/18784220.

71. Lee, Y., Marsden, C. D., Norris, L. C., Collier, T. C., Main, B. J., Fofana, A., et al. (2013). Spatiotemporal dynamics of gene flow and hybrid fitness between the M and S forms of the malaria mosquito, Anopheles gambiae. *Proc Natl Acad Sci U S A, 110*(49), 19854–19859. Retrieved from https://www.ncbi.nlm.nih.gov/pubmed/24248386. doi:10.1073/pnas.1316851110

72. Li, M., Yang, T., Kandul, N. P., Bui, M., Gamez, S., Raban, R., et al. (2020). Development of a confinable gene drive system in the human disease vector Aedes aegypti. *Elife, 9*. Retrieved from https://www.ncbi.nlm.nih.gov/pubmed/31960794. doi:10.7554/eLife.51701

73. Loiseau, C., Melo, M., Lee, Y., Pereira, H., Hanemaaijer, M. J., Lanzaro, G. C., et al. (2019). High endemism of mosquitoes on São Tomé and

Príncipe Islands: evaluating the general dynamic model in a worldwide island comparison. *Insect Conservation and Diversity, 12*(1), 69–79. Retrieved from https://onlinelibrary.wiley.com/doi/abs/10.1111/icad.12308. doi:https://doi.org/10.1111/icad.12308

74. Lopez Del Amo, V., Bishop, A. L., Sanchez, C. H., Bennett, J. B., Feng, X., Marshall, J. M., et al. (2020). A transcomplementing gene drive provides a flexible platform for laboratory investigation and potential field deployment. *Nat Commun, 11*(1), 352. Retrieved from https://www.ncbi.nlm.nih.gov/pubmed/31953404. doi:10.1038/s41467-019-13977-7
75. Losos, J. B., Ricklefs, R. E. (2009). Adaptation and diversification on islands. *Nature, 457* (7231), 830–836. Retrieved from https://www.ncbi.nlm.nih.gov/pubmed/19212401. doi:10.1038/nature07893
76. Lounibos, L. P. (2002). Invasions by insect vectors of human disease. *Annual Review of Entomology, 47*, 233–266. Retrieved from https://www.ncbi.nlm.nih.gov/pubmed/11729075. doi:10.1146/annurev.ento.47.091201.145206
77. Lounio, T. (2014). Population Dynamics and Livelihood Change on Ukara Island, Lake Victoria. (Master's Thesis), University of Helsinki,
78. Lukindu, M., Bergey, C. M., Wiltshire, R. M., Small, S. T., Bourke, B. P., Kayondo, J. K., et al. (2018). Spatio-temporal genetic structure of Anopheles gambiae in the Northwestern Lake Victoria Basin, Uganda: Implications for genetic control trials in malaria endemic regions. *Parasit Vectors, 11*(1), 246. Retrieved from https://www.ncbi.nlm.nih.gov/pubmed/29661226. doi:10.1186/s13071-018-2826-4
79. MacArthur, R. H., Wilson, E. O. (1967). The theory of island biogeography. Princeton, N.J., Princeton University Press.
80. Moreno, M., Cano, J., Nzambo, S., Bobuakasi, L., Buatiche, J. N., Ondo, M., et al. (2004). Malaria Panel Assay versus PCR: detection of naturally infected Anopheles melas in a coastal village of Equatorial Guinea. *Malar J, 3*, 20. Retrieved from https://www.ncbi.nlm.nih.gov/pubmed/15238168. doi:10.1186/1475-2875-3-20
81. Mosha, F. W., Petrarca, V. (1983). Ecological studies on Anopheles gambiae complex sibling species on the Kenya coast. *Transactions of the Royal Society of Tropical Medicine and Hygiene, 77*(3), 344–345. Retrieved from https://www.ncbi.nlm.nih.gov/pubmed/6623592. doi:10.1016/0035-9203(83)90161-x
82. Mugono, M., Konje, E., Kuhn, S., Mpogoro, F. J., Morona, D., Mazigo, H. D. (2014). Intestinal schistosomiasis and geohelminths of Ukara

Island, North-Western Tanzania: Prevalence, intensity of infection and associated risk factors among school children. *Parasit Vectors, 7*, 612. Retrieved from https://www.ncbi.nlm.nih.gov/pubmed/25533267. doi:10.1186/s13071-014-0612-5

83. Nambuya, A. S. C., Nkwiine, C., Wetala, P. (2013). Abundance and ecological functional categories of soil macrofauna as indicators of soil chemical properties status in oil palm plantations in Bugala island Kalangala District, Uganda. Paper presented at the Joint Proceedings of 27th Soil Science Society of East Africa & 6th African Soil Science Society, Nakuru, Kenya.
84. Nampijja, M., Webb, E. L., Kaweesa, J., Kizindo, R., Namutebi, M., Nakazibwe, E., et al. (2015). The Lake Victoria Island Intervention Study on Worms and Allergy-related diseases (LaVIISWA): study protocol for a randomized controlled trial. *Trials, 16*, 187. Retrieved from https://www.ncbi.nlm.nih.gov/pubmed/25902705. doi:10.1186/s13063-015-0702-5
85. Nash, A., Urdaneta, G. M., Beaghton, A. K., Hoermann, A., Papathanos, P. A., Christophides, G. K., et al. (2019). Integral gene drives for population replacement. *Biol Open, 8*(1). Retrieved from https://www.ncbi.nlm.nih.gov/pubmed/30498016. doi:10.1242/bio.037762
86. National Academies of Sciences Engineering and Medicine. (2016). Gene drives on the horizon: Advancing science, navigating uncertainty, and aligning research with public values. Washington, DC: The National Academies Press.
87. National Bureau of Statistics. (2012). 2012 PHC: Population distribution by administrative areas [Census report]. Retrieved from: https://www.nbs.go.tz/index.php/en/census-surveys/population-and-housing-census/162-2012-phc-population-distribution-by-administrative-areas
88. Neuhaus, C. P. (2018). Community engagement and field trials of genetically modified insects and animals. *Hastings Cent Rep., 48*(1), 25–36. Retrieved from https://www.ncbi.nlm.nih.gov/pubmed/29457234. doi:10.1002/hast.808
89. Norris, D. E., Shurtleff, A. C., Toure, Y. T., Lanzaro, G. C. (2001). Microsatellite DNA polymorphism and heterozygosity among field and laboratory populations of Anopheles gambiae ss (Diptera: Culicidae). *Journal of Medical Entomology, 38*(2), 336–340. Retrieved from https://www.ncbi.nlm.nih.gov/pubmed/11296845. doi:10.1603/0022-2585-38.2.336

90. Ogola, E., Villinger, J., Mabuka, D., Omondi, D., Orindi, B., Mutunga, J., et al. (2017). Composition of Anopheles mosquitoes, their blood-meal hosts, and Plasmodium falciparum infection rates in three islands with disparate bed net coverage in Lake Victoria, Kenya. *Malar J, 16*(1), 360. Retrieved from https://www.ncbi.nlm.nih.gov/pubmed/28886724. doi:10.1186/s12936-017-2015-5
91. Resnik, D. B. (2014). Ethical issues in field trials of genetically modified disease-resistant mosquitoes. *Dev World Bioeth, 14*(1), 37–46. Retrieved from https://www.ncbi.nlm.nih.gov/pubmed/23279283. doi:10.1111/dewb.12011
92. Ross, P. A., Endersby-Harshman, N. M., Hoffmann, A. A. (2019). A comprehensive assessment of inbreeding and laboratory adaptation in Aedes aegypti mosquitoes. *Evol Appl., 12*(3), 572–586. Retrieved from https://www.ncbi.nlm.nih.gov/pubmed/30828375. doi:10.1111/eva.12740
93. Salgueiro, P., Moreno, M., Simard, F., O'Brochta, D., Pinto, J. (2013). New insights into the population structure of Anopheles gambiae s.s. in the Gulf of Guinea Islands revealed by Herves transposable elements. *PLoS One, 8*(4), e62964. Retrieved from https://www.ncbi.nlm.nih.gov/pubmed/23638171. doi:10.1371/journal.pone.0062964
94. Sanford, M. R., Cornel, A. J., Nieman, C. C., Dinis, J., Marsden, C. D., Weakley, A. M., et al. (2014). Plasmodium falciparum infection rates for some Anopheles spp. from Guinea-Bissau, West Africa. *F1000Res, 3*, 243. Retrieved from https://www.ncbi.nlm.nih.gov/pubmed/25383188. doi:10.12688/f1000research.5485.2
95. Santos, A. M. C., Field, R., Ricklefs, R. E. (2016). New directions in island biogeography. *Global Ecology and Biogeography, 25*(7), 751–768. Retrieved from https://onlinelibrary.wiley.com/doi/abs/10.1111/geb.12477. doi:https://doi.org/10.1111/geb.12477
96. Schindler, D. W. (1998). Whole-ecosystem experiments: Replication versus realism: the need for ecosystem-scale experiments. *Ecosystems, 1*(4), 323–334. Retrieved from https://doi.org/10.1007/s10021 9900026. doi:10.1007/s100219900026
97. Schmidt, H., Collier, T. C., Hanemaaijer, M. J., Houston, P. D., Lee, Y., Lanzaro, G. C. (2020). Abundance of conserved CRISPR-Cas9 target sites within the highly polymorphic genomes of Anopheles and Aedes mosquitoes. *Nat Commun., 11*(1), 1425. Retrieved from https://www.ncbi.nlm.nih.gov/pubmed/32188851. doi:10.1038/s41467-020-15204-0

98. Schmidt, H., Lee, Y., Collier, T. C., Hanemaaijer, M. J., Kirstein, O. D., Ouledi, A., et al. (2019). Transcontinental dispersal of Anopheles gambiae occurred from West African origin via serial founder events. *Commun Biol., 2,* 473. Retrieved from https://www.ncbi.nlm.nih.gov/pubmed/31886413. doi:10.1038/s42003-019-0717-7
99. Scott, T. W., Takken, W., Knols, B. G., Boete, C. (2002). The ecology of genetically modified mosquitoes. *Science, 298*(5591), 117–119. Retrieved from https://www.ncbi.nlm.nih.gov/pubmed/12364785. doi:10.1126/science.298.5591.117
100. Service, M. W. (1997). Mosquito (Diptera: Culicidae) dispersal - the long and short of it. *Journal of Medical Entomology, 34*(6), 579–588. Retrieved from https://www.ncbi.nlm.nih.gov/pubmed/9439109. doi:10.1093/jmedent/34.6.579
101. Smith, A. (1955). The Transmission of Bancroftial Filariasis on Ukara Island, Tanganyika. I.—A Geographical and Ecological Description of the Island with an annotated List of Mosquitos and other Arthropods of medical Importance. *Bulletin of Entomological Research, 46*(2), 419–436. Retrieved from https://www.cambridge.org/core/article/transmission-of-bancroftial-filariasis-on-ukara-island-tanganyika-ia-geographical-and-ecological-description-of-the-island-with-an-annotated-list-of-mosquitos-and-other-arthropods-of-medical-importance/193FF09AACF861834482ACB9CF8AB3D6. doi:10.1017/S000748530003100X
102. Smith, A. C. I. (1984). International transportation of mosquitoes of public health importance. In M. Laird (Ed.), *Commerce and the spread of pests and disease vectors,* 1–21, New York: Praeger.
103. Song, M., Wang, B., Liu, J., Gratz, N. (2003). Insect vectors and rodents arriving in China aboard international transport. *Journal of Travel Medicine, 10*(4), 241–244. Retrieved from https://www.ncbi.nlm.nih.gov/pubmed/12946302. doi:10.2310/7060.2003.40603
104. Ssegawa, P., Nkuutu, D. N. (2006). Diversity of vascular plants on Ssese islands in Lake Victoria, central Uganda. *African Journal of Ecology, 44*(1), 22–29. Retrieved from https://onlinelibrary.wiley.com/doi/abs/10.1111/j.1365-2028.2006.00609.x. doi:https://doi.org/10.1111/j.1365-2028.2006.00609.x
105. Statistics Mauritius. (2020). *Population and Vital Statistics - Republic of Mauritius January – June 2020* [Census report]. Retrieved from: https://statsmauritius.govmu.org/Pages/Statistics/ESI/Population/Pop_Vital_Jan-Jun20.aspx

106. Takahashi, S. (1984). Survey on accidental introductions of insects entering Japan via aircraft. In M. Laird (Ed.), *Commerce and the Spread of Pests and Disease Vectors,* 65–79, New York: Praeger.

107. Tryon, C. A., Faith, J. T., Peppe, D. J., Beverly, E. J., Blegen, N., Blumenthal, S. A., et al. (2016). The Pleistocene prehistory of the Lake Victoria basin. *Quaternary International, 404,* 100–114. Retrieved from https://www.sciencedirect.com/science/article/pii/S1040618215011994. doi:https://doi.org/10.1016/j.quaint.2015.11.073

108. Tuhebwe, D., Bagonza, J., Kiracho, E. E., Yeka, A., Elliott, A. M., Nuwaha, F. (2015). Uptake of mass drug administration program for schistosomiasis control in Koome Islands, Central Uganda. *PLoS One, 10*(4), e0123673. Retrieved from https://www.ncbi.nlm.nih.gov/pubmed/25830917. doi:10.1371/journal.pone.0123673

109. Unckless, R. L., Clark, A. G., Messer, P. W. (2017). Evolution of resistance against CRISPR/Cas9 gene drive. *Genetics, 205*(2), 827–841. Retrieved from https://www.ncbi.nlm.nih.gov/pubmed/27941126. doi:10.1534/genetics.116.197285

110. United Nations Food and Agricultural Organization. (2001). *Global ecological zoning for the global forest resources assessment 2000.* Retrieved from Rome: http://www.fao.org/3/ad652e/ad652e00.htm

111. Weigelt, P., Jetz, W., Kreft, H. (2013). Bioclimatic and physical characterization of the world's islands. *Proceedings of the National Academy of Sciences, 110*(38), 15307–15312. Retrieved from https://www.pnas.org/content/pnas/110/38/15307.full.pdf. doi:10.1073/pnas.1306309110

112. White, G. B. (1971). Chromosomal evidence for natural interspecific hybridization by mosquitoes of the Anopheles gambiae complex. *Nature, 231*(5299), 184–185. Retrieved from https://www.ncbi.nlm.nih.gov/pubmed/4930676. doi:10.1038/231184a0

113. White, G. B. (1985). Anopheles bwambae sp.n., a malaria vector in the Semliki Valley, Uganda, and its relationships with other sibling species of the An.gambiae complex (Diptera: Culicidae). *Systematic Entomology, 10*(4), 501–522. Retrieved from https://onlinelibrary.wiley.com/doi/abs/10.1111/j.1365-3113.1985.tb00155.x. doi:https://doi.org/10.1111/j.1365-3113.1985.tb00155.x

114. WHO Expert Committee on Malaria & World Health Organization. (1979). *WHO Expert Committee on Malaria : Seventeenth report* (9241206403). Retrieved from Geneva: https://apps.who.int/iris/handle/10665/41359

115. WHO Vector Control Technical Expert Group Report to MPAC. (2013). *Capacity Building in Entomology and Vector Control.* Paper presented at the Malaria Policy Advisory Committee Meeting, Crowne Plaza Hotel, Geneva. https://www.who.int/malaria/mpac/mpac_sep13_vcteg_report.pdf?ua=1

116. WHO/TDR and FNIH. (2014). *The Guidance framework for testing genetically modified mosquitoes* [Framework]. Retrieved from https://www.who.int/tdr/publications/year/2014/Guidance_ framework_mosquitoes.pdf

117. World Health Organization. (2016). *Vector Surveillance and Control at Ports, Airports, and Ground Crossings.* Geneva, Switzerland: WHO Press.

118. World Health Organization. (2020). *World malaria report 2020: 20 years of global progress and challenges.* Retrieved from Geneva:

119. Wynn, G., Paradise, C. J. (2001). Effects of microcosm scaling and food resources on growth and survival of larval Culex pipiens. *BMC Ecology, 1,* 3. Retrieved from https://www.ncbi.nlm.nih.gov/pubmed/11527507. doi:10.1186/1472-6785-1-3

120. Zeemeijer, I. M. (2012). *Who gets What, When and How? New Corporate Land Acquisitions and the Impact on Local Livelihoods in Uganda.* (Master's Thesis), Utrecht University,

Chapter 7

Modeling Priorities as Gene Drive Mosquito Projects Transition from Lab to Field

John M. Marshall[a,*] **and Ace R. North**[b,*]
[a]*Divisions of Epidemiology & Biostatistics, School of Public Health, University of California, Berkeley, California, USA*
[b]*Department of Zoology, University of Oxford, Oxford, UK*
john.marshall@berkeley.edu, ace.north@zoo.ox.ac.uk

7.1 Introduction

Despite significant reductions in malaria incidence and prevalence over the last decade following the wide-spread distribution of long-lasting insecticide treated nets (LLINs), malaria is not expected to be eliminated with currently available tools - LLINs, indoor residual spraying with insecticides (IRS), and artemisinin combination therapy drugs (ACTs) (Walker et al., 2016). Consequently, there is interest in novel interventions that complement existing ones, such as attractive targeted sugar baits (ATSBs) (Traore et al., 2020),

*Equal contribution.

Mosquito Gene Drives and the Malaria Eradication Agenda
Edited by Rebeca Carballar-Lejarazú

ISBN 978-981-4968-33-1 (Hardcover), 978-1-003-30877-5 (eBook)
www.jennystanford.com

transmission-blocking vaccines (Coelho et al., 2019), and gene drive-modified mosquitoes. Gene drive mosquitoes have been described as playing a potentially transformative role toward malaria elimination as, provided that societal and regulatory approval can be gained, they are not hindered by human compliance issues that other interventions face and they should be capable of spreading beyond their site of application, potentially impacting disease transmission on a wide scale (James et al., 2018). These interventions are synergistic with currently available ones as, for example, reducing the malaria parasite population through the distribution of ACTs, and gene drive mosquitoes engineered with disease-refractory genes results in fewer opportunities for the evolution of parasite resistance to either intervention, more than would be the case for intervention in isolation (Marshall et al., 2019).

Mathematical modeling has a central role to play in determining the impact that gene drive systems could have, alongside other interventions, toward the goal of malaria elimination. Predicting the spread of genes through populations and their impacts on mosquito populations requires an understanding of the dynamics of mosquitoes, humans, and parasites, with specific attention to genetic inheritance, mosquito life history, and pathogen transmission. Mathematical models provide a means to integrate the knowledge of each of these components of vector-borne disease transmission, including a mechanistic understanding of the processes involved and data that could be used to validate these models. Given the potential irreversibility of a release of gene drive-modified organisms, models provide a means to address questions regarding safety and efficacy prior to a release based on the best available data. Such models may be used as a basis for risk assessment, field trial design, and to explore hypothetical intervention scenarios.

In this chapter, we survey modeling priorities as gene drive mosquito projects advance from the lab to the field. We begin by highlighting priorities in model building, namely (i) capturing nuances in the inheritance-biasing impacts of gene drive systems, (ii) incorporating data and insights on mosquito vector ecology, including life history, habitat distribution, and movement patterns, and (iii) aligning entomological models with detailed models of malaria transmission, including the impacts of currently available and novel interventions. We then highlight several priorities in the model application as gene drive products advance from the lab to the field. These include informing target product profiles (TPPs) for gene drive products to assess when they satisfy safety and efficacy criteria, and informing the design of cage trials, field trials, and eventually vector and disease control interventions. Other priorities include developing monitoring programs to assess the safety and efficacy of trials and interventions, developing surveillance programs to detect unintended spread, and addressing risk and regulatory questions requiring a quantitative analysis.

We focus on CRISPR-based homing gene drive systems, as these currently have the most promise to contribute to the malaria eradication agenda. Indeed, three of the most advanced malaria vector gene drive projects at present—Target Malaria, the UC Irvine Malaria Initiative, and Transmission Zero—are developing CRISPR-based homing gene drive systems to drive either a fitness load or disease-refractory gene into a mosquito population. Homing-based gene drive systems are able to spread through a population despite a fitness cost due to their overrepresentation in the gametes of a heterozygote. This is achieved by expressing an endonuclease which creates a double-stranded break at a highly specific site chosen as a target for the drive integration (Esvelt et al., 2014). Homology-directed repair (HDR) then copies the drive allele to the cut chromosome (Rong and Golic, 2003). If this occurs in the germline, it effectively converts a heterozygote into a homozygote in terms of inheritance.

Two general classes of CRISPR-based homing gene drive systems have been proposed for the control of malaria vectors: (i) constructs that aim to suppress vector populations by introducing fitness costs or a sex bias, and (ii) constructs that aim to modify vector populations by introducing traits that reduce disease transmission. Additionally, some constructs are intended to be self-sustaining, whereby they spread from one population to another with the goal of disease control over a wide area, and others are designed to be self-limiting, whereby their spread is limited in both space and time. Gene drive systems exist that fall into each pair of categories, and have been reviewed by others (Sinkins and Gould, 2006; Alphey, 2014; Burt, 2014; Godfray et al., 2017; Raban et al., 2020). Here, we provide an overview of modeling considerations that apply to the development of several gene drive designs, with an emphasis on homing-based systems intended for wide-scale malaria control.

7.2 Model Building

Good models should satisfy the principle of parsimony and be tailored to the questions they are designed for. In the words of physicist Albert Einstein, "Everything should be made as simple as possible, but no simpler," and in the words of statistician George Box, "All models are wrong, but some are useful." When gene drive systems were first proposed as a novel approach to control mosquitoes, the models used to explore the idea were appropriately abstract and generic (Burt, 2003; Deredec et al., 2008). These models focused on population genetics—considering changes in gene frequencies in mostly randomly mixing populations—and ignored most details of mosquito ecology and parasite transmission. Such models are also used for new technologies, such as CleaveR (Oberhofer et al., 2019) and TARE (Champer et al., 2020), which have only recently been proposed. But as the earlier technologies have advanced and data have become available, models with increasing levels of detail have been developed for specific gene drive systems in specific mosquito species intended for release in specific environments (North et al., 2019, 2020; Eckhoff et al., 2017). As potential field trials edge closer, models capable of addressing a myriad of logistical questions will be required, and increasing detail will be needed to address the nuances of biased inheritance, mosquito vector ecology, and local malaria transmission.

7.2.1 Population Genetics Models

The early conceptual models of homing-based gene drive preceded the discovery of CRISPR as a gene-editing tool and described the population genetics of a gene drive system, H, and a wildtype allele, W (Burt, 2003; Deredec et al., 2008). In particular, the model of Deredec et al. (2008) defined homing rate, e, to be the probability that a wildtype allele in the germline of a heterozygote is converted

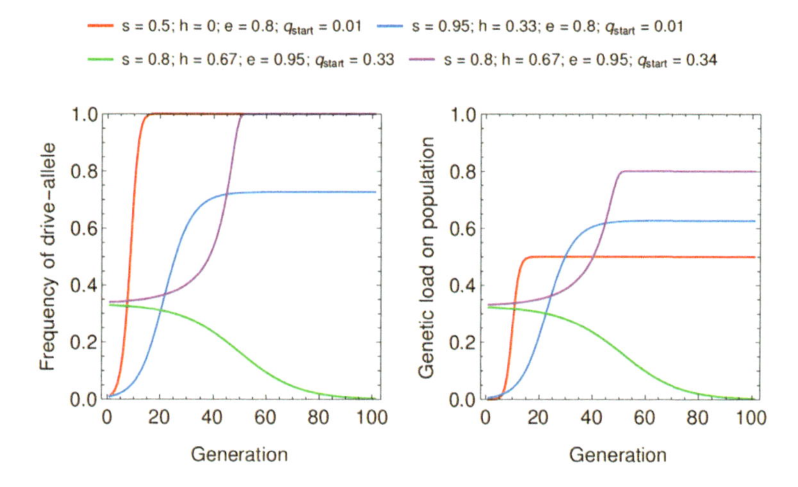

Figure 7.1 Population genetics of a homing drive allele. Time-series dynamics depending on the homing rate, *e*, fitness load, *s*, dominance of the fitness load, *h*, and initial drive allele frequency, q_{start}. The model, which assumes that homing occurs after gene expression so that costs of being homozygous are not experienced by individuals in which homing occurs, is described by Deredec et al. (2008).

to a gene drive allele, fitness load, *s*, to be the fitness penalty incurred by homozygotes (HH), and dominance, *h*, to be the extent to which this fitness cost is imposed on heterozygotes (HW). Depending on the parameter values, this model predicted a number of possible outcomes following the introduction of a gene drive system into a population: (i) the inheritance bias exceeds the fitness load and the drive allele spreads into the population from low initial frequencies, eventually eliminating the wildtype allele completely, (ii) the drive allele spreads to a stable equilibrium frequency at which both alleles coexist, (iii) the drive allele either spreads to fixation or is lost depending on its initial frequency in the population, and (iv) the drive allele is invariably lost from the population (Burt, 2003; Deredec et al., 2008) (Fig. 7.1).

These early models of homing-based gene drive also considered resistant alleles, R, which contain the coding frame and function of the target site but not the gene drive recognition site, and thus are reistant to homing (Deredec et al., 2008). These resistant alleles may already exist due to standing genetic variation (SGV), or may form through (i) error-prone repair mechanisms such as non-homologous end-joining (NHEJ) and microhomology-mediated end-joining (MMEJ), or (ii) *de novo* mutation after the drive allele has been introduced (Unckless et al., 2017). If the drive allele confers a fitness load while the drive-resistant allele does not, then the drive-resistant allele is expected to prevail in the end and the level of SGV, rate of resistant allele generation, and magnitude of fitness advantage of the drive-resistant allele over the drive allele will determine the timescale over which this occurs (Marshall et al., 2017; Noble et al., 2017).

Subsequent models of homing-based gene drive have incorporated two varieties of alleles conferring resistance to homing: (i) in-frame resistance alleles that preserve the function of the target gene, R (sometimes denoted as "R1"), and (ii) in-frame or out-of-frame costly resistant alleles that disrupt target gene function, B (sometimes denoted as "R2"). Some of these models also account for the different stages at which wildtype alleles are converted to intact drive or drive-resistant alleles, notably: (i) pre-fertilization, in which the allelic make-up of gametes in either parent is distorted, and (ii) post-fertilization, in which wildtype alleles in a fertilized embryo are converted to drive-resistant alleles following deposition of Cas9 by a mother having the intact drive allele (Deredec et al., 2008; Champer et al., 2017; Pham et al., 2019; Adolfi et al., 2020). Maternal deposition of Cas9 produces a mosaic marker phenotype in some offspring, as somatic allele conversions are manifest in some embryonic cells but not others, and are heritable when occurring in embryonic germ cells.

When incorporating these details of CRISPR-based homing drive into a model, parameters that need estimating include: (i) the proportion of W alleles that are cleaved in the gametes of HW heterozygotes, either female or male, (ii) the proportion of those cleaved that are subject to accurate HDR and become H alleles, either in females or males, (iii) the proportion of resistance alleles that are R versus B (or R1 versus R2), (iv) the proportion of W alleles among offspring of mothers having the Cas9 allele that is cleaved through maternal deposition in the fertilized embryo, and whether this depends on how many Cas9 alleles the mother has, and (v) the proportion of those maternally cleaved alleles that become R alleles versus B alleles in the fertilized embryo (Pham et al., 2019; Adolfi et al., 2020). Paternal deposition of Cas9 in the fertilized embryo is also possible and may be important. Fitness costs must be considered, including costs due to the H and B (or R2) alleles, especially when no copy of a functional target gene is present.

Laboratory cage studies are useful for estimating the parameters of population genetic models. These studies fall into two categories: (i) crossing experiments, in which mosquitoes having different genotypes are crossed and marker phenotypes (and corresponding genotypes) of offspring are tallied (Kyrou et al., 2018), and (ii) cage experiments, in which populations are monitored over several generations and the time-series of marker phenotypes (and hence genotypes) provides insights into their generating inheritance and fitness processes (Pollegioni et al., 2020; Adolfi et al., 2020). Crossing experiments may be used to directly estimate parameters such as cleavage and HDR rates, while large cage studies are useful to estimate fitness effects and to refine estimates of rates of resistance allele generation, including their fitness effects. Both approaches also allow 95% confidence or credible intervals to be estimated for model parameters.

In addition to single-locus homing-based gene drives, multi-locus gene drives are important to consider as many have properties that would enable them to serve as intermediate technologies in a phased release pipeline (Li et al., 2020), or as systems to remediate transgenes from the environment following a trial period (Xu et al., 2020). For instance, in split drive systems, transient drive activity is achieved by locating the Cas9 and guide RNA (gRNA) components at separate loci (Li et al., 2020). Drive occurs at the gRNA locus when the Cas9 and gRNA alleles co-occur in an organism; but this effect is transient as the Cas9 allele is gradually eliminated from the population due to a fitness cost, the two alleles segregate across generations, and the gRNA allele is also gradually eliminated due to a fitness cost. Modeling the population genetics of these systems requires many more possible genotypes and crosses to be considered, as the number of possible genotypes multiplies across loci. There are also additional biological features that sometimes emerge. For instance, a split drive system in *Drosophila melanogaster* displays "shadow drive" in addition to regular drive at the gRNA locus, in which maternally-deposited Cas9 biases inheritance of the gRNA allele for one extra generation, even if the Cas9 allele is not transmitted to the offspring (Terradas et al., 2021).

7.2.2 Mosquito Vector Models

Mosquito vector models come in a variety of shapes and sizes, reflecting differences in the population of interest and motivating questions. For instance, a model describing an island mosquito population will have different requirements to one describing spread across a region. Investigations requiring seasonality may include detailed consideration of rainfall (Lambert et al., 2018) and/or temperature (Beck-Johnson et al., 2013). Notwithstanding, there are basic ingredients that are common to models of mosquito vectors of malaria. Next, we discuss these commonalities while highlighting important points of divergence between models. We pay particular attention to how models address three major details of mosquito biology: (i) density dependence in the mosquito life cycle, (ii) mosquito movement behavior, and (iii) mosquito dry season biology.

7.2.2.1 Mosquito life cycle

The first step to building a mosquito population model is to decompose the mosquito life cycle into distinct life stages. While most models incorporate an aquatic juvenile and terrestrial adult stage, the inclusion of further life-history detail is more variable. For *Anopheles gambiae s.l.*, it is well documented that the juvenile stage is composed of eggs (2–3 days), followed by four larval instars (7–10 days in total), followed by pupae (2–3 days) (Silver, 2007). Some models treat these stages distinctly (Lunde et al., 2013; Eckhoff et al., 2017), while others distinguish eggs from larvae from pupae but lump all larval stages together (Deredec et al., 2011), or into two larval stages (White et al., 2011), or lump all the juvenile stages together (North and Godfray, 2018; Ermert et al., 2011). Adults are typically categorized by sex, and it is common to distinguish unmated from mated females (Deredec et al., 2011). Among mated females, some models decompose the stages of the gonotrophic cycle, for instance into host-seeking and ovipositing sub-stages (North et al., 2013), while others assume all mated females oviposit a number of eggs each day (Ermert et al., 2011; North and Godfray, 2018).

In addition to classifying mosquitoes by life stage, more detailed models also classify adult mosquitoes by chronological age, which allows consideration of age-dependent mortality, in particular "senescence," whereby the mortality rate increases with age. While senescence is often observed in laboratory conditions (Dawes et al., 2009; Benedict et al., 2009), there is less evidence that it is important in natural populations (Clements and Paterson, 1981). Senescence may be rare in wild mosquito populations because mosquitoes face many age-independent hazards, including predation, meaning that few individuals live long enough for age itself to impact mortality. Models of natural mosquito populations, therefore, tend to assume that adult mosquitoes have a constant mortality rate, which produces exponentially-distributed lifespans. Female mosquito lifespan is an important parameter in malaria epidemiology because it affects both vectorial capacity and the population growth rate (Macdonald, 1957). Several methods are available to estimate this parameter, including inference from decaying collection numbers in mark-release-recapture (MRR) studies, and dissections that are able to estimate the number of gonotrophic cycles that a female mosquito has completed (Detinova et al., 1945; Polovodova, 1949).

7.2.2.2 Spatial population structure

Spatially-explicit models of mosquito population dynamics are needed to model the spread of gene drives through spatially-structured populations. Two broad approaches that have been applied to study this are: (i) metapopulation models, which consist of a set of randomly mixing populations connected by migration, and (ii) models that describe the movement of individuals through continuous space, either using a reaction-diffusion or individual-based approach. The simplest metapopulation model is a two-patch model, whereby two distinct populations exchange migrants. The two-patch model has been used to investigate modifications that are not intended to spread beyond their release population. For instance, Sudweeks et al. (2019) used a two-patch model to investigate the potential of gene drives to suppress island populations of invasive rodents by targeting alleles that are fixed on the island but not present on the mainland. Various threshold-dependent drive systems have also been studied in this way (Dhole et al., 2019; Burt and Deredec, 2018; Marshall and Hay, 2012).

Metapopulation models generalize the two-patch model to two or more randomly mixing populations connected by migration. The most abstract metapopulation models assume the local populations can be defined by equilibrium states, such as wildtype, gene drive, or empty, without explicitly considering individuals within populations (Fig. 7.2A). These models can help to illustrate some abstract principles, for instance that spatial structure can result in wildtype and driving genes coexisting in a metapopulation, even when local

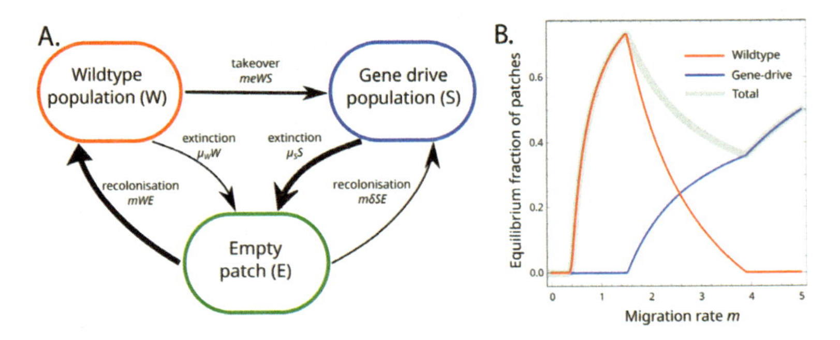

Figure 7.2 A population-based metapopulation model describing the spatial population dynamics of a suppression gene drive allele. **A.** It is assumed that the gene drive will rapidly spread to fixation in wildtype populations ("takeover" events, which occur at rate *meWS* for every wildtype patch), while gene drive populations become extinct more readily than wildtype populations ($\mu_S > \mu_W$) and are less efficient at recolonizing empty habitat ($\delta < 1$). Except for extinction, these processes are all mediated by the migration rate, *m*, among populations, and the predicted equilibrium state of the metapopulation depends critically upon this parameter. **B.** As migration rate increases from zero, the metapopulation transitions from extinct ($m < 0.4$) to exclusively wildtype (the gene drive cannot invade, $0.4 < m < 1.52$) to coexistence ($1.52 < m < 3.9$), to exclusively gene drive ($m > 3.9$). The model parameters are: $e = 1.5$, $\mu_S = 2$, $\mu_W = 0.4$, and $\delta = 0.8$. Full details of the simulation are provided in Bull et al. (2019).

coexistence is not possible (Bull et al., 2019). This outcome is predicted to occur for intermediate rates of migration between local populations. By contrast, low migration rates are predicted to prevent the gene drive from establishing, while high rates allow the gene drive to eliminate the wildtype (Fig. 7.2B). While these models can be instructive, metapopulation models that consider the internal dynamics of local populations reveal further complexity in how spatial structure may affect gene drive dynamics in real landscapes (North et al., 2019, 2020).

A variety of models have been created to describe both the metapopulation and local population dynamics of mosquito vectors. Over a decade ago, the Skeeter Buster model was created (Magori et al., 2009; Legros et al., 2012) by extending the CIMSiM model of spatial mosquito ecology (Focks et al., 1995) to model the spread of gene drive systems through spatially-explicit *Ae. aegypti* populations. Soon after, the EMOD malaria model (Eckhoff, 2011) was used to simulate the spread of homing-based gene drive systems through national-scale *An. gambiae* populations (Eckhoff et al., 2017). While both model frameworks describe mosquito populations on a grid, migration rates between populations may vary arbitrarily, meaning that the grid may represent a more complex geographical space. The MGDrivE framework (Sánchez et al., 2020; Wu et al., 2020) has recently been created to simulate the spread of user-defined gene drive systems through a network of populations arbitrarily arranged in space, and the framework of North and Godfray (2018) describes *An. gambiae* populations distributed in populations coinciding with human settlements on the basis that *An. gambiae* is anthropophilic. The decision on which approach is best to use depends on a number of factors, including mosquito species, knowledge of the study area, and available computational power.

Models that describe the movement of gene drive organisms through continuous space take either a reaction-diffusion or individual-based approach. A reaction-diffusion framework that describes the spatial spread of gene drives as traveling waves was proposed by Tanaka et al. (2017) as a spatial generalization of a population genetics model (Unckless et al., 2015). This work suggested that, for a drive system with a fitness cost within a certain range (i.e., $0.5 < s < 0.7$), the spread would occur for releases exceeding a critical population density, and a barrier conferring a selective disadvantage to gene drive organisms could potentially contain the spread. A reaction-diffusion model of a driving Y chromosome has been used to investigate additional characteristics of spatial spread, in particular the speed of spread (Beaghton et al., 2016). Finally, an individual-based model that describes the spread of gene drive organisms in continuous space, taking into account life history, was used by North et al. (2013) to investigate the effects of fine-scale stochasticity and larval and feeding site densities on gene drive spread. The computational requirements for these types of individual-based models mean that they are best suited to analyses at small spatial scales, while reaction-diffusion frameworks can explore less detailed models at larger spatial scales.

7.2.2.3 Density dependence

As mosquitoes develop from egg to larva to pupa to adult, there are some life-history processes that occur differently in sparse versus crowded populations. These are referred to as "density-dependent" processes, and are especially important to identify and characterize because they act to regulate populations, preventing them from growing indefinitely. In mosquitoes, there are two life-history processes that are widely accepted as being density-dependent: (i) larval development and mortality, as crowding of larvae in pools of water increases competition for space and resources, and (ii) mating, because the ease at which a newly-emerged female finds a mate depends on the density of the adult male population, and sometimes their ability to form swarms (Mozūraitis et al., 2020).

Larval density-dependence is widely regarded as the most significant factor regulating population size for anopheline mosquitoes, and a number of lab (Lyimo et al., 1992; Koenraadt and Takken, 2003) and field studies (Gimnig et al., 2002; Muriu et al., 2013) have sought to quantify this by measuring larval development and mortality at a range of densities. Based on these studies, three broad effects of larval crowding have been identified: reduced survival, prolonged development, and smaller adults. Uneven competition between the four instar stages has also been observed, with fourth-stage instars predating on first-stage instars under some circumstances (Koenraadt and Takken, 2003). Impacts on larval development time and resulting adult size were seen in two field studies, but the studies disagreed on whether larval density had an impact on mortality (Gimnig et al., 2002; Muriu et al., 2013).

Theoretical ecologists have classified crowding competition that affects survival along a spectrum from "contest" to "scramble" (Bellows, 1981). "Contest competition" arises when some contestants acquire their required resources at the expense of others who perish. "Scramble competition" arises when resources are equally shared. The limited empirical evidence suggests that contest competition is a better model for anopheline larvae (Muriu et al., 2013; White et al., 2011), and most models use a contest functional form to model larval survival. Under contest competition, populations converge to a stable size in a given environment, referred to as the "carrying capacity." Anopheline larvae tend to develop in temporary water bodies created by recent rainfall (Gimnig et al., 2001; Shililu et al., 2003), motivating some models to assume that larval competition reduces with recent rainfall (Ermert et al., 2011; Lambert et al., 2018; North and Godfray, 2018; Lunde et al., 2013; Wu et al., 2020). Some models include temperature effects in the contest competition function (Ermert et al., 2011; Lunde et al., 2013), or include additional effects such as asymmetric competition between larval instars at different stages (Lunde et al., 2013; Eckhoff et al., 2017). We are unaware of population models that incorporate density-dependent effects for larval development time or resulting adult size. The difficulty of including these factors is not so much building the model as estimating the parameters from limited data. We advocate for more research to deepen our understanding of these effects.

While larval competition is an example of negative density dependence, mating is an example of positive density dependence, as female mosquitoes find mates more easily at higher population densities (Mozūraitis et al., 2020). As such, density-dependent mating is unlikely to be important in large populations, where most females will mate within a day or two of emerging; however, it may be important in small populations that could be driven to extinction if males become so rare that most females fail to mate, an example of an Allee effect (Courchamp et al., 1999). This may be important in highly seasonal environments where population density becomes very low in the dry season, and in populations suppressed by suppression gene drive systems. Density-dependent mating can be incorporated into models by assuming females mate at slower rates in smaller populations leading to a significant risk of death before finding a mate (North and Godfray, 2018).

7.2.2.4 Movement ecology

Models of mosquito movement are crucial to gain an informed prediction of how transgenes may spread spatially. Understanding of mosquito movement patterns is limited due to the fact that mosquito flight is difficult to observe directly, and so movement patterns are inferred from indirect evidence, primarily MRR studies and genetic data. In MRR studies, dispersal of captured and released mosquitoes is inferred from their recapture locations. These studies are best suited to inferring small-scale movement patterns, and typically take place within a single village, although a handful of experiments have recaptured mosquitoes outside the release village, allowing the frequency of inter-village movements to be estimated (Thomson et al., 1995; Taylor et al., 2001; Costantini et al., 1996). These studies indicate that movements between neighboring villages occur at rates in the region of 0.5–3% per adult mosquito per day (North and Godfray, 2018), though caution is needed in extrapolating from so few data, and such rates will be highly dependent on the local geography and distance between villages.

Little is known about the limits of local mosquito flight distances (Verdonschot and Besse-Lototskaya, 2014), or about long-range movements that may occur if adult mosquitoes "hitchhike" in high-altitude seasonal winds (Huestis et al., 2019) or in human vehicles. These movements are not suited to MRR studies, however genetic or genomic data could be leveraged to obtain some estimates of their scale and rates. For instance, the *An. gambiae* 1000 Genomes Project has shown that insecticide-resistant haplotypes have, in recent decades, spread over large parts of sub-Saharan Africa from a small number of origins (Anopheles gambiae 1000 Genomes Consortium et al., 2017). Other approaches to infer intermediate and large-scale movement based on genetic data include Wright's fixation index, F_{ST}, using a variety of genetic markers such as single nucleotide polymorphisms (SNPs) and repeat sequences such as microsatellites (Marsden et al., 2013; Taylor et al., 2001), and novel methods utilizing the theory of identity-by-descent, which provide an estimate of effective dispersal averaged over several generations (Novembre and Slatkin, 2009). We hope that the growing interest in gene drive technology will spur further studies of anopheline genomic data, and further MRR experiments, to gain additional insights into both fine-scale and large-scale movement patterns.

7.2.2.5 Dry season ecology

Much of sub-Saharan Africa undergoes seasonal cycling between dry and rainy conditions; a duality that is particularly pronounced in the Sahel where the dry season may last up to ten months (Nicholson, 2013). Sahelian dry conditions are hostile to mosquito populations, which seem to disappear during these times only to reappear at the onset of each rainy season (Dao et al., 2014). How mosquito species achieve this remains the subject of debate, despite decades of research. The most prominent explanations are: (i) adult mosquitoes aestivate by hiding in shelters and re-emerge when rains begin (Omer and Cloudsley-Thompson, 1970; Dao et al., 2014), (ii) extinct areas are recolonized each year by adult mosquitoes dispersing from nearby locations with permanent larval habitat (Ramsdale et al., 1970; Jawara et al., 2008), and (iii) adult mosquitoes recolonize after being carried large distances by high-altitude seasonal winds (Garrett-Jones, 1962; Huestis et al., 2019). Evidence for each of these hypotheses, which are not mutually exclusive, is limited to indirect observations. For example, the recapture of a single *An. gambiae* female that was marked in the previous rainy season in a village in Mali is suggestive of aestivation (Lehmann et al., 2010), while the capture of small numbers of *An. coluzzii* females on sticky panels that were raised to 40–290 meters nightly in Mali are suggestive of wind-borne migration (Huestis et al., 2019). Genomic analyses offer hope for further insights.

To investigate how dry season ecology may influence a gene drive vector suppression program, North et al. implemented each of these hypotheses within a model of gene drive in West Africa (North et al., 2019, 2020). A number of subtle effects of dry season ecology were reported. For example, simulations suggest long-distance migration speeds up the spread of gene drives to remote locations, yet may also facilitate the recolonization of populations that have been extirpated by the drive allele in the case of a population suppression strategy. Aestivation was generally predicted to slow the rate of gene drive spread because it results in fewer active mosquito generations per year (it was assumed that mosquitoes aestivate in all locations, including those with permanent larval habitat). Additional modeling could explore the impact of dry season ecology on the use of different varieties of gene drive systems.

7.2.3 Malaria Transmission Models

Models of disease transmission are becoming increasingly relevant to models of gene drive, as: (i) the readiness of a gene drive system for field trials will be determined in part by its expected (i.e., modeled) epidemiological impact, and (ii) initial field trials are expected to have a measured entomological outcome alongside a modeled epidemiological outcome (James et al., 2020). Given the potential for a non-localized gene drive system to spread widely, it has been acknowledged that drive systems at the trial stage should be expected to cause a significant reduction in disease transmission. Therefore, readiness for field trials should be determined by alignment with a TPP that includes the expected impact on disease transmission (James et al., 2018). Models that incorporate both gene drive and epidemiological dynamics are also important for the design of monitoring protocols, so that epidemiological outcomes can be inferred from entomological field measurements.

Vector-borne disease models can be more complicated than those for diseases transmitted directly from human to human as they require consideration of the vector, host, and pathogen. However, the dynamics simplify when disease progression in humans and mosquitoes is considered in parallel, with the interaction between the two occurring according to "force of infection" (FOI) terms, the FOI in humans (per capita rate at which susceptible humans become infected), λ_H, and the FOI in vectors (per capita rate at which susceptible mosquitoes become infected), λ_V. Malaria infection in mosquitoes is reasonably described by an SEI model (susceptible-exposed-infectious), in which adult mosquitoes emerge from pupae in the susceptible state, become exposed and latently infected at a per capita rate equal to λ_V, and progress to infectiousness after an "extrinsic incubation period" (EIP) (Ross, 1910; Macdonald, 1957) (Fig. 7.3A). The FOI in mosquitoes is proportional to the fraction of humans that are infectious. Transmission parameters may be tied to specific mosquito genotypes; for instance, an antimalarial effector gene may be associated with a human-to-mosquito or mosquito-to-human transmission probability of zero (Fig. 7.3B).

The simplest model of human malaria is the Ross-Macdonald model (Ross, 1910; Macdonald, 1957), in which, alongside the SEI model in mosquitoes, human disease progression is described by an SIS model (Fig. 7.3A). Here, humans become infected at rate, λ_H, and recover at a rate, r. The FOI in humans, λ_H, is proportional to the size of the mosquito population and the fraction of mosquitoes that are infectious. This is a significant simplification of malaria transmission dynamics and hence may only be used to obtain ballpark estimates of transmission level. A number of more detailed models have been developed that concisely describe malaria transmission dynamics and may be used to obtain more accurate predictions. These include models developed by the malaria modeling group at Imperial College London (Griffin et al., 2010; Walker et al., 2016), and the Infectious Disease Modeling Unit at the Swiss Tropical and Public Health Institute. Important details that contribute to the accuracy of these models include acquired and maternal immunity, symptomatic and asymptomatic infection, variable parasite density and superinfection in humans, human age structure, mosquito biting heterogeneity, and antimalarial drug therapy and prophylaxis. It is important that these models are calibrated to the setting of interest, and that historical intervention coverage is accounted for so that modeled levels of immunity are accurate.

Alongside a detailed model of human malaria transmission, a detailed model of mosquito life history allows genetic control technologies to be assessed in combination with currently available

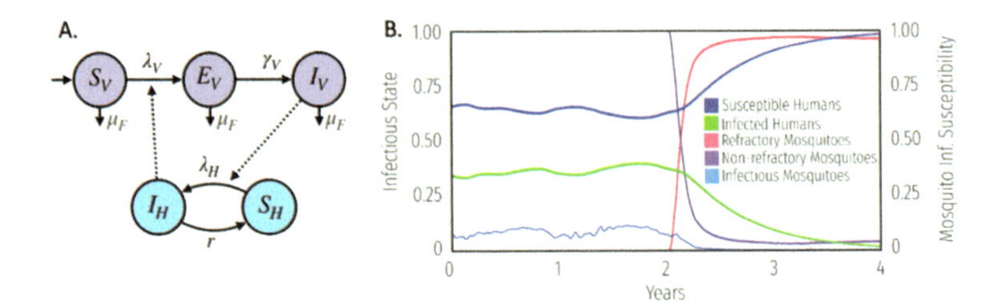

Figure 7.3 Integrating models of gene drive and malaria transmission. **A.** In the Ross-Macdonald model of malaria transmission, susceptible humans (S_H) become infected/infectious (I_H) at a rate equal to the force of infection in humans, λ_H, and recover at rate r, becoming susceptible again. Female mosquitoes emerge as susceptible adults (S_V), become exposed/latently infected (E_V) at a rate equal to the force of infection in mosquitoes, λ_V, and progress to infectiousness (I_V) through the extrinsic incubation period (EIP $= 1/\gamma_V$). The mortality rate, μ_F, is the same for female mosquitoes in each of these states. **B.** Example simulations from the MGDrivE 2 modeling framework of a split gene drive system and linked malaria-refractory gene introduced into an *Anopheles gambiae* population with implications for malaria transmission inferred from the Ross-Macdonald model. Following releases of the drive system at year two, the proportion of refractory female mosquitoes (solid red line) increases and the proportion of infectious mosquitoes (dotted light blue line) declines. As humans recover from infection and less develop new infections, the proportion of infected/infectious humans (solid green line) declines until it reaches near undetectable levels by year four. Full details of the simulation are provided in Wu et al. (2020).

vector control tools such as LLINs and IRS, as well as emerging technologies such as ATSBs. The elaborated feeding and gonotrophic cycle model proposed by Le Menach et al. (Le Menach et al., 2007) has been adapted and widely used (Griffin et al., 2010; White et al., 2011; Chitnis et al., 2010) to describe how LLINs and IRS reduce malaria transmission by: (i) increasing the death rate of adult female mosquitoes, (ii) increasing the proportion of bites taken on livestock rather than people, and (iii) decreasing the egg laying rate by increasing the time taken for female mosquitoes to obtain a blood meal. With growing awareness of the limitations of LLINs and IRS, interest has shifted to integrated vector management, and this model has been extended to include the impact of ATSBs, spatial repellents, spatial spraying, odor-baited traps, ovitraps, and livestock treated with systemic or topical insecticides (Marshall et al., 2013; Kiware et al., 2017).

Which malaria transmission model is most suited to the analysis will depend on a number of factors, including the data available to parameterize the model and the required specificity of the output. For analyses that seek to provide qualitative results for a generic setting, the Ross-Macdonald model may suffice; however, for models being applied to specific settings where interventions history is available, a parsimonious model such as the Imperial College London model (Griffin et al., 2010) or OpenMalaria may be well-suited. Models that need to include a high degree of spatial resolution and interventions such as reactive case detection or reactive spraying with insecticides may require an individual-based approach. Comparison of several models can help to build consensus where there is agreement, and to better understand the consequences of model assumptions when projections differ (Eaton et al., 2015).

7.3 Model Application

As gene drive mosquito projects transition from lab to field, a wide range of research questions arise that stand to benefit from modeling input (Fig. 7.4). Regulatory approval for environmental releases of gene drive mosquitoes will depend on a demonstration of product safety and efficacy, both of which have aspects suited to modeling analyses. Product efficacy may be apparent from laboratory and cage studies; however, modeling analyses allow us to extrapolate an expected impact to the population level, including impacts on human disease incidence. Likewise, many questions in an environmental risk assessment (ERA) may be addressed empirically by laboratory or cage studies, while other risks involve complex, interacting factors or become apparent at the population level are well suited to modeling analyses. Models will also be useful in planning field trials or interventions of gene drive mosquitoes. In these cases, models can explore possible outcomes prior to a release being performed based on data from laboratory or cage experiments and a quantitative understanding of local mosquito

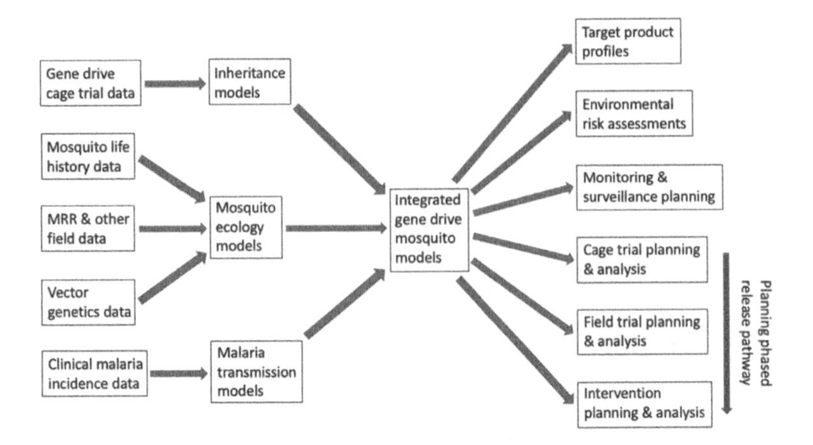

Figure 7.4 Models as a means of data integration. As gene drive mosquito projects progress from lab to field, models may be used to combine data from disparate sources including gene drive cage trials, field studies of mosquito vectors, and clinical malaria incidence from healthcare centers. These data inform models describing gene drive inheritance, mosquito ecology, and malaria transmission, which may be combined into an integrated modeling framework. This integrated framework, or components thereof, may be used to address a range of project-related modeling questions regarding gene drive product safety and efficacy, and planning and analysis of cage trials, field trials, and interventions.

ecology and malaria transmission. We explore these and other model applications in the following section.

7.3.1 Target Product Profiles

TPPs are valuable planning tools for determining whether a product is likely to have its desired effect in practice. For low-threshold gene drive systems, TPPs will likely play a role in deciding whether a product should be advanced from contained laboratory testing to semi-field or field testing, given the possibility that transgenes may become established in the local mosquito population following a release. While some aspects of TPPs for novel genetic technologies may be addressed empirically with data from molecular biologists, ecologists, and other specialists, other aspects require a quantitative approach that uses modeling to integrate data from multiple sources and make population-level inferences (Carballar-Lejarazú and James, 2017). Target outcomes will be both entomological, e.g., reducing the number of malaria-competent mosquitoes to less than 10% of their expected population size over a given time period, and epidemiological, e.g., reducing clinical malaria incidence by at least 20–50% within a given timeframe.

Addressing modeling questions related to TPPs entails: (i) defining the desired outcomes, and (ii) exploring regions of parameter space that achieve them. Primary outcomes of interest to regulatory agencies are likely to be epidemiological in nature. At a recent meeting organized by the Foundation for the National Institutes of Health (FNIH), it was discussed that a 20–50% reduction in the clinical incidence of malaria will likely be required prior to field testing of low-threshold gene drive systems in inhabited areas (James et al., 2020). The eventual incidence target is likely to represent a balance between demonstrating significant public health benefits, while also having an achievable goal that will enable the technology to be approved for use. This highlights the need for models to predict epidemiological impacts prior to a release based on available data.

In addition to the 20–50% reduction in the clinical incidence of malaria, other desired outcomes discussed at the FNIH meeting on TPPs include: (i) a minimum duration of epidemiological impact of three years, (ii) an entomological impact that would result in the required epidemiological impact, perhaps as measured by vectorial capacity, and (iii) a rate of spread that would produce the desired epidemiological impact within an acceptable time frame for a field trial (perhaps within two years) (James et al., 2020). Achievement of these outcomes depends on the field trial setting, which would be chosen, at least in part, for consistency with the objectives. TPPs may also include requirements regarding confinement and remediation, assessments of which benefit from modeling input.

Modeling analyses that support TPPs explore parameter value ranges that achieve these outcomes. To best scope such analyses, the parameter space capable of achieving these outcomes should be narrowed down as much as possible based on data from laboratory gene drive experiments, ecological and epidemiological characterization of the field site, and release schemes possible given feasible production capabilities. Construct parameters to be explored include: (i) rates of cleavage and accurate homology-directed repair, (ii) rates of resistant allele generation through NHEJ and other mechanisms, (iii) the proportion of resistant alleles that are in-frame and functional versus out-of-frame or otherwise costly, (iv) fitness costs of the various homing and resistant alleles, (v) efficacy of the anti-pathogen effector gene and its robustness to pathogen evolution (for population replacement strategies), and (vi) levels of standing genetic variation at the construct target site. Wide ranges of parameter space should be explored to determine those collectively satisfying TPP criteria. Sensitivity analyses should also be conducted where there is inadequate data to accurately specify input parameters. This activity can provide a basis for prioritizing field data to be collected by entomologists.

Until recently, TPPs for malaria vector control interventions have focused on mosquito-centric analyses and outcomes. For instance, TPPs for odor-baited traps and spatial repellents have utilized models of the mosquito life and feeding cycle and derived outcomes such as the impact on the entomological inoculation rate (EIR), which measures the number of infective bites that a person receives per unit time but does not require a detailed understanding of parasite transmission between mosquitoes and humans (Okumu et al., 2010; Killeen et al., 2011; Killeen and Moore, 2012). As models of malaria transmission have become more sophisticated, human-centric analyses and outcomes have become more common. For instance, the Imperial College London and OpenMalaria models of malaria transmission have been leveraged to explore the expected impact of vaccines and novel vector control tools such as ATSBs on clinical malaria incidence (Hogan et al., 2018; Fraser et al., 2021; Golumbeanu et al., 2021). Such analyses will be essential for gene drive-modified mosquitoes, given the interest in regulatory agencies in epidemiological end-points.

7.3.2 Monitoring and Surveillance

As several gene drive mosquito projects advance from contained laboratory testing to semi-field testing and/or small-scale field trials, there is an urgent need to assess both: (i) monitoring requirements to assess gene drive establishment and persistence at the field site, and (ii) surveillance requirements to detect the unintended spread of gene drive-modified mosquitoes beyond the testing or trial site. This is of particular importance as, for non-localized gene drive mosquito projects, the potential scale of intervention means that monitoring and surveillance are expected to be more costly than research, development, and deployment, perhaps even several times over.

Regarding surveillance of spread to unintended areas, open questions related to the optimal density and placement of traps, and frequency of checking traps, in order to detect gene drive-modified mosquitoes before they become too numerous to be remediated. Lessons may be learned from invasive species literature, in which early invasions may be halted through effective surveillance efforts, and the influence of geography on an organism's dispersal characteristics has been shown to be important (Koch et al., 2020). Surveillance program costs may be estimated based on unit costs for the purchase, distribution, and monitoring of traps and analysis of trapped mosquitoes. Trade-offs may be considered between the use of a larger number of cheaper BG-Sentinel traps, which may require more frequent visitation, or a smaller number of "smart traps," such as those being developed by Microsoft Premonition, which is more expensive but has increased functionality and may require less frequent visitation.

Regarding monitoring for gene drive establishment and persistence at field sites, lessons may be learned from the experiences of the World Mosquito Program with small-scale field trials of *Wolbachia*-transfected mosquitoes (Hoffmann et al., 2011). An important detail here is the incorporation of heterogeneity in spatial models, and the design of monitoring schemes capable of detecting geographic locations having low transgene or *Wolbachia* frequency. For low-threshold gene drive systems, monitoring requirements for detecting rare resistant alleles should also be assessed, as these could greatly compromise population suppression strategies. For population replacement strategies, a more pressing concern is the resistance of malaria parasites to the effector gene. Monitoring program costs may be estimated based on model parameters (e.g., density of traps, frequency of monitoring, number of mosquitoes analyzed) and unit costs for the purchase, distribution, and monitoring of traps and analysis of trapped mosquitoes.

7.3.3 Risk and Regulatory Considerations

The first gene drive products to be considered for malaria control will likely require stringent ERAs. Since field data on gene drives will not exist before releases begin, evaluation of many potential risks will benefit from evidence generated through modeling, in addition to evidence from laboratory studies, field data collection, pre-existing literature, and combinations thereof. One approach to conducting a rigorous ERA is the method of "problem formulation," in which a list of potential harms is generated, each of which is decomposed into a "risk hypothesis" that describes the causal chain of events by which it may occur (Raybould, 2006; Devos et al., 2019). This approach has recently been used to map the potential risks associated with the releases of mosquitoes with a population suppression gene drive (Connolly et al., 2021). Modeling can support this process by examining risk hypotheses that involve complex, interacting factors; for instance, those that are manifest at the population level.

Consider, for instance, the risk hypothesis that a population suppression gene drive system carried by female mosquitoes induces higher vectorial capacity, resulting in increased local malaria incidence following a release. A causal pathway by which this risk may occur is: (i) the transgene becomes common in the local mosquito population, (ii) transgenic females have a greater vectorial capacity for malaria, and (iii) elevated vectorial capacity causes a greater increase in malaria transmission than the reduction in transmission caused by mosquito population suppression. To assess the probability of this outcome, researchers would assess the probability of each step in the pathway. Laboratory studies would determine whether transgenic females do indeed have a higher vectorial capacity for malaria, and if this is true, modeling would be used to determine the implications of this for malaria transmission at the population level. Models may opt to explore the worst-case scenario by combining the largest reasonable increase in vector competence with the smallest reasonable suppression caused by the transgene. If malaria incidence is predicted to increase under this scenario, this will be an important concern for an ERA to consider further.

Many risk hypotheses that modeling may help to inform concern disease transmission outcomes, since these are generally extrapolated from pre-existing or generated data (Hosack et al., 2021). Two examples that may apply to either population suppression or population replacement gene drive strategies include: (i) the risk that female mosquitoes, if included in a release, could transmit and increase incidence of another vector-borne disease present in the community, such as o'nyong'nyong virus or lymphatic filariasis, in the months following a release, and (ii) the risk that a gene drive system only shows transient success, reducing malaria immunity in the human population and resulting in increased malaria susceptibility and incidence upon technology failure. These risks highlight the importance of developing versatile disease transmission models alongside models of mosquito ecology and gene drive inheritance.

7.3.4 Cage Trials

The World Health Organization (WHO) recommends a phased approach to the testing and release of gene drive-modified mosquitoes. Phase one comprises small-scale laboratory studies followed by testing in larger population cages in a laboratory setting, while phase two comprises confined field trials (Benedict et al., 2014). The intermediate step of large cage studies is recommended as this enables the observation of age-structured mosquito populations that undergo semi-natural population dynamics, albeit in highly simplified environments. Mosquitoes in cage studies live their entire lives in these populations, meaning fitness effects of transgenes may be revealed that are difficult to detect using small-scale laboratory studies alone. Following WHO guidelines, the Target Malaria project has used large cage experiments to investigate both self-limiting *An. gambiae* modifications (Valerio et al., 2016; Facchinelli et al., 2019; Pollegioni et al., 2020) and a self-sustaining gene drive system that targets the *doublesex* gene required for female fertility (Hammond et al., 2021). Modeling has been useful at multiple stages of the design and analysis of these cage trials. To assist in planning, models were constructed from pre-existing information, such as the results of smaller-scale studies, to predict how cage population dynamics might depend on the experimental set-up. After the experiments had run their course, the same models were fitted to the cage data and used to help estimate key model parameters.

7.3.5 Field Trial Design

By integrating a fitted model from cage trial data with models of mosquito ecology and malaria transmission tailored to a candidate field site, an integrated model can be used to explore potential release scenarios and to contribute to field trial design. Important considerations for an initial field trial of gene drive-modified mosquitoes are: (i) whether the measured outcome of interest is epidemiological, or entomological (with a modeled epidemiological outcome), and (ii) whether an intermediate drive system such as a split drive (Li et al., 2020) or an autosomal X-shredder (Facchinelli et al., 2019) be tested first, given concerns over confinement of non-localized gene drive systems to field sites. As such, key questions for models to explore are: (i) what release schemes are achievable and lead to the gene drive mosquito having the intended outcome within the timeframe of a field trial (1–3 years), and (ii) what checks and balances need to be put in place to ensure that the gene drive product being trialed is confined to the field site? Models may also be used to design remediation strategies following a trial and to help to inform the distribution of traps to monitor trial progress and confinement objectives.

The first field trial of a novel tool such as a gene drive-modified mosquito is likely to involve a single release site and accompanying control site. This was the case for initial trials of *Aedes aegypti* having the RIDL construct (releases of insects carrying a dominant lethal allele) (Harris et al., 2012), for trials of ATSBs to suppress *An. gambiae* in Mali (Müller et al., 2010), and for trials of *Wolbachia*-infected *Ae. aegypti* in Queensland, Australia (Hoffmann et al., 2011). In the event that a trial is successful at a single field site, trials involving a collection of intervention and control sites may be pursued. For instance, ATSBs were recently trialed in seven of 14 study villages in Mali (the other seven being control sites) where LLINs are in widespread use (Traore et al., 2020) and discussions are now proceeding toward randomized controlled trials (RCTs). For *Wolbachia*-infected *Ae. aegypti*, an RCT was recently conducted in Yogyakarta, Indonesia in which the intervention was shown to have a statistically significant impact on reducing dengue incidence (Indriani et al., 2020). Modeling and analysis of RCTs are statistical in nature, but may be paired with mathematical models, especially at the planning stage.

7.3.6 Intervention Design

In the event that a gene drive mosquito product is ultimately approved for use, a number of logistical questions will emerge regarding how it should be deployed. Past modeling has indicated the spread of gene drives will be affected by environmental conditions, including the distribution of mosquito resources across a landscape (North et al., 2013) and seasonality (Lambert et al., 2018; Eckhoff et al., 2017; North et al., 2020). For instance, North et al. (2013) found that the establishment of a population suppression gene drive system is more challenging in a spatially-clustered population as compared to a uniformly-distributed one because, in the presence of spatial clustering, the drive allele is at risk of locally extinguishing parts of the population and thereby becoming locally extinct itself. To prevent this from happening, population suppression gene drive mosquitoes should be released throughout a landscape rather than at a single site (North et al., 2013).

We anticipate that models will play a critical role in the planning and implementation of future gene drive mosquito interventions. Models will be useful in assessing the merits of a variety of release schemes, including: (i) the frequency, size, and number of releases, (ii) the spatial distribution of releases, (iii) the life stage of release, and (iv) whether to release males only or also females with a disease-refractory gene (Sánchez et al., 2020; Winskill et al., 2014). Additional factors to consider include the logistics of release infrastructure, e.g., the locations of lab rearing facilities, the rate at which gene drive mosquitoes can be produced, and transportation from rearing facilities to release sites. As low-threshold gene drive mosquitoes are self-sustaining, another option to consider, particularly for population replacement strategies, is sourcing gene drive mosquitoes from one of the release sites where they have become prevalent and distributing larvae from these sites widely.

In addition to release logistics, a large component of a gene drive mosquito intervention will be the accompanying monitoring and surveillance effort, and modeling will help to inform the distribution of mosquito traps and frequency of monitoring, as described earlier. Data collected from these traps will then be used to ensure that the release is proceeding as intended, and to parameterize and validate the model iteratively. For a large-scale intervention, rare events such as the generation of uncommon resistance alleles are more likely to occur, and so surveillance efforts should monitor for these, and models should be prepared to address their emergence. Models of remediation plans should also be iteratively updated as an intervention progresses, in the event of unwanted effects or a shift in public opinion.

7.4 Conclusion

As gene drive mosquito projects transition from lab to field in support of the wider malaria elimination agenda, mathematical and computational modeling is expected to play a growing role, in parallel with other project components such as community engagement, regulatory approval, and intervention delivery. In preparation for this role, it is important that models continue to be developed and iteratively refined based on the best available data regarding gene drive inheritance, vector ecology, malaria epidemiology, and other project components. Key data required to improve these models include those concerning mosquito life history, habitat distribution, and movement patterns, as well as those quantifying rare molecular events, such as the formation of drive-resistant alleles that may occur in large populations. Given the surging interest in the epidemiological implications of gene drive mosquitoes, transmission dynamics at locations of interest should be well studied.

Priorities regarding model application will evolve as these projects progress. Given the potential for low-threshold gene drive systems to spread, alignment with TPP criteria and an ERA is especially important prior to any open release. Some of these criteria and risks will require formal mathematical and computational analyses. Given that monitoring and surveillance are expected to be a significant cost driver for gene drive mosquito projects, models are needed to design cost-efficient monitoring strategies to ensure each project is having its intended effect, as well as cost-efficient surveillance strategies to detect unintended spread while it can still be remediated. Modeling priorities will then progress from designing and analyzing cage trials to designing and analyzing field trials, with the eventual goal of supporting a wide-scale intervention. We foresee the continued need for modeling efforts to ensure that gene drive mosquito technology can be deployed safely and effectively in line with regulatory requirements and the wishes of affected communities, and toward the eventual goal of malaria eradication.

Acknowledgments

The authors thank Dr. Héctor M. Sánchez C. for help in preparing Fig. 7.3.

References

Adolfi, Adriana, Valentino M. Gantz, Nijole Jasinskiene, Hsu-Feng Lee, Kristy Hwang, Gerard Terradas, Emily A. Bulger, et al. 2020. "Efficient Population Modification Gene-Drive Rescue System in the Malaria Mosquito Anopheles Stephensi." *Nature Communications* 11 (1): 5553.

Alphey, Luke. 2014. "Genetic Control of Mosquitoes." *Annual Review of Entomology* 59: 205–24.

Anopheles Gambiae 1000 Genomes Consortium, Data analysis group, Partner working group, Sample collections—Angola, Burkina Faso, Cameroon, Gabon, et al. 2017. "Genetic Diversity of the African Malaria Vector Anopheles Gambiae." *Nature* 552 (7683): 96–100.

Beaghton, Andrea, Pantelis John Beaghton, and Austin Burt. 2016. "Gene Drive through a Landscape: Reaction-Diffusion Models of Population Suppression and Elimination by a Sex Ratio Distorter." *Theoretical Population Biology* 108: 51–69.

Beck-Johnson, Lindsay M., William A. Nelson, Krijn P. Paaijmans, Andrew F. Read, Matthew B. Thomas, and Ottar N. Bjørnstad. 2013. "The Effect of Temperature on Anopheles Mosquito Population Dynamics and the Potential for Malaria Transmission." *PLoS ONE*. https://doi.org/10.1371/journal.pone.0079276.

Bellows, T. S. 1981. "The Descriptive Properties of Some Models for Density Dependence." *The Journal of Animal Ecology* 50 (1): 139–56.

Benedict, Mark Q., Rebecca C. Hood-Nowotny, Paul I. Howell, and Elien E. Wilkins. 2009. "Methylparaben in Anopheles Gambiae S.l. Sugar Meals Increases Longevity and Malaria Oocyst Abundance but Is Not a Preferred Diet." *Journal of Insect Physiology* 55 (3): 197–204.

Benedict, M., M. Bonsall, A. A. James, S. James, and J. Lavery. 2014. "Guidance Framework for Testing of Genetically Modified Mosquitoes." https://dspace2.flinders.edu.au/xmlui/handle/2328/37689.

Bull, James J., Christopher H. Remien, and Stephen M. Krone. 2019. "Gene-Drive-Mediated Extinction Is Thwarted by Population Structure and Evolution of Sib Mating." *Evolution, Medicine, and Public Health* 2019 (1): 66–81.

Burt, A., and A. Deredec. 2018. "Self-Limiting Population Genetic Control with Sex-Linked Genome Editors." *Proceedings of the Royal Society B*. https://royalsocietypublishing.org/doi/abs/10.1098/rspb.2018.0776.

Burt, A. 2003. "Site-Specific Selfish Genes as Tools for the Control and Genetic Engineering of Natural Populations." *Proceedings. Biological Sciences / The Royal Society* 270 (1518): 921–28.

Burt, A. 2014. "Heritable Strategies for Controlling Insect Vectors of Disease." *Philosophical Transactions of the Royal Society of London. Series B, Biological Sciences* 369 (1645): 20130432.

Carballar-Lejarazú, Rebeca, and Anthony A. James. 2017. "Population Modification of Anopheline Species to Control Malaria Transmission." *Pathogens and Global Health* 111 (8): 424–35.

Champer, Jackson, Esther Lee, Emily Yang, Chen Liu, Andrew G. Clark, and Philipp W. Messer. 2020. "A Toxin-Antidote CRISPR Gene Drive System for Regional Population Modification." *Nature Communications* 11 (1): 1082.

Champer, Jackson, Riona Reeves, Suh Yeon Oh, Chen Liu, Jingxian Liu, Andrew G. Clark, and Philipp W. Messer. 2017. "Novel CRISPR/Cas9 Gene Drive Constructs Reveal Insights into Mechanisms of Resistance Allele Formation and Drive Efficiency in Genetically Diverse Populations." *PLoS Genetics* 13 (7): e1006796.

Chitnis, Nakul, Allan Schapira, Thomas Smith, and Richard Steketee. 2010. "Comparing the Effectiveness of Malaria Vector-Control Interventions through a Mathematical Model." *The American Journal of Tropical Medicine and Hygiene* 83 (2): 230–40.

Clements, A. N., and G. D. Paterson. 1981. "The Analysis of Mortality and Survival Rates in Wild Populations of Mosquitoes." *The Journal of Applied Ecology* 18 (2): 373–99.

Coelho, Camila H., Rino Rappuoli, Peter J. Hotez, and Patrick E. Duffy. 2019. "Transmission-Blocking Vaccines for Malaria: Time to Talk about Vaccine Introduction." *Trends in Parasitology* 35 (7): 483–86.

Connolly, J.B., Mumford, J.D., Fuchs, S., Turner, G., Beech, C., North, A., and Burt, A. "Systematic Identification of Plausible Pathways to Potential Harm via Problem Formulation for Investigational Releases of a Population Suppression Gene Drive to Control the Human Malaria Vector *Anopheles Gambiae* in West Africa." *Malar J* **20,** 170 (2021). https://doi.org/10.1186/s12936-021-03674-6

Costantini, C., S. G. Li, A. Della Torre, N. Sagnon, M. Coluzzi, and C. E. Taylor. 1996. "Density, Survival and Dispersal of Anopheles Gambiae Complex Mosquitoes in a West African Sudan Savanna Village." *Medical and Veterinary Entomology* 10 (3): 203–19.

Courchamp, F., T. Clutton-Brock, and B. Grenfell. 1999. "Inverse Density Dependence and the Allee Effect." *Trends in Ecology & Evolution* 14 (10): 405–10.

Dao, A., A. S. Yaro, M. Diallo, S. Timbiné, D. L. Huestis, Y. Kassogué, A. I. Traoré, Z. L. Sanogo, D. Samaké, and T. Lehmann. 2014. "Signatures of Aestivation and Migration in Sahelian Malaria Mosquito Populations." *Nature* 516 (7531): 387–90.

Dawes, Emma J., Thomas S. Churcher, Shijie Zhuang, Robert E. Sinden, and María-Gloria Basáñez. 2009. "Anopheles Mortality Is Both Age- and Plasmodium-Density Dependent: Implications for Malaria Transmission." *Malaria Journal*. https://doi.org/10.1186/1475-2875-8-228.

Deredec, Anne, Austin Burt, and H. C. J. Godfray. 2008. "The Population Genetics of Using Homing Endonuclease Genes in Vector and Pest Management." *Genetics* 179 (4): 2013–26.

Deredec, Anne, H. Charles J. Godfray, and Austin Burt. 2011. "Requirements for Effective Malaria Control with Homing Endonuclease Genes." *Proceedings of the National Academy of Sciences of the United States of America* 108 (43): E874–80.

Detinova, T. S., et al. 1945. "Determination of the Physiological Age of the Females of Anopheles by the Changes in the Tracheal System of the Ovaries." *Medical Parasitology* 14 (2). https://www.cabdirect.org/cabdirect/abstract/19461000329.

Devos, Y., W. Craig, R. H. Devlin, and A. Ippolito. 2019. "Using Problem Formulation for Fit-for-purpose Pre-market Environmental Risk Assessments of Regulated Stressors." *Efsa*. https://efsa.onlinelibrary.wiley.com/doi/abs/10.2903/j.efsa.2019.e170708.

Dhole, S., A. L. Lloyd, and F. Gould. 2019. "Tethered Homing Gene Drives: A New Design for Spatially Restricted Population Replacement and Suppression." *Evolutionary Applications*. https://onlinelibrary.wiley.com/doi/abs/10.1111/eva.12827.

Eaton, Jeffrey W., Nicolas Bacaër, Anna Bershteyn, Valentina Cambiano, Anne Cori, Rob E. Dorrington, Christophe Fraser, et al. 2015. "Assessment of Epidemic Projections Using Recent HIV Survey Data in South Africa: A Validation Analysis of Ten Mathematical Models of HIV Epidemiology in the Antiretroviral Therapy Era." *The Lancet. Global Health* 3 (10): e598–608.

Eckhoff, Philip A. 2011. "A Malaria Transmission-Directed Model of Mosquito Life Cycle and Ecology." *Malaria Journal* 10 (October): 303.

Eckhoff, Philip A., Edward A. Wenger, H. Charles J. Godfray, and Austin Burt. 2017. "Impact of Mosquito Gene Drive on Malaria Elimination in a Computational Model with Explicit Spatial and Temporal Dynamics." *Proceedings of the National Academy of Sciences of the United States of America* 114 (2): E255–64.

Ermert, Volker, Andreas H. Fink, Anne E. Jones, and Andrew P. Morse. 2011. "Development of a New Version of the Liverpool Malaria Model. I. Refining the Parameter Settings and Mathematical Formulation of Basic Processes Based on a Literature Review." *Malaria Journal* 10 (February): 35.

Esvelt, Kevin M., Andrea L. Smidler, Flaminia Catteruccia, and George M. Church. 2014. "Concerning RNA-Guided Gene Drives for the Alteration of Wild Populations." *eLife* 3 (July). https://doi.org/10.7554/eLife.03401.

Facchinelli, Luca, Ace R. North, C. Matilda Collins, Miriam Menichelli, Tania Persampieri, Alessandro Bucci, Roberta Spaccapelo, Andrea Crisanti, and Mark Q. Benedict. 2019. "Large-Cage Assessment of a Transgenic Sex-Ratio Distortion Strain on Populations of an African Malaria Vector." *Parasites & Vectors* 12 (1): 70.

Focks, D. A., E. Daniels, D. G. Haile, and J. E. Keesling. 1995. "A Simulation Model of the Epidemiology of Urban Dengue Fever: Literature Analysis, Model Development, Preliminary Validation, and Samples of Simulation Results." *The American Journal of Tropical Medicine and Hygiene* 53 (5): 489–506.

Fraser, Keith J., Lazaro Mwandigha, Sekou F. Traore, Mohamed M. Traore, Seydou Doumbia, Amy Junnila, Edita Revay, et al. 2021. "Estimating the Potential Impact of Attractive Targeted Sugar Baits (ATSBs) as a New Vector Control Tool for Plasmodium Falciparum Malaria." *Malaria Journal* 20 (1). https://doi.org/10.1186/s12936-021-03684-4.

Garrett-Jones, C. 1962. "The Possibility of Active Long-Distance Migrations by Anopheles Pharoensis Theobald." *Bulletin of the World Health Organization* 27: 299–302.

Gimnig, J. E., M. Ombok, L. Kamau, and W. A. Hawley. 2001. "Characteristics of Larval Anopheline (Diptera: Culicidae) Habitats in Western Kenya." *Journal of Medical Entomology* 38 (2): 282–288.

Gimnig, John E., Maurice Ombok, Samson Otieno, Michael G. Kaufman, John M. Vulule, and Edward D. Walker. 2002. "Density-Dependent Development of Anopheles gambiae(Diptera: Culicidae) Larvae in Artificial Habitats." *Journal of Medical Entomology* 39(1): 162–172. https://doi.org/ 10.1603/0022-2585-39.1.162.

Godfray, H. Charles J., Ace North, and Austin Burt. 2017. "How Driving Endonuclease Genes Can Be Used to Combat Pests and Disease Vectors." *BMC Biology* 15 (1): 81.

Golumbeanu, Monica, Guojing Yang, Flavia Camponovo, Erin M. Stuckey, Nicholas Hamon, Mathias Mondy, Sarah Rees, Nakul Chitnis, Ewan Cameron, and Melissa A. Penny. 2021. "Combining Machine Learning and Mathematical Models of Disease Dynamics to Guide Development of Novel Disease Interventions." *medRxiv*. https://www.medrxiv.org/content/10.1101/2021.01.05.21249283v1.abstract.

Griffin, Jamie T., T. Deirdre Hollingsworth, Lucy C. Okell, Thomas S. Churcher, Michael White, Wes Hinsley, Teun Bousema, et al. 2010. "Reducing Plasmodium Falciparum Malaria Transmission in Africa: A Model-Based Evaluation of Intervention Strategies." *PLoS Medicine* 7 (8). https://doi.org/10.1371/journal.pmed.1000324.

Hammond, A., Pollegioni, P., Persampieri, T. et al. 2021. "Gene-drive suppression of mosquito populations in large cages as a bridge between lab and field." *Nat Commun* 12: 4589.

Harris, Angela F., Andrew R. McKemey, Derric Nimmo, Zoe Curtis, Isaac Black, Siân A. Morgan, Marco Neira Oviedo, et al. 2012. "Successful Suppression of a Field Mosquito Population by Sustained Release of Engineered Male Mosquitoes." *Nature Biotechnology* 30 (9): 828–30.

Hoffmann, A. A., B. L. Montgomery, J. Popovici, I. Iturbe-Ormaetxe, P. H. Johnson, F. Muzzi, M. Greenfield, et al. 2011. "Successful Establishment of Wolbachia in Aedes Populations to Suppress Dengue Transmission." *Nature* 476 (7361): 454–57.

Hogan, Alexandra B., Peter Winskill, Robert Verity, Jamie T. Griffin, and Azra C. Ghani. 2018. "Modeling Population-Level Impact to Inform Target Product Profiles for Childhood Malaria Vaccines." *BMC Medicine* 16 (1): 109.

Hosack, Geoffrey R., Adrien Ickowicz, and Keith R. Hayes. 2021. "Quantifying the Risk of Vector-Borne Disease Transmission Attributable to Genetically Modified Vectors." *Royal Society Open Science* 8 (3): 201525.

Huestis, Diana L., Adama Dao, Moussa Diallo, Zana L. Sanogo, Djibril Samake, Alpha S. Yaro, Yossi Ousman, et al. 2019. "Windborne Long-Distance Migration of Malaria Mosquitoes in the Sahel." *Nature* 574 (7778): 404–8.

Indriani, Citra, Warsito Tantowijoyo, Edwige Rancès, Bekti Andari, Equatori Prabowo, Dedik Yusdi, Muhammad Ridwan Ansari, et al. 2020. "Reduced Dengue Incidence Following Deployments of Infected in Yogyakarta, Indonesia: A Quasi-Experimental Trial Using Controlled Interrupted Time Series Analysis." *Gates Open Research* 4 (May): 50.

James, Stephanie, Frank H. Collins, Philip A. Welkhoff, Claudia Emerson, H. Charles J. Godfray, Michael Gottlieb, Brian Greenwood, et al. 2018. "Pathway to Deployment of Gene Drive Mosquitoes as a Potential Biocontrol Tool for Elimination of Malaria in Sub-Saharan Africa: Recommendations of a Scientific Working Group." *The American Journal of Tropical Medicine and Hygiene* 98 (6): 1–49.

James, Stephanie L., John M. Marshall, George K. Christophides, Fredros O. Okumu, and Tony Nolan. 2020. "Toward the Definition of Efficacy and Safety Criteria for Advancing Gene Drive-Modified Mosquitoes to Field Testing." *Vector Borne and Zoonotic Diseases* 20 (4): 237–51.

Jawara, Musa, Margaret Pinder, Chris J. Drakeley, Davis C. Nwakanma, Ebrima Jallow, Claus Bogh, Steve W. Lindsay, and David J. Conway. 2008.

"Dry Season Ecology of Anopheles Gambiae Complex Mosquitoes in The Gambia." *Malaria Journal*. https://doi.org/10.1186/1475-2875-7-156.

Killeen, Gerry F., Nakul Chitnis, Sarah J. Moore, and Fredros O. Okumu. 2011. "Target Product Profile Choices for Intra-Domiciliary Malaria Vector Control Pesticide Products: Repel or Kill?" *Malaria Journal* 10 (July): 207.

Killeen, Gerry F., and Sarah J. Moore. 2012. "Target Product Profiles for Protecting against Outdoor Malaria Transmission." *Malaria Journal* 11 (January): 17.

Kiware, Samson S., Nakul Chitnis, Allison Tatarsky, Sean Wu, Héctor Manuel Sánchez Castellanos, Roly Gosling, David Smith, and John M. Marshall. 2017. "Attacking the Mosquito on Multiple Fronts: Insights from the Vector Control Optimization Model (VCOM) for Malaria Elimination." *PloS One* 12 (12): e0187680.

Koch, Frank H., Denys Yemshanov, Robert G. Haight, Chris J. K. MacQuarrie, Ning Liu, Robert Venette, and Krista Ryall. 2020. "Optimal Invasive Species Surveillance in the Real World: Practical Advances from Research." *Emerging Topics in Life Sciences* 4 (5): 513–20.

Koenraadt, C. J. M., and W. Takken. 2003. "Cannibalism and Predation among Larvae of the Anopheles Gambiae Complex." *Medical and Veterinary Entomology* 17 (1): 61–66.

Kyrou, Kyros, Andrew M. Hammond, Roberto Galizi, Nace Kranjc, Austin Burt, Andrea K. Beaghton, Tony Nolan, and Andrea Crisanti. 2018. "A CRISPR–Cas9 Gene Drive Targeting Doublesex Causes Complete Population Suppression in Caged Anopheles Gambiae Mosquitoes." *Nature Biotechnology* 36 (11): 1062–66.

Lambert, Ben, Ace North, Austin Burt, and H. Charles J. Godfray. 2018. "The Use of Driving Endonuclease Genes to Suppress Mosquito Vectors of Malaria in Temporally Variable Environments." *Malaria Journal* 17 (1): 154.

Legros, Mathieu, Chonggang Xu, Kenichi Okamoto, Thomas W. Scott, Amy C. Morrison, Alun L. Lloyd, and Fred Gould. 2012. "Assessing the Feasibility of Controlling Aedes Aegypti with Transgenic Methods: A Model-Based Evaluation." *PloS One* 7 (12): e52235.

Lehmann, Tovi, Adama Dao, Alpha Seydou Yaro, Abdoulaye Adamou, Yaya Kassogue, Moussa Diallo, Traoré Sékou, and Cecilia Coscaron-Arias. 2010. "Aestivation of the African Malaria Mosquito, Anopheles Gambiae in the Sahel." *The American Journal of Tropical Medicine and Hygiene* 83 (3): 601–6.

Le Menach, Arnaud, Shannon Takala, F. Ellis McKenzie, Andre Perisse, Anthony Harris, Antoine Flahault, and David L. Smith. 2007. "An Elaborated Feeding Cycle Model for Reductions in Vectorial Capacity of Night-Biting Mosquitoes by Insecticide-Treated Nets." *Malaria Journal* 6 (January): 10.

Li, Ming, Ting Yang, Nikolay P. Kandul, Michelle Bui, Stephanie Gamez, Robyn Raban, Jared Bennett, et al. 2020. "Development of a Confinable Gene Drive System in the Human Disease Vector." *eLife* 9 (January). https://doi.org/10.7554/eLife.51701.

Lunde, Torleif Markussen, Diriba Korecha, Eskindir Loha, Asgeir Sorteberg, and Bernt Lindtjørn. 2013. "A Dynamic Model of Some Malaria-Transmitting Anopheline Mosquitoes of the Afrotropical Region. I. Model Description and Sensitivity Analysis." *Malaria Journal* 12 (January): 28.

Lyimo, E. O., W. Takken, and J. C. Koella. 1992. "Effect of Rearing Temperature and Larval Density on Larval Survival, Age at Pupation and Adult Size of Anopheles Gambiae." *Entomologia Experimentalis et Applicata* 63 (3): 265–71.

Macdonald, George. 1957. *The Epidemiology and Control of Malaria.*

Magori, Krisztian, Mathieu Legros, Molly E. Puente, Dana A. Focks, Thomas W. Scott, Alun L. Lloyd, and Fred Gould. 2009. "Skeeter Buster: A Stochastic, Spatially Explicit Modeling Tool for Studying Aedes Aegypti Population Replacement and Population Suppression Strategies." *PLoS Neglected Tropical Diseases* 3 (9): e508.

Marsden, Clare D., Anthony Cornel, Yoosook Lee, Michelle R. Sanford, Laura C. Norris, Parker B. Goodell, Catelyn C. Nieman, et al. 2013. "An Analysis of Two Island Groups as Potential Sites for Trials of Transgenic Mosquitoes for Malaria Control." *Evolutionary Applications* 6 (4): 706–20.

Marshall, John M., Anna Buchman, Héctor M. Sánchez C, and Omar S. Akbari. 2017. "Overcoming Evolved Resistance to Population-Suppressing Homing-Based Gene Drives." *Scientific Reports* 7 (1): 3776.

Marshall, John M., and Bruce A. Hay. 2012. "Confinement of Gene Drive Systems to Local Populations: A Comparative Analysis." *Journal of Theoretical Biology* 294 (February): 153–71.

Marshall, John M., Robyn R. Raban, Nikolay P. Kandul, Jyotheeswara R. Edula, Tomás M. León, and Omar S. Akbari. 2019. "Winning the Tug-of-War Between Effector Gene Design and Pathogen Evolution in Vector Population Replacement Strategies." *Frontiers in Genetics* 10 (October): 1072.

Marshall, John M., Michael T. White, Azra C. Ghani, Yosef Schlein, Gunter C. Muller, and John C. Beier. 2013. "Quantifying the Mosquito's Sweet Tooth: Modeling the Effectiveness of Attractive Toxic Sugar Baits (ATSB) for Malaria Vector Control." *Malaria Journal* 12 (August): 291.

Mozūraitis, Raimondas, Melika Hajkazemian, Jacek W. Zawada, Joanna Szymczak, Katinka Pålsson, Vaishnovi Sekar, Inna Biryukova, et al. 2020. "Male Swarming Aggregation Pheromones Increase Female Attraction and Mating Success among Multiple African Malaria Vector Mosquito Species." *Nature Ecology & Evolution* 4 (10): 1395–1401.

Müller, Günter C., John C. Beier, Sekou F. Traore, Mahamadou B. Toure, Mohamed M. Traore, Sekou Bah, Seydou Doumbia, and Yosef Schlein. 2010. "Successful Field Trial of Attractive Toxic Sugar Bait (ATSB) Plant-Spraying Methods against Malaria Vectors in the Anopheles Gambiae Complex in Mali, West Africa." *Malaria Journal* 9 (July): 210.

Muriu, Simon M., Tim Coulson, Charles M. Mbogo, and H. Charles J. Godfray. 2013. "Larval Density Dependence in Anopheles Gambiae S.s., the Major African Vector of Malaria." *The Journal of Animal Ecology* 82 (1): 166–74.

Nicholson, Sharon E. 2013. "The West African Sahel: A Review of Recent Studies on the Rainfall Regime and Its Interannual Variability." *International Scholarly Research Notices* 2013. https://www.hindawi.com/archive/2013/453521/abs/.

Noble, Charleston, Jason Olejarz, Kevin M. Esvelt, George M. Church, and Martin A. Nowak. 2017. "Evolutionary Dynamics of CRISPR Gene Drives." *Science Advances* 3 (4): e1601964.

North, Ace, Austin Burt, and H. Charles J. Godfray. 2013. "Modeling the Spatial Spread of a Homing Endonuclease Gene in a Mosquito Population." *Journal of Applied Ecology*. https://doi.org/10.1111/1365-2664.12133.

North, Ace R., Austin Burt, and H. Charles J. Godfray. 2019. "Modeling the Potential of Genetic Control of Malaria Mosquitoes at National Scale." *BMC Biology* 17 (1): 26.

North, Ace R., Austin Burt and H. Charles J. Godfray. 2020. "Modeling the Suppression of a Malaria Vector Using a CRISPR-Cas9 Gene Drive to Reduce Female Fertility." *BMC Biology* 18 (1): 98.

North, Ace R., and H. Charles J. Godfray. 2018. "Modeling the Persistence of Mosquito Vectors of Malaria in Burkina Faso." *Malaria Journal* 17 (1): 140.

Novembre, John, and Montgomery Slatkin. 2009. "Likelihood-Based Inference in Isolation-by-Distance Models Using the Spatial Distribution of Low-Frequency Alleles." *Evolution; International Journal of Organic Evolution* 63 (11): 2914–25.

Oberhofer, Georg, Tobin Ivy, and Bruce A. Hay. 2019. "Cleave and Rescue, a Novel Selfish Genetic Element and General Strategy for Gene Drive." *Proceedings of the National Academy of Sciences of the United States of America* 116 (13): 6250–59.

Okumu, Fredros O., Nicodem J. Govella, Sarah J. Moore, Nakul Chitnis, and Gerry F. Killeen. 2010. "Potential Benefits, Limitations and Target Product-Profiles of Odor-Baited Mosquito Traps for Malaria Control in Africa." *PloS One* 5 (7): e11573.

Omer, S. M., and J. L. Cloudsley-Thompson. 1970. "Survival of Female Anopheles Gambiae Giles through a 9-Month Dry Season in Sudan." *Bulletin of the World Health Organization* 42 (2): 319–30.

Pham, Thai Binh, Celine Hien Phong, Jared B. Bennett, Kristy Hwang, Nijole Jasinskiene, Kiona Parker, Drusilla Stillinger, John M. Marshall, Rebeca Carballar-Lejarazú, and Anthony A. James. 2019. "Experimental Population Modification of the Malaria Vector Mosquito, Anopheles Stephensi." *PLoS Genetics* 15 (12): e1008440.

Pollegioni, Paola, Ace R. North, Tania Persampieri, Alessandro Bucci, Roxana L. Minuz, David Alexander Groneberg, Tony Nolan, Philippos-Aris Papathanos, Andrea Crisanti, and Ruth Müller. 2020. "Detecting the Population Dynamics of an Autosomal Sex Ratio Distorter Transgene in Malaria Vector Mosquitoes." *The Journal of Applied Ecology* 57 (10): 2086–96.

Polovodova, P. V. 1949. "The Determination of the Physiological Age of Female Anopheles by the Number of Gonotrophic Cycles Completed." *Medskaya. Parazit.* 18: 352–55.

Raban, Robyn R., John M. Marshall, and Omar S. Akbari. 2020. "Progress towards Engineering Gene Drives for Population Control." *The Journal of Experimental Biology* 223 (Pt Suppl 1). https://doi.org/10.1242/jeb.208181.

Ramsdale, C. D., Russell E. Fontaine, and World Health Organization. 1970. "Ecological Investigations of Anopheles Gambiae and Anopheles Funestus." WHO/MAL/70.735. World Health Organization. https://apps.who.int/iris/handle/10665/65589.

Raybould, Alan. 2006. "Problem Formulation and Hypothesis Testing for Environmental Risk Assessments of Genetically Modified Crops." *Environmental Biosafety Research* 5 (3): 119–25.

Rong, Yikang S., and Kent G. Golic. 2003. "The Homologous Chromosome Is an Effective Template for the Repair of Mitotic DNA Double-Strand Breaks in Drosophila." *Genetics* 165 (4): 1831–42.

Ross, Sir Ronald. 1910. *The Prevention of Malaria*.

Sánchez C, Héctor M., Jared B. Bennett, Sean L. Wu, Gordana Rašić, Omar S. Akbari, and John M. Marshall. 2020. "Modeling Confinement and Reversibility of Threshold-Dependent Gene Drive Systems in Spatially-Explicit Aedes Aegypti Populations." *BMC Biology* 18 (1): 50.

Sánchez C., H. M., Wu, Sean L., Bennett, J. B., Marshall, J.M. 2020. "MGDrivE: A modular simulation framework for the spread of gene drives through spatially-explicit mosquito populations." *Methods in Ecology and Evolution* 11 (2): 229–239.

Shililu, Josephat, Tewolde Ghebremeskel, Fessahaye Seulu, Solomon Mengistu, Helen Fekadu, Mehari Zerom, Asmelash Ghebregziabiher, et al. 2003. "Larval Habitat Diversity and Ecology of Anopheline Larvae in Eritrea." *Journal of Medical Entomology* 40 (6): 921–29.

Silver, John B. 2007. *Mosquito Ecology: Field Sampling Methods*. Springer Science & Business Media.

Sinkins, Steven P., and Fred Gould. 2006. "Gene Drive Systems for Insect Disease Vectors." *Nature Reviews. Genetics* 7 (6): 427–35.

Sudweeks, Jaye, Brandon Hollingsworth, Dimitri V. Blondel, Karl J. Campbell, Sumit Dhole, John D. Eisemann, Owain Edwards, et al. 2019. "Locally Fixed Alleles: A Method to Localize Gene Drive to Island Populations." *Scientific Reports* 9 (1): 15821.

Tanaka, Hidenori, Howard A. Stone, and David R. Nelson. 2017. "Spatial Gene Drives and Pushed Genetic Waves." *Proceedings of the National Academy of Sciences of the United States of America* 114 (32): 8452–57.

Taylor, C., Y. T. Touré, J. Carnahan, D. E. Norris, G. Dolo, S. F. Traoré, F. E. Edillo, and G. C. Lanzaro. 2001. "Gene Flow among Populations of the Malaria Vector, Anopheles Gambiae, in Mali, West Africa." *Genetics* 157 (2): 743–50.

Terradas, Gerard, Anna B. Buchman, Jared B. Bennett, Isaiah Shriner, John M. Marshall, Omar S. Akbari, and Ethan Bier. 2021. "Inherently Confinable Split-Drive Systems in Drosophila." *Nature Communications* 12 (1): 1480.

Thomson, M. C., S. J. Connor, and M. L. Quinones. 1995. "Movement of Anopheles Gambiae Sl Malaria Vectors between Villages in The Gambia." *Medical and Veterinary Entomology* 9(4):413–419. https://

onlinelibrary.wiley.com/doi/abs/10.1111/j.1365-2915.1995.tb00015.x?casa_token=wj8G00TcsjQAAAAA:s0y0ExXOmshGwMQ-mhRooT_nFSFPgSjc-E8bifnoYQs-tEcDBIOd5vlngFKWDC2ke8jKEd1U2MEfBg.

Traore, Mohamad M., Amy Junnila, Sekou F. Traore, Seydou Doumbia, Edita E. Revay, Vasiliy D. Kravchenko, Yosef Schlein, et al. 2020. "Large-Scale Field Trial of Attractive Toxic Sugar Baits (ATSB) for the Control of Malaria Vector Mosquitoes in Mali, West Africa." *Malaria Journal* 19 (1): 72.

Unckless, Robert L., Andrew G. Clark, and Philipp W. Messer. 2017. "Evolution of Resistance Against CRISPR/Cas9 Gene Drive." *Genetics* 205 (2): 827–41.

Unckless, Robert L., Philipp W. Messer, Tim Connallon, and Andrew G. Clark. 2015. "Modeling the Manipulation of Natural Populations by the Mutagenic Chain Reaction." *Genetics* 201 (2): 425–31.

Valerio, Laura, Ace North, C. Matilda Collins, John D. Mumford, Luca Facchinelli, Roberta Spaccapelo, and Mark Q. Benedict. 2016. "Comparison of Model Predictions and Laboratory Observations of Transgene Frequencies in Continuously-Breeding Mosquito Populations." *Insects* 7 (4). https://doi.org/10.3390/insects7040047.

Verdonschot, Piet F. M., and Anna A. Besse-Lototskaya. 2014. "Flight Distance of Mosquitoes (Culicidae): A Metadata Analysis to Support the Management of Barrier Zones around Rewetted and Newly Constructed Wetlands." *Limnologica* 45 (March): 69–79.

Walker, Patrick G. T., Jamie T. Griffin, Neil M. Ferguson, and Azra C. Ghani. 2016. "Estimating the Most Efficient Allocation of Interventions to Achieve Reductions in Plasmodium Falciparum Malaria Burden and Transmission in Africa: A Modeling Study." *The Lancet. Global Health* 4 (7): e474–84.

White, Michael T., Jamie T. Griffin, Thomas S. Churcher, Neil M. Ferguson, María-Gloria Basáñez, and Azra C. Ghani. 2011. "Modeling the Impact of Vector Control Interventions on Anopheles Gambiae Population Dynamics." *Parasites & Vectors* 4 (July): 153.

Winskill, Peter, Angela F. Harris, Siân A. Morgan, Jessica Stevenson, Norzahira Raduan, Luke Alphey, Andrew R. McKemey, and Christl A. Donnelly. 2014. "Genetic Control of Aedes Aegypti: Data-Driven Modeling to Assess the Effect of Releasing Different Life Stages and the Potential for Long-Term Suppression." *Parasites & Vectors* 7 (February): 68.

Wu, S. L., J. B. Bennett, A. J. Dolgert, T. M. Leon, and J. M. Marshall. 2020. "MGDrivE 2: A Simulation Framework for Gene Drive Systems Incorporating Seasonality and Epidemiological Dynamics." *bioRxiv*. https://www.biorxiv.org/content/10.1101/2020.10.16.343376v1. abstract.

Xu, Xiang-Ru Shannon, Emily A. Bulger, Valentino M. Gantz, Carissa Klanseck, Stephanie R. Heimler, Ankush Auradkar, Jared B. Bennett, et al. 2020. "Active Genetic Neutralizing Elements for Halting or Deleting Gene Drives." *Molecular Cell* 80 (2): 246–62.e4.

Section IV

Risk Assessment and Community Engagement

Chapter 8

Probabilistic Ecological Risk Assessment: An Overview of the Process

Keith R. Hayes, Geoffrey R. Hosack, and Adrien Ickowicz
CSIRO Data61, Hobart, Tasmania
keith.hayes@csiro.au

8.1 Introduction

Ecological risk assessment has been broadly characterized as a process that entails three phases - problem formulation, risk analysis, and risk characterization - preceded by a planning stage where stakeholders and risk professionals agree on the purpose, scope and technical approach of the assessment, and accompanied throughout by data acquisition, verification, and monitoring (Suter, 2008; National Academies of Sciences Engineering and Medicine, 2016).

More detailed descriptions portray the process as a series of steps that may differ according to the specific context (see, for example, Hayes, 2007). For genetically modified organisms (GMOs), the Secretariat of the Convention on Biological Diversity (Secretariat of the Convention on Biological Diversity (SCBD), 2016), and the

Mosquito Gene Drives and the Malaria Eradication Agenda
Edited by Rebeca Carballar-Lejarazú

ISBN 978-981-4968-33-1 (Hardcover), 978-1-003-30877-5 (eBook)
www.jennystanford.com

European Food Safety Authority (European Food Safety Authority (EFSA) Panel on Genetically Modified Organisms (GMO), 2013) describe a five-step process that addresses the characteristics of the GMO and the environment in which it will be used (step 1), the likelihood and consequence of adverse effects (steps 2 and 3), risk calculations (step 4), and risk management and acceptability (step 5). In the SCBD guidance these steps are again preceded by a planning phase that includes problem formulation and choice of comparator organisms, whereas in the EFSA guidance problem formulation is incorporated into the first step. The EFSA guidance also includes a sixth step where the overall (with management) risk is estimated.

This chapter briefly examines these steps, together with the preceding planning phase; from the perspective of risk practitioners who wish to conduct probabilistic risk assessments for gene drive modified mosquitoes (GDMMs). Typically, the GMO risk assessment guidance provided by national biosafety authorities is either agnostic to methodology or recommends qualitative risk assessments (see summary by Hayes et al., 2014). For gene drive modified organisms (GDMOs), however, the National Academy of Sciences, Engineering and Medicine recommends quantitative, probabilistic risk assessments, supported by modeling of off-target and non-target effects from the genome level through to the ecosystem level (National Academies of Sciences Engineering and Medicine, 2016). Similarly, the Australian Academy of Sciences recommends a comprehensive risk assessment, which includes ecological and evolutionary modeling (Australian Academy of Sciences, 2017).

This chapter aims to provide an overview of the overall risk assessment process and explain why probabilistic approaches to risk characterization have been advocated as best practices for synthetic gene drive systems, including GDMMs. Moreover by identifying examples of the methods and models that enable probabilistic risk estimates, we hope that this chapter will help others conduct their own quantitative assessments. Wherever possible the references and examples highlighted here are drawn from risk assessments for GDMOs or at least GMOs. In many instances, however, this is not possible because relatively few GDMO risk assessments, or probabilistic GMO risk assessments, have been completed to date. In these instances, examples and references from environmental risk assessments more generally are provided.

8.2 Stakeholders, Planning, and Problem Formulation

It is unethical to undertake a GDMM field trial or any form of environmental release without first conducting an honest and transparent public engagement (Long et al., 2020; Annas et al., 2021). There are also important epistemic and practical reasons for engaging affected communities and stakeholders throughout the ecological risk assessment process (National Academies of Sciences Engineering and Medicine, 2016, 2017; Kuzma, 2019). The large areas over which gene-drive based interventions might act present practical challenges in this context, but guidance on how to address these challenges is available and gradually improving (Kolopack et al., 2015; Singh, 2019; Thizy et al., 2019, 2021).

To date, however, very few risk assessments for GMOs have tried to engage stakeholders throughout the entire risk assessment. The only example that we are aware of suggests that stakeholders are most readily engaged in the early planning and problem formulation phase, particularly in the collation of relevant information, identification of the risk assessment endpoints, and description of conceptual models and hazards that this phase entails (Dana et al., 2012). The risk assessment endpoints are the environmental, human health, and other socio-economic values that are potentially threatened by the risk source, and which the risk assessment process seeks to protect. It is imperative that all relevant parties - officials, scientists, analysts, and stakeholders - are involved in the choice of endpoints to include in any particular assessment to ensure that the risk assessment identifies and addresses all "significant concerns" (National Research Council, 1996).

The values that are, or are perceived to be, threatened by GDMMs will likely include human and animal (e.g., livestock) health, together with a variety of environmental concerns, but may also include other socio-economic values. The values captured by the assessment endpoints, and importantly how GDMMs might adversely affect these values, should be described in the hazard analysis component of the problem formulation phase. This part of the analysis seeks to answer the question "What and how might things go wrong?" as comprehensively and systematically as possible, within the context of carefully specified field release proposal.

A variety of hazards associated with genetically modified mosquitoes have been identified by a number of international and national organizations (American Society of Tropical Medicine and Hygeine, 2003; European Food Safety Authority (EFSA) Panel on Genetically Modified Organisms (GMO), 2013; World Health Organisation Special Programme for Research and Training in Tropical Diseases, 2014; Secretariat of the Convention on Biological Diversity (SCBD), 2016; National Academies of Sciences Engineering and Medicine, 2016; Australian Academy of Sciences, 2017; European Food Safety Authority (EFSA) Panel on Genetically Modified Organisms (GMO), 2020), and individual scientists or scientific working groups (Benedict et al., 2008; David et al., 2013; Roberts et al., 2017; Hayes et al., 2018a; James et al., 2018; Rode et al., 2019; Teem et al., 2019; James et al., 2020; Romeis et al., 2020). Community engagement activities in places such as Africa have also started (Hartley et al., 2019) and some of these provide additional examples of the significant concerns and hazards that the community associates with GDMMs (see, for example, Finda et al., 2021). Taken together these publications and emerging community concerns provide a checklist of issues and specific hazards against which a proposal for a field trial or environmental release of a GDMM could be evaluated. It is important, however, that any such proposal is preceded by its own dedicated community engagement and hazard analysis because risk assessments for GMOs should be conducted on a case-by-case basis.

When identifying hazards associated with any particular proposal, it is important to describe the circumstances by which exposure to the GDMM leads to adverse outcomes on the environmental values identified previously. In the GMO context, these exposure scenarios (Wolt et al., 2010), adverse outcomes pathways (Ankley et al., 2010), or pathways to harm (Devos et al., 2019) have been graphically portrayed either as simple influence diagrams (Dana et al., 2012) or more commonly as block diagrams (Connolly et al., 2021), possibly augmented by more sophisticated methods such as Signed Digraphs (Hayes et al., 2014), Fault Trees (Kapuscinski et al., 2007), or Bayes Nets (Landis et al., 2020) that can also provide a basis for risk calculations at subsequent steps in the process.

These pathways to harm are effectively conceptual models of how things may go wrong, and their graphical presentation makes them amenable to stakeholder input. They also provide an opportunity to characterize the weight of evidence, and identify risk measurement endpoints (Suter, 1990), which maybe interim steps in the pathway where risk predictions are made that are most easily compared with laboratory or field observations. Furthermore, minimally acceptable efficacy or safety outcomes might be identified for one or more of the events in these pathways, and hence form part of a formally defined Target Product Profile (Carballar-Lejarazú and James, 2017; James et al., 2020), with the expectation that transgenic strains that fail to meet these criteria would not be pursued further.

8.3 The GMO and Receiving Environment

The probability of adverse events following the release of a GDMM will be strongly influenced by the type of gene drive system, and when and where it is released. It is important that the target mosquito species, the genetic construct, and the release conditions are carefully defined prior to the risk calculation steps so that uncertainty about what is being proposed (the genetic intervention) does not add to, and potentially confound, the other sources of variability and epistemic uncertainty in the risk assessment process (World Health Organisation, 2008; Hayes, 2011).

A variety of synthetic gene drive systems have been described in the literature and new systems are continually being developed and tested. Standardized nomenclature and definitions for these systems are available (Alphey et al., 2020) and this will help remove linguistic uncertainty from the early stages of the risk assessment by removing ambiguity, reducing the chances of misinterpretation, and assisting with translations that will be necessary when engaging stakeholders (Chemonges Wanyama et al., 2021).

Localized drives are deliberately designed with molecular confinement systems to limit their spread and persistence (see, for example, Nash et al., 2019; Zapletal et al., 2020). Molecular confinement systems can also be implemented in conjunction with physical containment and/or geographical isolation allowing different developmental phases to be implemented within a phased testing pathway (World Health Organisation Special Programme for Research and Training in Tropical Diseases, 2014). The gradual relaxation of the molecular confinement strategy and the physical/geographical containment strategy allows data to be gathered on the behavior of the GDMM and environmental outcomes, whilst maintaining (gradually expanding) limits on the spread and persistence of the GDMM. This approach is essential to the iterative risk calculation methodology described in Section 8.4.

One of the most important aspects of gene drives, from a risk assessment perspective, is that ultimately they are designed to spread throughout wild-type populations. This increases the spatial and temporal scope of a risk assessment, as compared to genetically modified crops, for example. This relatively larger spatio-temporal context can present challenges when defining the receiving environment. For non-localized drives, the spatial extent of the receiving environment may extend to an entire continent and any adjacent islands, whilst the temporal scope could be tens of years for a suppression drive and possibly longer for a population replacement drive.

For GDMMs, the important characteristics of the receiving environment include the amount and seasonality of precipitation, annual temperature regimes, humidity, wind, land use, the extent and location of permanent, ephemeral or intermittent water bodies, and the location and population density of humans and associated livestock. These physical and socio-economic characteristics, together with the distribution of sympatric species that compete with the target mosquito species for resources, influence its dispersal and population dynamics and will therefore have a strong bearing on a risk assessment. Fortunately several global, spatially explicit, often longitudinal datasets (Table 8.1) are now available to support the environmental characterization and risk modeling.

Table 8.1 Global, spatially explicit data sets that can be used to characterize important characteristics of the receiving environment to support environmental risk assessment of gene drive modified mosquitoes

Characteristic	Data sources
Rivers and inland water bodies	http://www.diva-gis.org/gdata
Non-perennial rivers and streams	https://doi.org/10.6084/m9.figshare.14633022.v1
Hydrological & environment attributes	https://hydrosheds.org/page/hydroatlas
Precipitation, humidity and temperature	https://www.worldclim.org/data/monthlywth.html
Human population density	http://www.diva-gis.org/gdata https://sedac.ciesin.columbia.edu/data/collection/gpw-v4
Human settlements	https://data.humdata.org/
Livestock distribution	http://www.fao.org/livestock-systems/en/
Climate and human population metrics	https://worldview.earthdata.nasa.gov/

8.4 Estimating the Probability and Consequences of Adverse Outcomes

8.4.1 Models and Risk Assessment

Probabilistic risk calculations are often based on the past performance of a risk source. Information on the number of failures, accidents, or fatalities per hours of operation or workplace activity, for example, are routinely collected to estimate risks in socio-economic settings (Jonkman et al., 2003), and the risks associated with various financial instruments are often based on the profits and losses they have previously incurred (McNeil et al., 2015).

In the absence of operating experience, however, risk assessments for new technologies must initially be based on a conceptual model. This conceptual model maybe informed by experience gained in similar circumstances. For example, the ecosystem effects of removing a mosquito species using a population suppression drive maybe informed by experience gained with Sterile Insect Technology (SIT) (Nagel and Peveling, 2005) or other biological control methods (Romeis et al., 2020), but the case-specific risks are nonetheless new and therefore conceptual in the first instance.

Conceptual models informed by beliefs, perceptions of, and experiences with, relevant situations provide the basis for qualitative risk assessments (Fig. 8.1). In the context of GMOs, qualitative risk assessments have several advantages (OGTR, 2013) but they remain fundamentally unscientific because they do not make predictions that can be empirically verified (Hayes et al., 2014). For this reason, they are also an unsuitable basis for a problem formulation, risk hypothesis-driven paradigm for GMO risk assessment that requires risk to be defined in terms of the probability of unacceptable outcomes (Raybould, 2006).

The process of moving from a conceptual model to probabilistic risk calculations is facilitated by elicitation (Fig. 8.1). Guidance on how to prepare for, structure, and instigate formal elicitation procedures are available from EFSA (European Food Safety Authority (EFSA), 2014) and in various scientific publications (Garthwaite et al., 2005; Martin et al., 2012; Morgan, 2014). Elicitation can target the structure and types of models used for quantitative risk assessments and/or the parameters in these models. Effects on ecosystems and their services that arise directly or indirectly through changes to predator-prey or competitive interactions, for example, can be investigated by elicitation and analysis of Signed Digraphs Graphs, through a process known as qualitative mathematical modeling (Dambacher et al., 2003; Hayes et al., 2014). Signed digraphs (and other graphical representations) can provide the structural basis for process-based models, which together with statistical models, comprise the two major types of quantitative models that support probabilistic risk assessments (Fig. 8.1).

Conducting the elicitations and modeling for probabilistic risk calculations increases the time and resources necessary to conduct an environmental risk assessment. Elicitation is not a low-cost, low-effort alternative (Morgan, 2014) and should be treated with the same care and attention to detail as any other scientific data gathering exercise. The principal advantage of this approach, over the simpler and less resource-intensive qualitative approach, is that

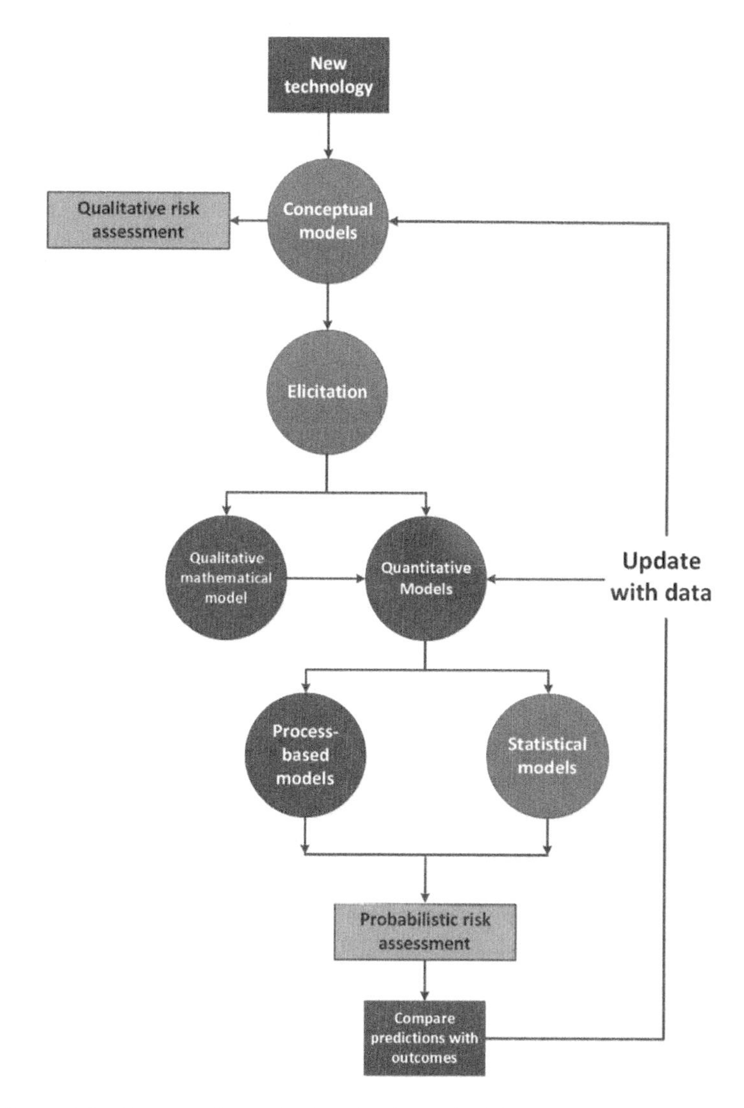

Figure 8.1 Simple schematic showing how all risk assessments for a new technology are based on a model. For qualitative risk assessment the model is a conceptual one. Elicitation of the structure and/or parameters of qualitative mathematical models or quantitative models (process-based or statistical) enable probabilistic risk assessments. The key advantage of probabilistic risk assessment is that its predictions can be compared to outcomes and these observations can be used to update all of the previous models.

it allows the risk analyst to make testable predictions. Moreover, with an appropriate statistical model to describe the probability of the data given these prior predictions (the likelihood), these prior (before the data) predictions themselves can be updated to posterior predictions using Bayes theory:

$$\underbrace{p\left(s|y\right)}_{\text{Posterior}} \propto \underbrace{p\left(y|s\right)}_{\text{Likelihood}} \underbrace{\pi\left(s\right)}_{\text{Prior}} \tag{8.1}$$

where s is an uncertain state of the world, represented by the uncertain parameters of a quantitative model, and y is the observation of the relevant processes concerned.

Bayesian methods provide a powerful suite of tools for statistical inference that enable uncertainty in models and their parameters to be coherently revised in light of observations (Lindley, 2000; van de Schoot et al., 2021), and their utility to risk assessment for novel technologies and rare events have been recognized for many decades (Rasmussen, 1981; Apostolakis, 1990).

8.4.2 Probabilistic Risk Assessment Methods

Statistical models and process-based models provide the basis for quantitative risk predictions. Although imperfect, further classification is nonetheless useful to identify important relationships between the two approaches and highlight methodological approaches to, and constraints on, probabilistic risk calculations (Fig. 8.2).

Process-based models mathematically describe how (for example) biological entities and systems change over time and space. These types of models can be categorized as deterministic or stochastic, and as linear or non-linear. The complexity of these models can vary from just a few parameters, to very complex, whole-of-ecosystem, end-to-end models, with thousands of parameters (Holsman et al., 2017; Audzijonyte et al., 2019). Probabilistic risk assessments often use direct elicitation to determine point estimates or probability density functions for the parameters in deterministic and stochastic models respectively (see, for example, O'Hagan, 2012), and a range of graphical techniques can be used to qualitatively describe the key parts of the system and hence the model structure (Jackson et al., 2000; Burgman, 2005).

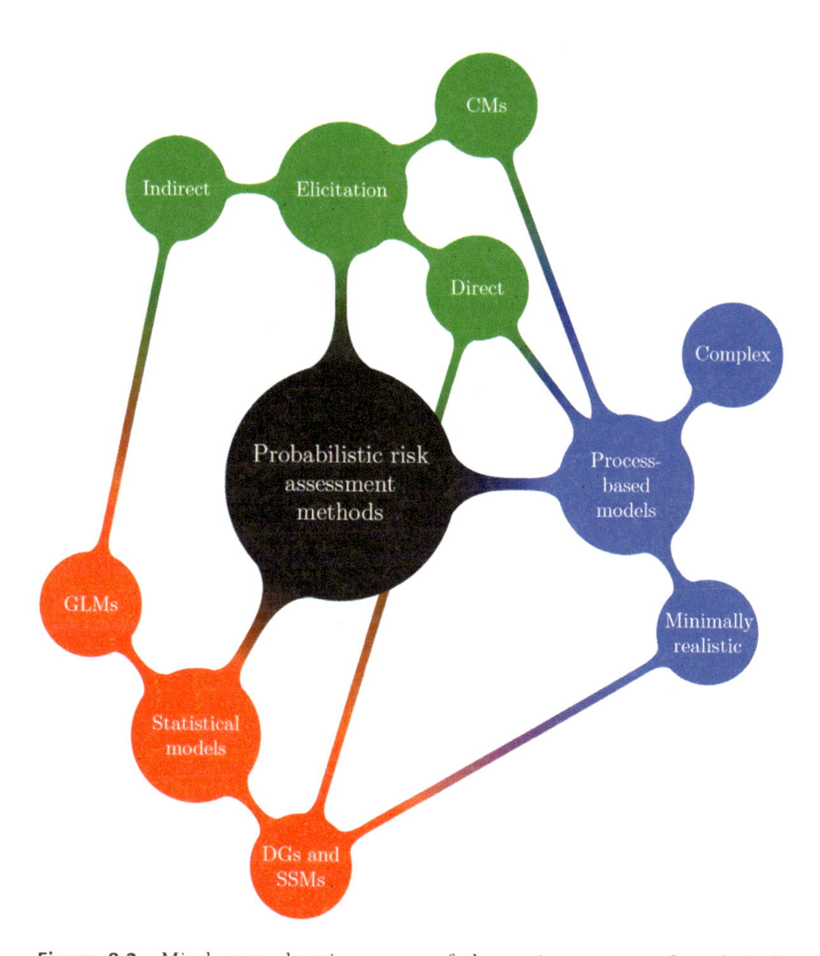

Figure 8.2 Mind map showing some of the major groups of analytical methods that can be used in a probabilistic risk assessment. Color schema follows that of Fig. 8.1 showing conceptual models in green, process-based models in blue and statistical models in red. Important connections between these three model classes are shown as links with merging colors. Abbreviations are: Conceptual models (CMs), Generalized Linear Models (GLMs), Directed Graphs (DGs) and State Space Models (SSMs).

Statistical models are mathematical descriptions of data generating processes. Statistical inference amends these descriptions once the outcomes of these processes have been observed in time and space. In this context, important classes of statistical models include directed graphs, such as Bayes Nets and Fault Trees, Generalized

Linear Models, and State Space Models (Fig. 8.2). We have used Fault Trees to estimate risks associated with Horizontal Gene Transfer (Hayes et al., 2015), and Landis et al. (2020) has proposed Bayes Nets as a means to estimate synthetic biology risks. The conditional probability relationships encoded by Fault Trees and Bayes Nets can be determined through direct elicitation, whereas indirect elicitation can be used to parameterize the important class of Generalized Linear Models (see, for example, Hosack et al., 2021).

State Space Models separate descriptions of the data collection process from the latent (hidden) process that generates the data, with the latter described by a stochastic or deterministic process-based model. This approach allows statistical inference to be performed on the parameters (and sometimes structure) of the process model. Hidden Markov Models and the Kalman filter, for example, enable inference over linear process models (Meinhold and Singpurwalla, 1983), whereas more sophisticated techniques such as the extended Kalman filter and particle filters enable inference over a wider class of non-linear models (Doucet et al., 2001). There are analytical constraints to these techniques, which to some extent can be alleviated through the use of informative prior distributions. They can also entail significant computational costs but this has increasingly been offset by the growing availability of high-performance computing facilities. Bayesian inference and State Space Models are now widely used in ecological contexts (McClintock et al., 2020), including applications directly relevant to GDMM risk assessment (see White et al., 2011; Morris et al., 2021, for example).

Statistical inference over complex process models is at the cutting edge of statistical research, and a balance must be struck between the desire to place all interesting processes within a model and our ability to meaningfully update and validate the model parameters within the Bayesian paradigm described above. In practice, this places limits on the complexity of the process model, and has motivated a modeling strategy that seeks to retain enough complexity to be "minimally realistic" for the purposes of managing environmental systems, whilst being simple enough to quantify parameter uncertainty using standard Monte Carlo methods (Plagányi et al., 2014). We suspect a similar strategy will be useful for GDMM risk assessment and that further research into the requisite minimally realistic models is warranted.

8.5 Risk Calculations

The definition of risk and the calculations used to quantify it vary considerably across different domains. We prefer the following definitions: risk is *expected loss* or *the probability of loss*. Losses occur through adverse changes to the values that we care about - that is the values we seek to protect by conducting the risk assessment. These values are the risk assessment's endpoints, represented by its measurement endpoints (Section 8.2), which form all or part of the uncertain quantities s in the risk assessment models (Section 8.4). These quantities are random variables in the model, hence the loss function is a function of one or more random variables denoted $L(s)$.

In decision-theoretic settings, risk is typically defined as the expected loss, and the results of a probabilistic risk assessment are represented as a scalar:

$$E\left[L(s)\right] = \int_{S} L(s)\, p\,(s\,|y)\, ds \qquad (8.2)$$

where prior to collection of data, the prior probability $\pi\,(s)$ can be substituted for the posterior $p\,(s\,|y)$. This definition also permits the risk result to be expressed as any quantile s_q of the distribution function of loss $P\,(s \le s_q)$ due to the equivalence between probability and the expected value of an indicator function:

$$L(s_q) = E\left[\mathrm{I}\{s \le s_q\}\right] = P\,(s \le s_q) \qquad (8.3)$$

Some authors, however, recommend that the results of the assessment are not reduced to a scalar but rather presented as the entire distribution function of loss, i.e., the probability of loss, sometimes referred to as the "risk curve" (Kaplan and Garrick, 1981). In our experience, the results of quantitative risk calculations are commonly reported as some scalar characteristic of the distribution function, such as the expected value (mean) or as the probability of being above or below a specified quantile s_q.

In our analyses, we summarize the results in this manner, but often also provide the entire risk curve.

Loss functions describe the harmful consequence component of risk, and importantly can be tailored to different endpoints in a manner that helps decision makers distinguish acceptable from unacceptable changes, whilst also contextualizing the risk predictions to help stakeholders understand the results. For example, Hosack et al. (2021) proposed a loss function for pathogen transmission risks that quantifies the difference between the basic reproduction number of genetically engineered (GE) and wild type (WT) mosquitoes

$$L(s) = \log_{10}\left[\frac{R_0^{GE}(s)}{R_0^{WT}(s)}\right] \tag{8.4}$$

We have used this loss function to estimate the probability that the vectorial capacity of female mosquitoes carrying a dominant sterile male construct is higher than wild type mosquitoes of the same genetic background, whilst accounting for the uncertainty in the vectorial capacity parameters s, such as the daily mortality rate of the different mosquito strains (see also Hayes et al., 2015, 2018b). Here the risk is calculated as the probability that the vectorial capacity of genetically engineered mosquitoes is higher than wild type such that $P(L(s) \geq 0)$ from Eq. 8.4. Hosack et al. (2021) examined this risk for different combinations of laboratory generations and backcrosses, and quantifies the effect of different vector-host ratios on the risk calculations.

This characterization of disease transmission risk is more generalizable than an estimate of the basic reproduction number of a cohort of released GDMMs as it focuses on relative attributes of the GMM versus wild-type vector rather than the particularities of the deployment locale. It is also is easier to communicate and reflects the recommendation that GDMM risk assessment decisions be based on “go/no go” safety criteria designed to ensure that genetically engineered mosquitoes do no more harm to human health than wild type mosquitoes of the same genetic background (James et al., 2018).

8.6 Monitoring, Management, and Acceptability

8.6.1 Risk Acceptance

The previous vectorial capacity example demonstrates how loss functions can be developed to assist decision makers to identify unacceptable changes to risk assessment endpoints within an overarching philosophy of "do no more harm". Similar loss functions could be defined for other GDMM relevant endpoints such as the probability of enhanced insecticide resistance, changes in fitness parameters or changes to the allergenic/toxicological properties, and appropriate experimental procedures established as part of the GDMM's target product profile (Carballar-Lejarazú and James, 2017; James et al., 2020).

Loss functions for what might be considered a complete list of GDMM risk endpoints requires more research, and consultation with stakeholders to help identify ways to define harms in a meaningful manner. An important challenge in this context is distinguishing change from harm. This is a value judgment and different stakeholder groups will likely have quite divergent views on what they consider to be harmful change and thereby different risk acceptance criteria. For changes to environmental components/processes and ecosystem services, the "do no more harm" approach necessitates some understanding of the ecosystem-wide impacts of the larvicides and adulticides currently used to control mosquito vector populations, distinguished from the impacts of insecticides that are used for other agricultural purposes. This is a challenging proposition.

Ideally, risk acceptance criteria for GDMMs are identified early in the product development cycle and reflected as specific, measurable changes in the target product profile, following extensive consultation and discussions with relevant stakeholders. Loss functions can then be used to contextualize changes in a manner that helps stakeholders express unacceptable outcomes, and risk assessments conducted to calculate the probability of unacceptable change, against which decisions makers can balance the expected benefits of the new technology. In practice, the process maybe iterative as successful laboratory trials of new products tend to motivate stakeholder engagement and risk assessment activities, from which risk acceptance criteria start to emerge sometime after the product has been tested in the laboratory.

8.6.2 Monitoring and Management

Monitoring outcomes and comparing risk predictions to outcomes is the distinguishing feature of a scientific risk assessment. Hayes et al. (2018a) discuss some of the challenges in this context, most notably those associated with rare events and the resourcing required to monitor over large spatio-temporal scales, and identify strategies such as spatially-balanced sample designs to help improve survey efficiency. Monitoring epidemiological outcomes at large spatial scales will likely leverage the infrastructure of the existing national and trans-national initiatives that gather data on the incidence of diseases and the causes of mortality and morbidity. Initiatives currently exist in some parts of the world to develop equivalent infrastructure that could provide a similar basis for large scale, potentially continent wide, monitoring of entomological, or environmental outcomes (Karan et al., 2016), but in malaria endemic areas such as Africa, this would likely require significant additional resources, planning, and coordination.

8.7 Concluding Remarks

The National Academies of Sciences Engineering and Medicine (2016) identifies quantitative, probabilistic methods as the appropriate means for assessing the potential risks associated with the deployment of synthetic gene drive systems. We are sympathetic to this point of view because: (a) probability is the natural and most widely used language for quantifying uncertainty; and, (b) risk predictions cannot be scientifically tested when they are expressed qualitatively.

For novel technologies, quantitative risk assessments are harder, and typically take longer to complete, than qualitative assessments. We also suspect that many regulatory authorities will not have sufficient resources to routinely conduct quantitative assessments. In Australia, for example, the federal regulator provides (on average) a risk assessment and risk management plan for a proposed GMO release about once every six weeks (http://www.ogtr.gov.au/internet/ogtr/publishing.nsf/Content/ir-1). This workload likely precludes lengthy elicitation and modeling exercises.

Gene drives, however, are significantly different from the class of currently regulated GMOs, and a thorough evaluation of their risks is warranted, within the phased development/release strategy recommended by the World Health Organization (World Health Organisation Special Programme for Research and Training in Tropical Diseases, 2014). Quantitative, scientifically tested, probabilistic predictions are the only way to confidently demonstrate that GDMMs meet the biosafety standards (ideally encapsulated within their target product profile) that are necessary to progress through each phase, hence we believe probabilistic risk assessments will be essential for any "first in class" GDMMs that progress through such a strategy.

As the operating experience with GDMOs grows, then quantitative risk assessments completed for first in class products maybe able to support case-specific qualitative risk assessments for very similar products. Qualitative assessments for the current class of regulated GMOs are often supported by underlying quantitative models and analysis, and this body of evidence provides the basis for the qualitative descriptions risk and uncertainty that are used by regulators in what are now routine settings. At the moment, however, there is nothing routine about gene drive releases and time and effort is needed to generate their biosafety evidence base.

Acknowledgments

The authors gratefully acknowledge the funding and support provided by the GeneConvene Global Collaborative at the Foundation for the National Institutes of Health (https://fnih.org/our-programs/geneconvene), and the CSIRO Data61 and Health and Biosecurity business units.

References

1. Alphey, L. S., Crisanti, A., Randazzo, F. and Akbari, O. S. (2020). Opinion: Standardizing the definition of gene drive, *Proceedings of the National Academy of Sciences* **117**, 49, pp. 30864–30867, doi:10.1073/pnas.2020417117.
2. American Society of Tropical Medicine and Hygeine (2003). Arthropod containment guidelines (version 3.1): Risk assessment for arthropod vectors, *Vector-Borne and Zoonotic Diseases* **3**, 2, pp. 69–73, URL https://doi.org/10.1089/153036603322163466.
3. Ankley, G. T., Bennett, R. S., Erickson, R. J., Hoff, D. J., Hornung, M. W., Johnson, R. D., Mount, D. R., Nichols, J. W., Russom, C. L., Schmieder, P. K., Serrrano, J. A., Tietge, J. E. and Villeneuve, D. L. (2010). Adverse outcome pathways: A conceptual framework to support ecotoxicology research and risk assessment, *Environmental Toxicology and Chemistry* **29**, 3, pp. 730–741, doi:https://doi.org/10.1002/etc.34, URL https://setac.onlinelibrary.wiley.com/doi/abs/10.1002/etc.34.
4. Annas, G. J., Beisel, C. L., Clement, K., Crisanti, A., Francis, S., Galardini, M., Galizi, R., Grünewald, J., Immobile, G., Khalil, A. S., Muller, R., Pattanayak, V., Petri, K., Paul, L., Pinello, L., Simoni, A., Taxiarchi, C. and Joung, J. K. (2021). A code of ethics for gene drive research, *The CRISPR Journal* **4**, 1, pp. 19–24, URL https://doi.org/10.1089/crispr.2020.0096.
5. Apostolakis, G. (1990). The concept of probability in safety assessments of technological systems, *Science* **250**, 4986, pp. 1359–1364, doi:10.1126/science.2255906, URL https://science.sciencemag.org/content/250/4986/1359.
6. Audzijonyte, A., Pethybridge, H., Porobic, J., Gorton, R., Kaplan, I. and Fulton, E. A. (2019). Atlantis: A spatially explicit end-to-end marine ecosystem model with dynamically integrated physics, ecology and socio-economic modules, *Methods in Ecology and Evolution* **10**, 10, pp. 1814–1819, doi:https://doi.org/10.1111/2041-210X.13272, URL https://besjournals.onlinelibrary.wiley.com/doi/abs/10.1111/2041-210X.13272.
7. Australian Academy of Sciences (2017). Synthetic gene drives in Australia: Implications of emerging technologies, Tech. rep., Australian Academy of Sciences, Canberra, Australia, 24 pp.
8. Benedict, M., D'Abbs, P., Dobson, S., Gottlieb, M., Harrington, L., Higgs, S., James, A. A., James, S., Knols, B., Lavery, J., O'Neill, S., Scott, T., Takken, W. and Toure, Y. (2008). Guidance for contained field trials of vector mosquitoes engineered to contain a gene drive system:

Recommendations of a scientific working group. *Vector Borne and Zoonotic Diseases* **8**, pp. 127–166, doi:10.1089/vbz.2007.0273.

9. Burgman, M. (2005). *Risks and Decisions for Conservation and Environmental Management* (Cambridge University Press, Cambridge, England).
10. Carballar-Lejarazú, R. and James, A. A. (2017). Population modification of Anopheline species to control malaria transmission, *Pathogens and Global Health* **111**, 8, pp. 424–435, doi:10.1080/20477724.2018.1427192, https://doi.org/10.1080/20477724.2018.1427192, pMID: 29385893.
11. Chemonges Wanyama, E., Dicko, B., Pare Toe, L., Coulibaly, M. B., Barry, N., Bayala Traore, K., Diabate, A., Drabo, M., Kayondo, J. K., Kekele, S., Kodio, S., Ky, A. D., Linga, R. R., Magala, E., Meda, W. I., Mukwaya, S., Namukwaya, A., Robinson, B., Samoura, H., Sanogo, K., Thizy, D. and Traoré, F. (2021). Co-developing a common glossary with stakeholders for engagement on new genetic approaches for malaria control in a local African setting, *Malaria Journal* **20**, 1, p. 53, URL https://doi.org/10.1186/s12936-020-03577-y.
12. Connolly, J. B., Mumford, J. D., Fuchs, S., Turner, G., Beech, C., North, A. R. and Burt, A. (2021). Systematic identification of plausible pathways to potential harm via problem formulation for investigational releases of a population suppression gene drive to control the human malaria vector *Anopheles gambiae* in west africa, *Malaria Journal* **20**, 1, p. 170, URL https://doi.org/10.1186/s12936-021-03674-6.
13. Dambacher, J. M., Li, H. W. and Rossignol, P. A. (2003). Qualitative predictions in model ecosystems, *Ecological Modelling* **161**, pp. 79–93.
14. Dana, G. V., Kapuscinski, A. R. and Donaldson, J. S. (2012). Integrating diverse scientific and practitioner knowledge in ecological risk analysis: A case study of biodiversity risk assessment in South Africa, *Journal of Environmental Management* **98**, pp. 134–146, doi:https://doi.org/10.1016/j.jenvman.2011.12.021, URL https://www.sciencedirect.com/science/article/pii/S0301479711004749.
15. David, A. S., Kaser, J. M., Morey, A. C., Roth, A. M. and Andow, D. A. (2013). Release of genetically engineered insects: A framework to identify potential ecological effects, *Ecology and Evolution* **3**, 11, pp. 4000–4015, doi:https://doi.org/10.1002/ece3.737, URL https://onlinelibrary.wiley.com/doi/abs/10.1002/ece3.737.
16. Devos, Y., Craig, W., Devlin, R. H., Ippolito, A., Leggatt, R. A., Romeis, J., Shaw, R., Svendsen, C. and Topping, C. J. (2019). Using problem formulation for fit-for-purpose pre-market environmental risk assessments

of regulated stressors, *EFSA Journal* **17**, S1, p. e170708, doi:https://doi.org/10.2903/j.efsa.2019.e170708, URL https://efsa.onlinelibrary.wiley.com/doi/abs/10.2903/j.efsa.2019.e170708.

17. Doucet, A., de Freitas, N. and Gordon, N. (2001). *An Introduction to Sequential Monte Carlo Methods* (Springer New York, New York, NY), ISBN 978-1-4757-3437-9, pp. 3–14, doi:10.1007/978-1-4757-3437-9_1, URL https://doi.org/10.1007/978-1-4757-3437-9_1.
18. European Food Safety Authority (EFSA) (2014). Guidance on expert knowledge elicitation in food and feed safety risk assessment, *EFSA Journal* **12**, 6, p. 3734, URL https://doi.org/10.2903/j.efsa.2014.3734.
19. European Food Safety Authority (EFSA) Panel on Genetically Modified Organisms (GMO) (2013). Guidance on the environmental risk assessment of genetically modified animals, *EFSA Journal* **11**, 5, p. 3200, doi:https://doi.org/10.2903/j.efsa.2013.3200, URL https://efsa.onlinelibrary.wiley.com/doi/abs/10.2903/j.efsa.2013.3200.
20. European Food Safety Authority (EFSA) Panel on Genetically Modified Organisms (GMO) (2020). Adequacy and sufficiency evaluation of existing EFSA guidelines for the molecular characterisation, environmental risk assessment and post-market environmental monitoring of genetically modified insects containing engineered gene drives, *EFSA Journal* **18**, 11, p. e06297, doi:https://doi.org/10.2903/j.efsa.2020.6297, URL https://efsa.onlinelibrary.wiley.com/doi/abs/10.2903/j.efsa.2020.6297.
21. Finda, M. F., Okumu, F. O., Minja, E., Njalambaha, R., Mponzi, W., Tarimo, B. B., Chaki, P., Lezaun, J., Kelly, A. H. and Christofides, N. (2021). Hybrid mosquitoes? Evidence from rural Tanzania on how local communities conceptualize and respond to modified mosquitoes as a tool for malaria control, *Malaria Journal* **20**, 1, p. 134, URL https://doi.org/10.1186/s12936-021-03663-9.
22. Garthwaite, P. H., Kadane, J. B. and O'Hagan, A. (2005). Statistical methods for eliciting probability distributions, *Journal of the American Statistical Association* **100**, 470, pp. 680–701, doi:10.1198/016214505000000105, https://doi.org/10.1198/016214505000000105.
23. Hartley, S., Thizy, D., Ledingham, K., Coulibaly, M., Diabaté, A., Dicko, B., Diop, S., Kayondo, J., Namukwaya, A. and Nourou, L., Paré Toé (2019). Knowledge engagement in gene drive research for malaria control, *PLoS Neglected Tropical Diseases* **13**, 4, p. e0007233, URL https://doi.org/10.1371/journal.pntd.0007233.
24. Hayes, K. R. (2007). *Environmental Risk Assessment of Genetically Modified Organisms Volume 3: Methodologies for Transgenic Fish*, chap.

Introduction to Environmental Risk Assessment for Transgenic Fish (CAB International, Wallingford, UK), pp. 1–28.

25. Hayes, K. R. (2011). Uncertainty and uncertainty analysis methods: Issues in quantitative and qualitative risk modeling with application to import risk assessment. ACERA project (0705). Report Number: EP102467, Tech. rep., CSIRO, Hobart, Australia, URL https://doi.org/10.4225/08/585189e5f2360.
26. Hayes, K. R., Barry, S. C., Beebe, N., Dambacher, J. M., De Barro, P., Ferson, S., Ford, J., Foster, S., Goncalves da Silva, A., Hosack, G. R., Peel, D. and Thresher, R. (2015). Risk assessment for controlling mosquito vectors with engineered nucleases, sterile male construct: Final report. Tech. rep., CSIRO Biosecurity Flagship, Hobart, Australia.
27. Hayes, K. R., Hosack, G. R., Dana, G. V., Foster, S. D., Ford, J. H., Thresher, R., Ickowicz, A., Peel, D., Tizard, M., Barro, P. D., Strive, T. and Dambacher, J. M. (2018a). Identifying and detecting potentially adverse ecological outcomes associated with the release of gene-drive modified organisms, *Journal of Responsible Innovation* **5**, sup1, pp. S139–S158, doi:10.1080/23299460.2017.1415585, URL https://doi.org/10.1080/23299460.2017.1415585.
28. Hayes, K. R., Hosack, G. R., Ickowicz, A., Foster, S., Peel, D., Ford, J. and Thresher, R. (2018b). Risk assessment for controlling mosquito vectors with engineered nucleases: Controlled field release for sterile male construct risk assessment final report, Tech. rep., CSIRO Biosecurity Flagship, Hobart, Australia, 152 pp.
29. Hayes, K. R., Leung, B., Thresher, R. E., Dambacher, J. M. and Hosack, G. R. (2014). Assessing the risks of genetic control techniques with reference to the common carp (*Cyprinus carpio*) in Australia. *Biological Invasions* **16**, pp. 1273–1288.
30. Holsman, K., Samhouri, J., Cook, G., Hazen, E., Olsen, E., Dillard, M., Kasperski, S., Gaichas, S., Kelble, C. R., Fogarty, M. and Andrews, K. (2017). An ecosystem-based approach to marine risk assessment, *Ecosystem Health and Sustainability* **3**, 1, p. e01256, doi:https://doi.org/10.1002/ehs2.1256, URL https://esajournals.onlinelibrary.wiley.com/doi/abs/10.1002/ehs2.1256.
31. Hosack, G. R., Ickowicz, A. and Hayes, K. R. (2021). Quantifying the risk of vector-borne disease transmission attributable to genetically modified vectors, *Royal Society Open Science* **8**, p. 201525, URL http://doi.org/10.1098/rsos.201525.
32. Jackson, L. J., Trebitz, A. S. and Cottingham, K. L. (2000). An Introduction to the practice of ecological modeling, *BioScience* **50**, 8, pp. 694–706,

doi:10.1641/0006-3568(2000)050[0694:AITTPO]2.0.CO;2, URL https://doi.org/10.1641/0006-3568(2000)050[0694:AITTPO]2.0.CO;2.

33. James, S., Collins, F. H., Welkhoff, P. A., Emerson, C., Godfray, H., Gottlieb, M., Greenwood, B., Lindsay, S. W., Mbogo, C. M., Okumu, F. O., Quemada, H., Savadogo, M., Singh, J. A., Tountas, K. H. and Touré, Y. T. (2018). Pathway to deployment of gene drive mosquitoes as a potential biocontrol tool for elimination of malaria in Sub-Saharan Africa: Recommendations of a scientific working group, *The American Journal of Tropical Medicine and Hygiene* **98**, pp. 1–49, URL https://doi.org/10.4269/ajtmh.18-0083.
34. James, S. L., Marshall, J. M., Christophides, G. K., Okumu, F. O. and Nolan, T. (2020). Toward the definition of efficacy and safety criteria for advancing gene drive-modified mosquitoes to field testing, *Vector-Borne and Zoonotic Diseases* **20**, 4, pp. 237–251, doi:10.1089/vbz.2019.2606, URL https://doi.org/10.1089/vbz.2019.2606.
35. Jonkman, S. N., van Gelder, P. H. A. J. M. and Vrijling, J. K. (2003). An overview of quantitative risk measures for loss of life and economic damage, *Journal of Hazardous Materials* **99**, 1, pp. 1–30, doi: https://doi.org/10.1016/S0304-3894(02)00283-2, URL https://www.sciencedirect.com/science/article/pii/S0304389402002832.
36. Kaplan, S. and Garrick, J. B. (1981). On the quantitative definition of risk, *Risk Analysis* **1**, 1, pp. 11–27.
37. Kapuscinski, A. R., Hard, J. J., Paulson, K. M., Neira, A., Ponniah, A., Kamonrat, W., Mwanja, W., Fleming, I. A., Gallardo, J., Devlin, R. H. and Trisak, J. (2007). *Environmental Risk Assessment of Genetically Modified Organisms Volume 3: Methodologies for Transgenic Fish*, chap. Approaches to Assessing Gene Flow (CAB International, Wallingford, UK), pp. 112–150.
38. Karan, M., Liddell, M., Prober, S. M., Arndt, S., Beringer, J., Boer, M., Cleverly, J., Eamus, D., Grace, P., Van Gorsel, E., Hero, J.-M., Hutley, L., Macfarlane, C., Metcalfe, D., Meyer, W., Pendall, E., Sebastian, A. and Wardlaw, T. (2016). The Australian SuperSite Network: A continental, long-term terrestrial ecosystem observatory, *Science of The Total Environment* **568**, pp. 1263–1274, doi:https://doi.org/10.1016/j.scitotenv.2016.05.170, URL https://www.sciencedirect.com/science/article/pii/S0048969716311007.
39. Kolopack, P. A., Parsons, J. A. and Lavery, J. V. (2015). What makes community engagement effective? Lessons from the Eliminate Dengue Program in Queensland Australia, *PLoS Neglected Tropical Diseases* **9**, 25875485, pp. e0003713–e0003713, URL https://www.ncbi.nlm.nih.gov/pmc/articles/PMC4395388/.

40. Kuzma, J. (2019). Procedurally robust risk assessment framework for novel genetically engineered organisms and gene drives, *Regulation & Governance* **n/a**, n/a, doi:https://doi.org/10.1111/rego.12245, URL https://www.onlinelibrary.wiley.com/doi/abs/10.1111/rego.12245.
41. Landis, W. G., Brown, E. A. and Eikenbary, S. (2020). *An Initial Framework for the Environmental Risk Assessment of Synthetic Biology-Derived Organisms with a Focus on Gene Drives* (Springer International Publishing, Cham), ISBN 978-3-030-27264-7, pp. 257–268, doi:10.1007/978-3-030-27264-7_11, URL https://doi.org/10.1007/978-3-030-27264-7_11.
42. Lindley, D. V. (2000). The Philosophy of Statistics, *Journal of the Royal Statistical Society. Series D (The Statistician)* **49**, 3, pp. 293–337, URL http://www.jstor.org/stable/2681060.
43. Long, K. C., Alphey, L., Annas, G. J., Bloss, C. S., Campbell, K. J., Champer, J., Chen, C.-H., Choudhary, A., Church, G. M., Collins, J. P., Cooper, K. L., Delborne, J. A., Edwards, O. R., Emerson, C. I., Esvelt, K., Evans, S. W., Friedman, R. M., Gantz, V. M., Gould, F., Hartley, S., Heitman, E., Hemingway, J., Kanuka, H., Kuzma, J., Lavery, J. V., Lee, Y., Lorenzen, M., Lunshof, J. E., Marshall, J. M., Messer, P. W., Montell, C., Oye, K. A., Palmer, M. J., Papathanos, P. A., Paradkar, P. N., Piaggio, A. J., Rasgon, J. L., Rašić, G., Rudenko, L., Saah, J. R., Scott, M. J., Sutton, J. T., Vorsino, A. E. and Akbari, O. S. (2020). Core commitments for field trials of gene drive organisms, *Science* **370**, 6523, pp. 1417–1419, doi:10.1126/science.abd1908, URL https://science.sciencemag.org/content/370/6523/1417.
44. Martin, T. G., Burgman, M. A., Fidler, F., Kuhnert, P. M., Low-Choy, S., Mcbride, M. and Mengersen, K. (2012). Eliciting expert knowledge in conservation science, *Conservation Biology* **26**, 1, pp. 29–38, doi:https://doi.org/10.1111/j.1523-1739.2011.01806.x, URL https://conbio.onlinelibrary.wiley.com/doi/abs/10.1111/j.1523-1739.2011.01806.x.
45. McClintock, B. T., Langrock, R., Gimenez, O., Cam, E., Borchers, D. L., Glennie, R. and Patterson, T. A. (2020). Uncovering ecological state dynamics with hidden Markov models, *Ecology Letters* **23**, 12, pp. 1878–1903, doi:https://doi.org/10.1111/ele.13610, URL https://onlinelibrary.wiley.com/doi/abs/10.1111/ele.13610.
46. McNeil, A. J., Frey, R. and Embrechts, P. (2015). *Quantitative Risk Management: Concepts, Techniques and Tools* (Princeton University Press, Princeton, New Jersey, USA).
47. Meinhold, R. J. and Singpurwalla, N. D. (1983). Understanding the Kalman Filter, *The American Statistician* **37**, 2, pp. 123–127, doi:10.1080/00031305.1983.10482723, URL https://www.tandfonline.com/doi/abs/10.1080/00031305.1983.10482723.

48. Morgan, M. G. (2014). Use (and abuse) of expert elicitation in support of decision making for public policy, *Proceedings of the National Academy of Sciences* **111**, 20, pp. 7176–7184, doi:10.1073/pnas.1319946111, URL https://www.pnas.org/content/111/20/7176.

49. Morris, A. L., Ghani, A. and Ferguson, N. (2021). Fine-scale estimation of key life-history parameters of malaria vectors: Implications for next-generation vector control technologies, *Parasites and Vectors* **14**, 1, p. 311, doi:10.1186/s13071-021-04789-0, URL https://europepmc.org/articles/PMC8188720.

50. Nagel, P. and Peveling, R. (2005). *Environment and the Sterile Insect Technique* (Springer Netherlands, Dordrecht), ISBN 978-1-4020-4051-1, pp. 499–524, doi:10.1007/1-4020-4051-2_19, URL https://doi.org/10.1007/1-4020-4051-2_19.

51. Nash, A., Urdaneta, G. M., Beaghton, A. K., Hoermann, A., Papathanos, P. A., Christophides, G. K. and Windbichler, N. (2019). Integral gene drives for population replacement, *Biology Open* **8**, 1, doi:10.1242/bio.037762, URL https://doi.org/10.1242/bio.037762, bio037762.

52. National Academies of Sciences Engineering and Medicine (2016). Gene drives on the horizon: Advancing science, navigating uncertainty, and aligning research with public values, Tech. rep., The National Academies Press, Washington DC, USA, doi:10.17226/23405.

53. National Academies of Sciences Engineering and Medicine (2017). Preparing for future products of biotechnology, Tech. rep., The National Academies Press, Washington DC, USA, doi:10.17226/24605.

54. National Research Council (1996). Understanding risk: Informing decisions in a democratic society, Tech. rep., National Academy Press, Washington DC, USA., URL http://www.nap.edu/catalog/5138.html.

55. OGTR (2013). Risk analysis framework, Tech. rep., Office of the Gene Technology Regulator (OGTR), Canberra, Australia, URL www.ogtr.gov.au/internet/ogtr/publishing.nsf/Content/risk-analysis-framework.

56. O'Hagan, A. (2012). Probabilistic uncertainty specification: Overview, elaboration techniques and their application to a mechanistic model of carbon flux, *Environmental Modelling and Software* **36**, pp. 35–48, doi:https://doi.org/10.1016/j.envsoft.2011.03.003, URL https://www.sciencedirect.com/science/article/pii/S1364815211000624, thematic issue on Expert Opinion in Environmental Modelling and Management.

57. Plagányi, E. E., Punt, A. E., Hillary, R., Morello, E. B., Thébaud, O., Hutton, T., Pillans, R. D., Thorson, J. T., Fulton, E. A., Smith, A. D. M., Smith, F., Bayliss, P., Haywood, M., Lyne, V. and Rothlisberg, P. C.

(2014). Multispecies fisheries management and conservation: Tactical applications using models of intermediate complexity, *Fish and Fisheries* **15**, 1, pp. 1–22, doi:https://doi.org/10.1111/j.1467-2979.2012.00488.x, URL https://onlinelibrary.wiley.com/doi/abs/10.1111/j.1467-2979.2012.00488.x.

58. Rasmussen, N. C. (1981). The application of probabilistic risk assessment techniques to energy technologies, *Annual Reviews in Energy* **6**, pp. 123–138.
59. Raybould, A. (2006). Problem formulation and hypothesis testing for environmental risk assessments of genetically modified crops, *Environmental Biosafety Research* **5**, 3, pp. 119–125, doi:10.1051/ebr:2007004.
60. Roberts, A., de Andrade, P. P., Okumu, F., Quemada, H., Savadogo, M., Singh, J. A. and James, S. (2017). Results from the workshop "Problem Formulation for the Use of Gene Drive in Mosquitoes", *The American Journal of Tropical Medicine and Hygiene* **96**, pp. 530–533, URL http://doi.org/10.4269/ajtmh.16-0726.
61. Rode, N. O., Estoup, A., Bourguet, D., Courtier-Orgogozo, V. and Débarre, F. (2019). Population management using gene drive: Molecular design, models of spread dynamics and assessment of ecological risks, *Conservation Genetics* **20**, 4, pp. 671–690, URL https://doi.org/10.1007/s10592-019-01165-5.
62. Romeis, J., Collatz, J., Glandorf, D. C. M. and Bonsall, M. B. (2020). The value of existing regulatory frameworks for the environmental risk assessment of agricultural pest control using gene drives, *Environmental Science and Policy* **108**, pp. 19–36, doi:https://doi.org/10.1016/j.envsci.2020.02.016, URL https://www.sciencedirect.com/science/article/pii/S1462901119311098.
63. Secretariat of the Convention on Biological Diversity (SCBD) (2016). Guidance on risk assessment of living modified organisms and monitoring in the context of risk assessment, Tech. rep., Secretariat of the Convention on Biological Diversity, URL https://www.cbd.int/doc/meetings/bs/bsrarm-03/official/bsrarm-03-03-en.pdf.
64. Singh, J. A. (2019). Informed consent and community engagement in open field research: Lessons for gene drive science, *BMC Medical Ethics* **20**, 1, p. 54, URL https://doi.org/10.1186/s12910-019-0389-3.
65. Suter, G. W. (1990). Endpoints for regional ecological risk assessments, *Environmental Management* **14**, 1, pp. 9–23, URL https://doi.org/10.1007/BF02394015.

66. Suter, G. W. (2008). Ecological risk assessment in the United States environmental protection agency: A historical overview, *Integrated Environmental Assessment and Management* **4**, 3, pp. 285–289, doi:https://doi.org/10.1897/IEAM_2007-062.1, URL https://setac.onlinelibrary.wiley.com/doi/abs/10.1897/IEAM_2007-062.1.

67. Teem, J. L., Ambali, A., Glover, B., Ouedraogo, J., Makinde, D. and Roberts, A. (2019). Problem formulation for gene drive mosquitoes designed to reduce malaria transmission in Africa: Results from four regional consultations 2016-2018, *Malaria Journal* **18**, 1, p. 347, URL https://doi.org/10.1186/s12936-019-2978-5.

68. Thizy, D., Emerson, C., Gibbs, J., Hartley, S., Kapiriri, L., Lavery, J., Lunshof, J., Ramsey, J., Shapiro, J., Singh, J. A., Toe, L. P., Coche, I. and Robinson, B. (2019). Guidance on stakeholder engagement practices to inform the development of area-wide vector control methods, *PLoS Neglected Tropical Diseases* **13**, 4, p. e0007286, URL https://doi.org/10.1371/journal.pntd.0007286.

69. Thizy, D., Pare Toe, L., Mbogo, C., Matoke-Muhia, D., Alibu, V. P., Barnhill-Dilling, S. K., Chantler, T., Chongwe, G., Delborne, J., Kapiriri, L., Nassonko Kavuma, E., Koloi-Keaikitse, S., Kormos, A., Littler, K., Lwetoijera, D., Vargas de Moraes, R., Mumba, N., Mutengu, L., Mwichuli, S., Nabukenya, S. E., Nakigudde, J., Ndebele, P., Ngara, C., Ochomo, E., Odiwuor Ondiek, S., Rivera, S., Roberts, A. J., Robinson, B., Sambakunsi, R., Saxena, A., Sykes, N., Tarimo, B. B., Tiffin, N. and Tountas, K. H. (2021). Proceedings of an expert workshop on community agreement for gene drive research in Africa - Co-organised by KEMRI, PAMCA and Target Malaria [version 2; peer review: 2 approved], *Gates Open Research* **5**, 19, doi:10.12688/gatesopenres.13221.2.

70. van de Schoot, R., Depaoli, S., King, R., Kramer, B., Märtens, K., Tadesse, M. G., Vannucci, M., Gelman, A., Veen, D., Willemsen, J. and Yau, C. (2021). Bayesian statistics and modelling, *Nature Reviews Methods Primers* **1**, 1, p. 1, URL https://doi.org/10.1038/s43586-020-00001-2.

71. White, M. T., Griffin, J. T., Churcher, T. S., Ferguson, N. M., Basáñez, M.-G. and Ghani, A. C. (2011). Modelling the impact of vector control interventions on *Anopheles gambiae* population dynamics, *Parasites & Vectors* **4**, 1, p. 153, URL https://doi.org/10.1186/1756-3305-4-153.

72. Wolt, J. D., Keese, P., Raybould, A., Fitzpatrick, J. W., Burachik, M., Gray, A., Olin, S. S., Schiemann, J., Sears, M. and Wu, F. (2010). Problem formulation in the environmental risk assessment for genetically modified plants, *Transgenic Research* **19**, 3, pp. 425–436, URL https://doi.org/10.1007/s11248-009-9321-9.

73. World Health Organisation (2008). Uncertainty and data quality in exposure assessment, Tech. rep., World Health Organisation (WHO), Geneva, Switzerland, URL https://apps.who.int/iris/handle/10665/44017.

74. World Health Organisation Special Programme for Research and Training in Tropical Diseases (2014). *Guidance framework for testing of genetically modified mosquitoes* (World Health Organization, Geneva, Switzerland), URL https://www.who.int/tdr/publications/year/2014/guide-fmrk-gm-mosquit/en/.

75. Zapletal, J., Najmitabrizi, N., Erraguntla, M., Lawley, M. A., Myles, K. M. and Adelman, Z. N. (2020). Making gene drive biodegradable, *Philosophical Transactions of the Royal Society B: Biological Sciences* **376**, 1818, URL https://doi.org/10.1098/rstb.2019.0804.

Chapter 9

Community Engagement and Mosquito Gene Drives

Ana Kormos
Vector Genetics Laboratory, Department of Pathology, Microbiology and Immunology, School of Veterinary Medicine, University of California, Davis, California, USA
akormos@ucdavis.edu

9.1 Introduction

Engagement is a critical and challenging component in the science of public health. The work of promoting health and wellness, disease detection and prevention, and the development of novel strategies to prevent infectious diseases, requires multiple communication methods and an environment that facilitates open dialog and exchange of knowledge, perspective, and preference by diverse groups of people.

In 2006, the National Institutes of Health began emphasizing translational research, which included a community engagement component [11]. The community engagement has become increasingly viewed as the keystone to translational science [18]. Emerging technologies like gene drives for public health have accelerated the

Mosquito Gene Drives and the Malaria Eradication Agenda
Edited by Rebeca Carballar-Lejarazú

ISBN 978-981-4968-33-1 (Hardcover), 978-1-003-30877-5 (eBook)
www.jennystanford.com

need for effective community engagement practices and credible, independent risk assessment to ensure safe, and ethical and scientifically rigorous trials of the technologies. Unlike drug or vaccine development, no industry standard yet exists to evaluate gene drive vector control methods [3], and the scope and impact of these methods may differ greatly, suggesting that a different set of research ethics and engagement considerations will be required.

There is an existing body of knowledge on engagement practices across many public health sectors [5, 7, 14, 16, 17, 20, 26, 29, 30], including recommendations specific to the testing of novel technologies for vector control [21, 24, 26, 28]. All of these recommendations state (or imply) the importance of communication and development of trusting relationships in successful engagement activities. While there are existing foundations on which to build an engagement program, attention has been drawn to specific questions surrounding the testing and introduction of new gene drive technologies. These include scale of application, participation and consent, existing regulatory and safety mechanisms, and long-term effects and accountability. Novel technologies present novel questions about best practices for engagement. Genetically engineered mosquitoes with gene drive offer the benefit of providing sustainable, long-term vector control over large areas without requiring individual behavior modification, or bias related to socio-economic determinants. While this offers obvious advantages, it also presents challenges, which include individual participation/involvement in decision-making and the determination of definitions related to engagement (e.g., communities, approval, acceptance, and consent). Additional challenges include the lack of existing regulation, ethical standards, and policies related to a technology that has the potential to affect human communities over large geographic areas, which may increase in scale over time. The changing public perceptions and attitudes about science and technology, and the increasing expectation that communities should be involved early in the decision-making surrounding the science, continue to drive the evolution of engagement strategies.

9.2 Defining Engagement

We define **engagement** as the facilitation and support of collaboration, open communication, and dialog for the exchange of knowledge, perspective, opinion, and preference among diverse groups of people. We apply this definition for audience (communities, stakeholders, and public) outlined by the National Academies of Sciences, Engineering, and Medicine (NASEM) [21], and thus, also reference the definitions they have provided for these groups (Box 9.1; taken from Kormos *et al*, 2020 [33]).

Box 1. Engagement audience definitions: definitions as provided in *Gene Drives on the Horizon: Advancing Science, Navigating Uncertainty, and Aligning Research with Public Values*, published by the National Academies of Sciences, Engineering, and Medicine [20].

- **Communities:** Groups of people who live within the geographical location or biologically-relevant proximity to a potential site where research is taking place or where field releases may take place such that they have tangible and immediate interests in the research project. Communities are included within the broader category of 'stakeholders'.
- **Stakeholders:** Organizations, groups, or persons with professional or personal interests sufficient to justify engagement, but who may, or may not, have geographic proximity to potential intervention sites for the research project.
- **Publics:** Groups who lack the direct connection to a project that stakeholders and communities have but nonetheless have interests, concerns, hopes, fears, and values that can contribute and influence decision-making about the research and possible use of the vector control intervention.

We define **regulatory engagement** as the engagement process with relevant stakeholders, community members, and public that will be involved in the assessment of risk, benefit, and regulatory oversight of the technology at the field trial site. The regulatory engagement process applies to the development of the risk assessment processes, and new regulatory frameworks/pathways required for activities related to testing gene drive technology.

Audience definition and identification are critical and iterative parts of the engagement process. The creation of meaningful pathways for open dialog and exchange of ideas and information require a deep understanding of the important differences and distinctions between audience, sub-groups that exist within them, and the ways in which these groups intersect and communicate with one another. The audience definitions provided by NASEM [21] are represented on a continuum; individual members of one group may also be members of another group. The process of identifying audience for engagement is an ongoing process that may continually evolve as the research and engagement activities evolve. The audience may change and expand in scope and definition, and new audience members or groups may emerge, requiring constant evaluation and assessment over the duration of the project.

Engagement and communication strategies and activities should be tailored for each specific audience, ensuring that the engagement efforts are meeting the individuals "where they are" with respect to experience, understanding, beliefs, values, and perceptions. For this reason, it is critical that early engagement be focused on the development of an understanding of **who** the program audience is that should be considered.

9.3 Importance of Engagement

The application of gene drive technology for public health requires effective community engagement practices and credible, independent risk assessment to ensure safe and ethical translation of the laboratory research to the field. Engagement is critically important to the technology in order to establish pathways for sharing information, improved understanding, determining acceptance and consent, considering ethics and safety, and actively encouraging community and stakeholder participation in the project (Fig. 9.1).

Engagement activities allow for an open exchange of information between the research program and the engagement audience. Engagement encourages sharing of information about the science and technology, program research goals, and proposed communication methods or strategies with stakeholders and community members. They also encourage stakeholders and community members to share information about their history, culture, values, and priorities; important social, political, economic, and environmental policies and governance; public health priorities and strategic plans; and current public health data and information systems with the

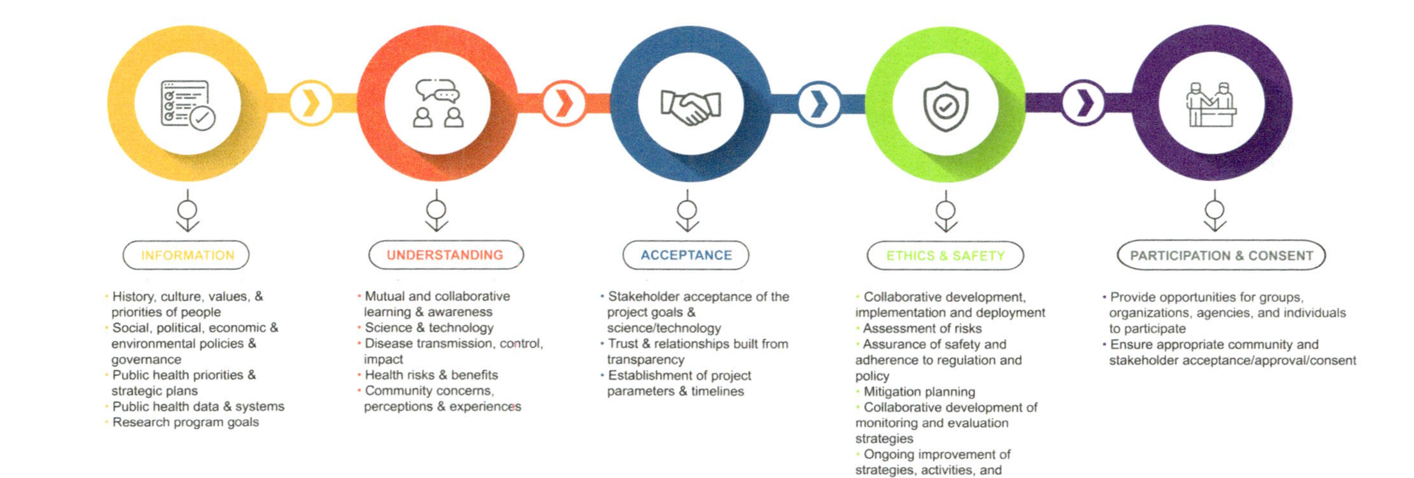

Figure 9.1 Importance of engagement.

research program. Information shared between groups in the early phase of engagement helps build trust and informs all other phases of the engagement and research processes.

Through the process of sharing information, engagement activities support understanding, not only regarding the complexities of the science and technology, but also about local public health concerns, disease transmission, control and management, potential community health risks and benefits associated with the technology, and community concerns, perceptions, and experiences. Shared understanding of opportunities for mutual and collaborative learning and awareness, and community communication preference are also established through meaningful engagement activities.

Trust and relationships established through shared information and understanding may positively impact the stakeholder and community perceptions of the program. Engagement activities help establish trust through transparency and build understanding of the history, concepts, and mechanisms of the proposed science and technology. The establishment of trust and advanced understanding is important in establishing overall acceptance of the science and project goals among stakeholders and will help establish parameters for determining whether or not the technology should be applied.

Assessment of risk and assurance of project safety and adherence to regulation and policy is a critically important contribution of engagement activities. Information shared by communities and stakeholders can inform researchers, regulators, and policymakers about risks and safety concerns that need to be considered. Engagement provides opportunities for the collaborative development of mitigation planning, monitoring and evaluation, and improvement of strategies, activities, and communication in response to community and stakeholder concerns and questions.

Perhaps of the greatest importance, engagement provides a framework for diverse audience participation in the processes leading up to the determination of whether or not gene drive technology should be applied in field trials leading to downstream broader application for public health interventions. Opportunities for participation in decisions about the development and application of the technology play an essential role in ensuring that appropriate stakeholder and community approval and/or acceptance/consent has been defined, determined, and considered prior to conducting a field trial. The importance of an early, robust, and iterative engagement program is an essential component of any research program that intersects with humans and the environment that we live in.

9.4 General Engagement Considerations

9.4.1 Funding

Early availability of funding and support from funders to conduct engagement activities is essential for the success of the program. Engagement should begin at the onset of the program, and in order to successfully engage with stakeholders and community members, it is critical that a sufficient part of the program budget is allocated to these efforts over the entire course of the program.

Most funding agencies who support field-based research understand the need for flexibility, given that this type of research is prone to timeline disruptions, due to both human and environmental dynamics in the system. Of great importance in conducting engagement over extended space and time is the establishment of shared understanding between research program leaders and funders that flexibility will be necessary regarding project goals and timelines. Engagement, by definition, involves dynamic human populations that cannot be easily confined to pre-determined definitions, models, or formulas. Thus, establishing acceptable program reporting mechanisms with funders that describe successes, challenges, and adjustments to activities or approaches is important, as well as providing a detailed description of the core concepts that guide engagement at each phase of research.

9.4.2 Timeline

Engagement should begin at the time that the engineered gene drive mosquito is developed and continue through the duration of the project and beyond. Early engagement can contribute to project research and development, providing important information about the efficacy of existing vector control methods, vector biology, geographic dispersal, patterns of transmission, and important social, political, and environmental considerations.

Early engagement with stakeholders is critical in the establishment of a project timeline. Engagement activities throughout the duration of the project should be guided and developed by project stakeholders so that their concerns, expectations, and perspective can help shape future project activities and outcomes. Given that field research is often subject to timeline disruptions, due to both human and environmental dynamics in the system, it is important to establish early on that flexibility will be necessary regarding project goals and timelines. Establishing acceptable program reporting mechanisms with funders, stakeholders, and project team members is important, as well as determining the process or model that will be used to guide program decision-making processes and project timeline adjustment at each phase of research.

9.4.3 Risks

Engagement activities are often subject to risks to project team members and the audience who are engaging in the project. Risks include establishing conflict of interest by inadvertently impacting or influencing the functionality or perception of stakeholder groups, creating unrealistic stakeholder and community expectations, perceptions of privileging specific groups over others, creating negative stigma of individuals or groups who may or may not be associated with project engagement, and the creation of an imbalance in power or authority among groups. These risks can have a profoundly negative impact not only on the project, but also on the people and the environment where the project is being conducted. It is important to have an early, open dialog with stakeholders to collaboratively identify and mitigate these potential risks, and establish shared, as opposed to conflicting, interests within the engagement framework or model to be applied. This subject is addressed in greater detail in the model for engagement section of this chapter.

9.5 Implementing Engagement

Engagement can be considered as a series of phases over the lifetime of a project; each phase informing the next, a series of iterative and evolving responses to the dynamic audience and environment surrounding the project. Thizy, et al. [26] describe preliminary considerations for engagement for area-wide vector control methods by emphasizing the importance of the project team's evaluation of requirements needed to ensure a successful engagement process. These considerations, which comprise the initial phase of engagement, are expanded here and defined by three primary activities: stakeholder identification, engagement model/framework identification, and understanding program responsibilities within the complex system where the project is being conducted. These initial activities will be important in the development of a more detailed, phased approach, to operationalizing engagement within specific communities as the project evolves.

9.5.1 Identifying Stakeholders

A clear understanding of the individuals and groups who are central to decision-making and determining appropriate responses to malaria control, treatment, and surveillance at the field site is vital in the development of an effective engagement strategy in support of the application of genetically engineered mosquitoes (GEM) with gene drive for malaria eradication. Government and health agencies responsible for malaria control and treatment will likely determine whether a field trial of GEM should be considered and will be instrumental in assisting the research team in the identification/mapping of additional stakeholders who will need to be engaged throughout the project. Stakeholder identification should also take into account other institutions, non-governmental organizations, funding agencies, and international partners involved in the national malaria control strategy. This subject is more extensively explored in the *application of the* Relationship-Based Model (RBM) section of this chapter.

9.5.2 Identifying a Model/Strategy for Engagement

A formalized strategy or identification of a model for engagement that reflects the research program objectives, mission, and values should be determined prior to initiating project activities. An engagement model or strategy should consider how the overall project research, development, and deployment timeline will be informed by, and coordinated with, engagement activities. The model/strategy should also provide a framework to help guide project team members, funders, and research institutions in how they approach important decisions about personnel, project communication, media engagement, training and capacity building, and development of definitions (communities, consent, acceptance, etc.) for the project.

9.5.3 Understanding Responsibilities

It is important that the research program team has the ability to remain responsive to project concerns, questions, and grievances, work within existing governance and accountability processes, and consider recommendations and guidelines from local authorities and agencies. The research program should seek and obtain appropriate project approval from relevant national and/or regional regulatory agencies and ethics committees to ensure accountability prior to conducting engagement activities. It is important that project engagement strategies are designed to work within established systems; they do not supersede existing and accepted laws, policy, regulation and governance, and respect existing international decision-making hierarchies or those within the country or community where applied. Developing a shared understanding of project expectations and mechanisms of existing systems is essential in the earliest phase of the project engagement.

9.6 RBM for Engagement

Notably, questions have arisen surrounding engagement activities and of the need for examples of models or frameworks that can be applied to guide engagement. Large conversations about engagement strategies and the development of engagement frameworks for GEM are initiated and led by outside experts and academics working to support specific research programs and goals [31]. The RBM provides a framework for GEM engagement that places stakeholders and community members at the center of decision-making processes rather than as recipients of pre-determined strategies, methods, and definitions. Successful RBM application in the transformation of healthcare delivery has demonstrated the importance of open dialog and relationship development in establishing an environment where individuals are actively engaged in decision-making processes regarding their health. While guidelines and recommendations for engagement for gene drives have previously been described, the RBM is unique in that and it provides a framework that supports decentralized decision-making and emphasizes the importance of stakeholder and community member leadership in the development and implementation of community and regulatory engagement strategies, definitions, and decisions (Fig. 9.2). The RBM provides a new approach to the development of ethical, transparent, and effective engagement strategies for field trials of GEM with a gene drive that reflects the important biological and social dynamics unique to the place and population where the technology may be applied.

9.7 RBM, Context, and Concepts

RBMs, developed for transforming healthcare delivery, have evolved from a "patient-centered" concept that was originally elaborated in 1969 [2]. There have been many conceptualizations described since then, using different terms including relationship-centered care and person-centered care. The relationship-based care model emerged in 2004 [15] and is widely known for providing a framework and concepts designed to focus on the specific needs of the patient and family. In 2006, relationship-centered care described a similar framework expanding the focus to the community and has since been adapted for successful population health management strategies [22].

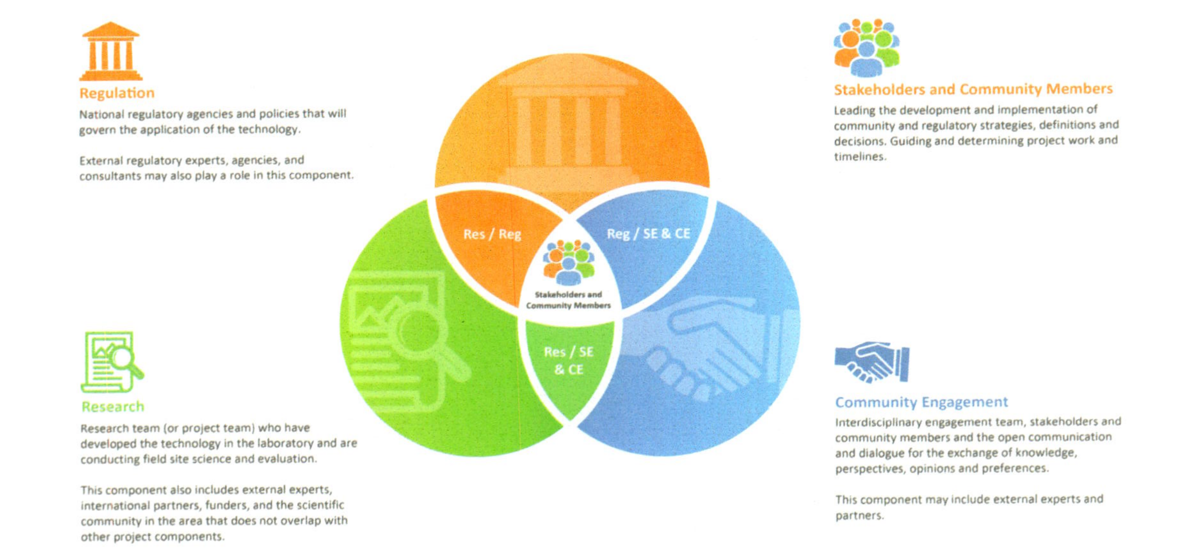

Figure 9.2 Relationship-Based Model (RBM) for engagement. Venn Diagram is showing interactions among key components applying an RBM. Stakeholders and community members are at the center and drive decision-making processes. RES/REG: Dialog among researchers and regulators is critical in the assessment of risk, application of the technology, and monitoring and surveillance. REG/SE & CE: Dialog among regulators, stakeholders, and community members informs and guides the risk-assessment and regulatory development processes, monitoring, and surveillance. RES/SE & CE: Dialog among researchers, stakeholders, and community will direct the scientific research and timelines, and will determine if, when, where, and how the technology is applied.

The RBM provides applications for building and strengthening relationships and establishes an environment in which individuals are actively engaged in decision-making processes regarding their health and how they receive care, rather than as recipients of messaging campaigns and prescriptive guidelines. RBMs recognize that the variables a public health system can positively affect or influence, such as personal habits, choices, options, and understanding, require collaborative, trusting relationships. Population health outcomes and acceptance of public health interventions are ultimately controlled by the decisions, values, beliefs, and circumstances of unique individuals who do not respond in a "one size fits all" environment.

Focusing on the value of relationships has resulted in successful, innovative health systems that have shown improved population and community health outcomes [6, 8, 9, 13, 19, 23, 27]. An expanded relationship between health delivery systems and the community promotes support of larger health improvement initiatives and traditional public health responsibilities of disease control and prevention that can be designed to meet the specific needs of the community/population. The World Health Organization (WHO) applied this concept to malaria, recently reporting that 'conditions and opportunities must be created to support the co-production of health in a way that places people at the center of malaria eradication considerations', and that if engagement for malaria control is to be successful 'one particularly important part of the broader process of engagement is the interaction between communities at risk of malaria and the health system with which they need to collaborate for a successful outcome' [31]. The RBM provides a salient framework for developing an effective, collaborative engagement strategy for malaria eradication programs.

The University of California Irvine Malaria Initiative (UCIMI) has developed gene drive-based systems for population modification of the African malaria vector mosquito, *Anopheles gambiae,* with the goal of contributing to the eradication of malaria [4]. Our program is concurrently conducting field research, work required **prior** to consideration of a release of GEM, to identify candidate sites for possible confined field trials. The field research includes conducting extensive collections and analyses of local mosquito populations, and engagement with stakeholders and community leaders. UCIMI has adopted an RBM approach for engagement, applying the core concepts of the RBM to our program (Box 9.2; taken from Kormos *et al*, 2020 [33]).

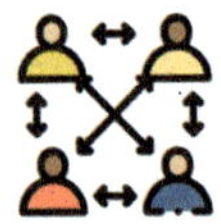

Box 2. Relationship-based model core concepts.

- Individuals drive decisions that determine their health and health outcomes.
- Relationships are important to understand the history, values, beliefs, and circumstances of communities and community members.
- Relationships of trust support collaboration, shared decision-making, and active participation.
- Optimal conditions for decision-making in the development of public health programs/interventions include the recipients at the point of care.
- Communication, collaboration and coordination between individuals/ communities and the programs, agents, health systems, and services involved in the delivery of health and public health interventions are needed.
- In complex, dynamic human systems, one size does not fit all.

While the application of RBM acknowledges the importance of stakeholder and community involvement, it does not ignore existing international decision-making hierarchies or those within the country or community where it is applied. On the contrary, it complements the WHO guidance for the evaluation of GEM by adapting to current governance rather than replacing it [32]. The RBM ensures that communities have a central role in directing program activities and strategies within existing systems. A research program like UCIMI, applying an RBM to engagement, is conducted in an attempt to reduce asymmetry between the program, external experts, academics, and the recipients of the technology. Applying an RBM acknowledges that stakeholders and individuals who are considering the technology should be leading conversations that inform decisions, policies, and frameworks for the application of the technology.

9.8 Applying the RBM

The established RBM outlines specific concepts for the successful implementation of the model [15]. We have adopted these concepts as a set of guidelines for the application of engagement emphasizing the importance of collaboration, relationship development, and full participation of people and their leaders in malaria-endemic countries (Box 9.3; taken from Kormos *et al*, 2020 [33]).

Box 3. Guidelines for applying the relationship-based model to engagement strategies.

- Commitment from program/institutional leadership to the model.
- Team-based approach to engagement.
- Build on existing strengths and resources.
- Develop an environment that supports the model: communication and engagement training, safe spaces and places for open communication, pathways for information sharing among groups.
- Engagement planning, development and implementation is led by site stakeholders and communities.
- Continuous evaluation and improvement.

9.8.1 Commitment to the Model

Program commitment to the RBM model is critical, and this commitment should be reflected in the mission, vision, values, and actions that program leaders demonstrate with colleagues, staff, collaborators, and community members. Decision-making is decentralized in an RBM. This changes who is making decisions, moving from an entirely position-based decision-making process to one that is knowledge-based. Decentralized decision-making creates conditions in which decision-making authority is given to those who are in the best position to determine the adequacy and efficacy of the decisions being made. It often involves active participation from individuals at the point of service/intervention in the development of specific strategies, timelines, and activities.

Within this approach, researchers expand their considerations when developing a project strategy and timeline to anticipate that there will be regular adjustments and revisions, and with the flexibility to respond to dynamic environments. Timelines for phases of development to deployment will be determined not only by scientific feasibility, but also by community and stakeholder groups driving decisions regarding engagement, and regulation and acceptance of the technology. The UCIMI program applied this concept by sharing a proposed project plan and timeline early in the initial engagement phase with stakeholders from potential field sites in order to address specific needs, requirements, and concerns unique to the site. From these early discussions, a final site-specific program timeline was determined by its stakeholders. The timeline and project work plans were then shared with existing and potential funders in conjunction with a written description of the RBM for engagement. At each stage of the project work, the work plan and timeline are revisited and reviewed with site stakeholders and partners, adjusting as needed to meet the requirements and priorities of the field site.

An RBM requires system-wide commitment, and this begins with the ability of program leadership to effectively communicate with funders and research institutions to which they are held accountable, benefits, challenges, and philosophy behind the application of the model so that there is a shared understanding about how program goals will be met over time. Commitment from funders and research institutions is of critical importance in a model that requires flexibility in the ability of the program to respond to community feedback. Program leadership commitment to the model also will require a commitment to establishing collaborative environments that are supportive in the development of relationships of trust and respect.

9.8.2 Interdisciplinary Approach

Community engagement

Every member of the research program is an integral part of the community engagement team, as they come into contact with public, stakeholder, or community groups by virtue of their participation in the program. In addition to the program research team, the RBM calls for a local, site-specific, interdisciplinary team that represents all entities contributing directly or indirectly to the health of the community at the field site to lead community engagement efforts. Applying this concept for novel malaria control strategies requires the inclusion of those affected by malaria and those who are central to decision-making and determining appropriate responses to malaria transmission, including control, treatment, and surveillance at the field site. Determination of the appropriate agencies and programs that need to be involved in engagement efforts should be guided by stakeholders at the field site where the research is being conducted.

The UCIMI research program initiated engagement with the Ministries of Health in the prospective field site of countries to begin conversations about the research and to develop a relationship with the government agencies responsible for malaria control, and who ultimately will determine whether a field trial of GEM with gene drive can/should be considered. This relationship has proven beneficial in connecting the program to the appropriate people in leadership roles and has provided the research program with credibility for having already passed through the required channels to allow for the initiation of conversation and collaboration. Additionally, this relationship assisted in providing the research program with an understanding of the national health system and the important intersections among community members, the health system, and malaria-related agencies and programs that will lead to the development of the site-specific engagement team. Focusing initial engagement efforts on assessing **who** is involved in malaria control efforts, **where** community members receive information about malaria, **what** sources of information are trusted, and **how** they intersect with the communities is essential in the development of the interdisciplinary engagement team.

Regulatory engagement

Just as a site-specific interdisciplinary team is important for community engagement, so is it important for regulatory engagement. The regulatory team, working within existing international and national regulations and policies, would determine appropriate regulation of this novel new technology. They also would determine how an assessment of risk is completed, who is involved, and if and when there is a need for external experts, international regulators, and neighboring countries to consult in the assessment and evaluation processes. Risk identification exercises can be carried out by the research team (comprising project and local collaborators) and information developed passed to regulators for consideration in the statutory process. The regulatory team composition and participation are determined by site stakeholders and may consist of national regulators, biosafety committees, environmental health experts, policymakers, and community leaders. Potential conflicts of interest must be declared and taken into consideration when assembling the team. The role of the research program in the development of this team is to provide support, information, and resources to stakeholders who are driving decisions to ensure that the overall regulatory process reflects and respects national goals and legislation.

9.8.3 Build on Existing Strengths and Resources

The RBM transforms healthcare delivery within an existing system, building on the strengths, resources, and expertise already present. How well individuals are able to engage with other professionals, sectors, agencies, and local communities influence? How trust is built effectiveness and coordination of functions? How health problems and issues are defined and addressed? Malaria research programs applying gene drive technology at a field site can learn and benefit from existing relationships and collaborations, education and awareness campaigns, social media and website access, and other tools and resources that are part of the existing malaria control programs and/or health system. Additionally, understanding the existing regulations, policies, and governing agencies is a critical first step in discussions about the regulation of gene drive organisms at a field site.

Recognizing and respecting the history, development, functionality, success, and challenges of local efforts and infrastructure are critical in the development of engagement strategies. UCIMI, in collaboration with the Ministries of Health at their candidate field sites, hosted workshops with stakeholders, community leaders, educators, non-governmental organizations (NGOs), and other agencies in an effort to assess existing malaria resources and outreach activities. Through these workshops and guidance from the Ministries, existing programs and initiatives were identified as important partners and collaborators for the research program in guiding best practices and strategies for engaging local communities.

The RBM places field site stakeholders in an important position of advising how to allocate research program resources to effectively and sustainably engage communities about each phase of the research they may consider advancing. Investing resources and support to strengthen existing structures, services, and initiatives may enhance the effectiveness and coordination of malaria-related functions of the larger public health and regulatory systems within the field site, particularly in early phases of the engagement. For example, UCIMI was advised by stakeholders to partner with existing malaria health education organizations in the initial phases of engagement to assess community understanding of malaria transmission and current controls being used. These partnerships offer opportunities for the program to advance their understanding of community knowledge through trusted sources, while providing resources and support to enhance existing community education efforts. In the advanced phases of engagement that require the delivery of information about the technology, stakeholders will provide direction to the program regarding appropriate engagement strategies. The goal is to establish shared as opposed to conflicting interests.

Here again, conflict of interest is an important consideration when determining with site stakeholders how and when to partner with external resources, as opposed to existing internal resources to support program engagement. UCIMI works with guidance from trusted local experts and agencies to advance the ability of the research program in assessing risk and community acceptance of the technology. Application of the RBM places these local experts and agencies as drivers of these activities in an effort to minimize the possibility that the program inadvertently impacts or influences how these existing resources function and are perceived within their communities. Guidance from stakeholders is essential in addressing these challenges.

9.8.4 Environment for Building/Strengthening Relationships

Developing an environment that supports the building of strong, trusting relationships among program team members, stakeholders, and community members is a foundational component of the RBM, and an important responsibility of the program. Support for good communication and engagement practices begins with participatory training and capacity building for the program team and field site collaborators that strengthen team-based problem solving and responsiveness, individual communication and engagement skills, and a shared understanding of how these skills assist in the development of relationships. Research program collaboration with expert(s) in communications and engagement are important to successfully guide the development and implementation of these opportunities for program team members and collaborators.

Relationship-based communication skills include active listening, open-ended questions, understanding individual communication styles and roles in relationships, identifying assumptions and communication roadblocks, reflecting (paraphrasing and restating both the feelings and words of the speaker), and developing shared language and definitions associated with the program. Group trainings and activities for the application of these skills help build capacity for good communication and relationship development.

Providing an environment that supports and encourages active community involvement in decision-making processes involves identifying/developing venues and mechanisms that allow people to engage comfortably in a dialog. This will require a direction from stakeholders and community leaders about appropriate places, safe spaces, times, and processes to encourage the active participation of community members. Additionally, the establishment of pathways that determine who, what, where, and when for consistent program communication and information sharing among key program components is an important part of developing an environment for relationships and the successful application of the RBM (Fig. 9.2).

9.8.5 Relationship-Based Engagement Planning, Development, and Implementation

In considering the who, how, what, where, and when questions that arise when developing community and regulatory engagement activities, the RBM points to the field site stakeholders and communities who will be directly impacted by the program for answers. There is active discussion among researchers, academics, regulators, and engagement experts about how communities are defined, how communities define themselves, what role the community wants to play/should play in the research, who influences the views and beliefs of the community, and how to assess, define, and determine risk [1, 10, 12, 21, 26]. These are critically important questions for global public health interventions, and specifically for a gene drive intervention that involves a field trial over a large geographical area.

Again, there are no "one size fits all" answers to these complex questions. Each country considering technologies, like gene drive for malaria control, has its own unique set of demographics, resources, infrastructure, governance, and its own unique geographic and political position in the larger global framework that will influence how these questions are answered. External research teams, social scientists, regulators, and other experts may present a framework or a set of guidelines and suggested definitions for thinking about these questions, but at the national level, the RBM places the stakeholders and community leaders in a position for determining the answers specific to their country. Working within an RBM, the responsibility of the research program includes sharing of knowledge and information about global and regional guidelines and frameworks and encouraging open dialog about how the national decisions will work within them, influence them, and may require some accountability to them.

Research program responsibilities also include support and resources (including external experts if requested) to encourage capacity building, and provide insight, advice, and guidance on the development of site-specific engagement and regulatory processes.

9.8.6 Continuous Evaluation and Improvement

Continuous and ongoing evaluation and improvement of activities, methods, and processes are key components of the RBM. If the model is to be effective, the program team at the point-of-service delivery will be involved in the outcome indicators being measured [15] and the process by which risk is assessed. The continuous improvement process within the RBM does not supersede independent assessments of program activities; rather it is established to allow the local program teams to facilitate appropriate responses to changing community needs, concerns, and priorities.

Evaluation and improvement for community engagement activities

The local community engagement team will direct what engagement outcomes will be measured and evaluated, how measures are defined, and how results from engagement will be shared with the larger community. Their direct involvement in evaluation begins with the collection of baseline data/information that will help inform the development of engagement activities. Activity outcome measures will guide and inform improvements and modifications to current and future engagement practices to improve effectiveness. Communication strategies and activities may need to be adapted in order to meet the needs, questions, and concerns of the community. Results also may inform the field site risk-assessment processes for the program, and provide valuable feedback and information to program stakeholders, funders, and the broader network of researchers who are interested in engagement practices.

Evaluation and improvement for regulatory engagementactivities

The WHO states that phased development and evaluation of a GEM will include continuous consideration of product safety and quality, as well as efficacy, which is consistent in the application of an RBM [32]. Risk assessment activities typically begin with an identification of protection goals followed by identification of potential hazards and the construction of pathways to harm in order to estimate likelihood and impact [25]. This evaluation is usually performed using qualitative and quantitative data derived from experts. As indicated previously, this process is initiated by the site team (researchers and site collaborators) and the information is provided to the statutory agencies. The agencies are expected to ask for additional information and analysis in their considerations of product safety and quality. This current and established process whereby field site regulators and experts determine how and when the risk assessment process will be conducted, what safety considerations need to be addressed, and if any outside experts and partners participate in this activity, complements the RBM. As with previous activities, potential conflicts of interest are declared and weighted accordingly.

9.8.7 Capacity Building

Active participation in the development of the new technologies, engagement strategies, regulatory pathways, and risk assessment processes will likely require different levels of capacity building for stakeholders and community partners. It is important to provide thoughtful consideration about where and how this training/capacity building may be conducted, and who is capable of providing it, so that conversations with field site stakeholders about training and capacity building opportunities and options can happen early in the program engagement efforts. The research program may consider collaborating with regulatory and public health agencies, research institutions, and NGOs that have expertise in specific areas, has advanced language or cultural competency, and/or previous experience at the prospective field site. Providing field site stakeholders with a menu of options at an early stage will help in the overall development of program timelines and activities. Capacity-building activities will help guarantee the continuity of transmission of important and fundamental knowledge among the participants involved in the project, guaranteeing both long-term technical capacity and sustainability, and facilitating the establishment of relationships of common interest among all parties involved, which strengthens and validates the RBM.

9.9 Conclusion

Engagement is an essential part of a GEM program for malaria research. Engagement activities can facilitate shared learning and decision making, assist in the identification of potential benefits and harms related to the technology, support transparency, and provide an opportunity for important dialog between researchers and diverse audience to explore questions regarding when, why, how, and if the technology is to be applied. Engagement provides a mechanism for those most affected to be involved, respected, and empowered in the decisions surrounding the use of the technology in relation to their health and the health of their environment.

Application of the RBM for the engagement of field trials of GEM for malaria control provides a framework for investigators who wish to establish meaningful and effective dialog, collaboration, and relationships of trust in the communities where their research is conducted. The core concepts of the model not only reflect and complement recently published guidelines and frameworks for community and regulatory engagement, but they also apply previously tested and effective methods for developing a public health strategy that is determined by the community. The core concepts of the model offer a new approach to engagement by emphasizing the importance of stakeholder and community member leadership in the development and implementation of engagement strategies, definitions, and decisions. This means that conversations, debates, and decisions surrounding important questions about who to engage, how to engage, and when to engage involve stakeholders and community members from field sites where research is being conducted. It is critically important to have these voices represented where these conversations take place and this model emphasizes this need. The RBM provides guidance for integrating field site communities in the development and implementation of novel malaria control interventions. It is our hope that consideration of this model for global public health interventions will contribute to the long-term goal of developing ethical, effective, and relationship-centered practices in engagement research.

References

1. AUDA-NEPAD. (2018) Gene drives for malaria control and elimination in Africa. Available at: https://www.nepad.org/publication/gene-drives-malaria-control-and-elimination-africa-0. Accessed June 1, 2019.
2. Balint E. (1969). The possibilities of patient-centered medicine. *J R Coll Gen Pract;17(82)*: 269–276. PMID: 5770926
3. Carballar-Lejarazú R, James AA. (2017) Population modification of Anopheline species to control malaria transmission. *Pathog Glob Health*; 111(8): 424–435. doi: 10.1080/20477724.2018.1427192
4. Carballar-Lejarazú R, Ogaugwu C, Tushar T, Kelsey A, Pham TB, Murphy J, et al. (2020) Next-generation gene drive for population modification of the malaria vector mosquito, *Anopheles gambiae. PNAS*. 117: 22805–22814. DOI: 10.1073/pnas.2010214117
5. Clancy CM. (2011) Patient engagement in health care. *Health Serv Res; 46(2)*: 389–393. doi:10.1111/j.1475-6773.2011.01254.x
6. Clever SL, Ford DE, Rubenstein LV, Rost KM, Meredith LS, Sherbourne CD, Wang N, Arbelaez JJ, Cooper LA. (2006) Primary care patients' involvement in decision-making is associated with improvement in depression. *Med Care; 44(5)*: 398–405. doi:10.1097/01.mlr.0000208117.15531.da
7. Delany-Moretlwe S, Stadler J, Mayaud P, Rees H. (2011) Investing in the future: Lessons learnt from communicating the results of HSV/HIV intervention trials in South Africa. *Health Res Policy Syst; 9(Suppl 1)*: S8. Available from: https://doi.org/10.1186/1478-4505/9/S1/S8.
8. Gluyas H. (2015) Patient-centered care: Improving healthcare outcomes. *Nurs Stand; 30(4)*: 50–57. doi:10.7748/ns.30.4.50.e10186
9. Gottlieb K. (2013) The Nuka System of Care: Improving health through ownership and relationships. *Int Jo Circumpolar Health; 72*. doi: 10.3402/ijch.v72i0.21118
10. Hartley S, Thizy D, Ledingham K, Coulibaly M, Diabate'A, Dicko B, et al. (2019) Knowledge engagement in gene drive research for malaria control. *PLoS Negl Trop Dis*; 13(4): e0007233. https://doi.org/10.1371/journal. pntd.0007233
11. Hood NE, Brewer T, Jackson R, Wewers ME. (2010) Survey of community engagement in NIH-funded research. *Clin Transl*; *3(1)*:19–22. doi:10.1111/j.1752-8062.2010.00179.x

12. James S, Collins FH, Welkhoff PA, Emerson C, Godfray HC, Gottlieb M, et al. (2018) Pathway to deployment of gene drive mosquitoes as a potential biocontrol tool for elimination of malaria in sub-Saharan Africa: Recommendations of a scientific working group. *Am J Trop Med Hyg;* 98(6 suppl):1–49. doi: 10.4269/ajtmh.18-0083
13. Johnston JM, Smith JJ, Hiratsuka VY, Dillard DA, Szafran QN, Driscoll DL. (2013) Tribal implementation of a patient-centered medical home model in Alaska accompanied by decreased hospital use. *Int J Circumpolar Health; 72.* doi: 10.3402/ijch.v72i0.20960
14. Kolopack PA, Parsons JA, Lavery JV. (2015) What makes community engagement effective? Lessons from the Eliminate Dengue program in Queensland Australia? *PLoS Negl Trop Dis; 9(4)*: e0003713. doi: 10.1371/journal.pntd.0003713
15. Koloroutis M. (2004) Relationship-based care: A model for transforming practice. 1st ed. Minneapolis, MN: Creative Health Care Management.
16. Lavery JM, Tindana PO, Scott TW, Harrington LC, Ramsey JM, Ytuarte-Nunez C, James AA. (2010) Towards a framework for community engagement in global health research. *Trends Parasitol; 26(6)*: 279–283. doi: 10.1016/j.pt.2010.02.009
17. MacQueen KM, Harlan SV, Slevin KW, Hannah S, Bass E, Moffet J. (2012) Stakeholder engagement toolkit for HIV prevention trials [Internet]. Durham: FHI360; [cited 2020 May 25]. Available from: https://www.fhi360.org/resource/stakeholder-engagement-toolkit-hiv-prevention-trials.
18. Michener L, Cook J, Ahmed SM, Yonas MA, Coyne-Beasley T, Aguilar-Gaxiola S. (2012) Aligning the goals of community-engaged research: Why and how academic health centers can successfully engage with communities to improve health? *Acad Med; 87(3)*:285-291. doi:10.1097/ACM.0b013e3182441680
19. Milstein A, Gilbertson E. (2009) American medical home runs. *Health Aff (Millwood); 28(5)*:1317–1326. doi: 10.1377/hlthaff.28.5.1317.
20. Nakibinge S, Maher D, Katende J, Kamali A, Grosskurth H, Seeley J. (2009) Community engagement in health research: Two decades of experience from a research project on HIV in rural Uganda. *Trop Med Int Health; 14(2)*: 190–195. doi:10.1111/j.1365-3156.2008.02207.x.
21. National Academies of Sciences, Engineering, and Medicine. (2016) Gene Drives on the horizon: Advancing science, navigating uncertainty, and aligning research with public values. 1st ed. Washington DC: The National Academies Press.

22. Nundy S, Oswald J. (2014) Relationship-centered care: A new paradigm for population health management. *Health; 2(4)*:216–219. doi: 10.1016/j.hjdsi.2014.09.003
23. Platchek T, Rebitzer R, Zulman D, Milstein A. (2014) Better health, less spending: Stanford University's Clinical Excellence Research Center. *Health Management, Policy and Innovation; 2(1)*:10–17. Available from: https://med.stanford.edu/content/dam/sm/cerc/documents/HMPI—Platcheck-Rebitzer-Zulman-Milstein-Better-Health-Less-Spending-2014.pdf
24. Ramsey JM, Bond JG, Macotela ME, Facchinelli L, Valerio L, Brown DM, Scott TW, James AA. (2014) A regulatory structure for working with genetically-modified mosquitoes: Lessons from Mexico. *PLoS Negl Trop Dis; 8(3)*: e2623. doi:10.1371/journal.pntd.0002623 PMID: 24626164
25. Roberts A, Andrade PP, Okumu F, Quemada H, Savadogo M, Singh JA, James S. (2017) Results from the workshop "problem formulation for the use of gene drive in mosquitoes". *Am J Trop Med Hyg; 96(3)*:530–533. doi: 10.4269/ajtmh.16-0726. Epub 2017 Apr 6.PMID: 27895273
26. Thizy D, Emerson C, Gibbs J, Hartley S, Kapiriri L, Lavery J, Lunshof J, Ramsey J, Shapiro J, Singh JA, et al. (2019) Guidance on stakeholder engagement practices to inform the development of area-wide vector control methods. *PLoS Negl Trop Dis; 13(4)*: e0007286. doi: 10.1371/journal.pntd.0007286
27. Weiner SJ, Schwartz A, Sharma G, Binns-Calvey A, Ashley N, Kelly B, Dayal A, Patel S, Weaver FM, Harris I. (2013) Patient-centered decision-making and health care outcomes. *Ann Intern Med; 158(8)*: 573–579. doi: 10.7236/0003-4819-158-8-201304160-00001
28. WHO Special Program for Research and Training in Tropical Diseases. (2014) Guidance framework for testing of genetically modified mosquitoes. Geneva: World Health Organization. Available from: https://apps.who.int/iris/handle/10665/127889
29. World Health Organization. (2016) Patient engagement: Technical series on safer primary care. Geneva: WHO. License: CC BY-NC-SA 3.0 IGO
30. World Health Organization. (2017) WHO community engagement framework for quality, people-centered and resilient health services. Geneva: World Health Organization (WHO/HIS/SDS/2017.15). License: CC BY-NC-SA 3.0 IGO.
31. World Health Organization. (2020) Malaria eradication: Benefits, future scenarios and feasibility. Geneva: World Health Organization. License: CC BY-NC-SA 3.0 IGO.

32. World Health Organization. (2020) Evaluation of genetically modified mosquitoes for the control of vector-borne diseases. Geneva: World Health Organization. License: CC-BY-NC-SA 3.0 IGO. ISBN: 978-92-4-001315-5 (electronic version).

33. Kormos A, Lanzaro GC, Bier E et al. (2020). Application of the Relationship-Based Model to Engagement for Field Trials of Genetically Engineered Malaria Vectors. *Am J Trop Med Hyg*; 104(3): 805–811.

Section V

Policy, Regulatory, and Ethical Considerations

Chapter 10

Review of International Regulatory Instruments and Processes

Felicity Keiper[a] and Ana Atanassova[b]

[a]*BASF Australia Ltd, Southbank, VIC, Australia*
[b]*BASF Belgium Coordination Center, Technologiepark-Zwijnaarde, Ghent, Belgium*
felicity.keiper@basf.com

10.1 Introduction

Regulatory oversight for the research and development (in containment) and eventual deployment (into the environment) of organisms containing engineered gene drives[1] is a topic of active debate involving many actors: governments, regulatory authorities, technology developers, the scientific community, non-government organizations (NGOs), and indigenous people and local communities (IPLCs). Internationally, this debate is predominantly occurring

[1]The term "organisms containing engineered gene drives" is used throughout this chapter to refer to organisms that have been genetically modified to introduce an engineered (also known as "synthetic") gene drive system. These organisms are genetically modified organisms (GMOs) developed using a specific application of genome editing technology. Gene/genome editing is another area of active regulatory debate but this is not examined in this chapter.

Mosquito Gene Drives and the Malaria Eradication Agenda
Edited by Rebeca Carballar-Lejarazú

ISBN 978-981-4968-33-1 (Hardcover), 978-1-003-30877-5 (eBook)
www.jennystanford.com

under the United Nations (UN) Convention on Biological Diversity (CBD),[2] which is a broad environmental treaty that requires its member countries (parties) to regulate the risks associated with the use and release of living modified organisms (LMOs)[3] resulting from biotechnology. New developments in biotechnology, including gene drive technology more broadly, have been under discussion since 2010 within the long-running CBD agenda item of "synthetic biology" [33].

The CBD discussions – as is often the case with new technologies – are emotive and polarized, fuelled by hype and sensational language on what might be possible. These discussions are largely driven by concerns regarding uncertainties as to the potential uncontrolled spread and serious or irreversible impacts of these LMOs on species and ecosystems [19],[4] and the existence/adequacy/preparedness of current LMO regulatory mechanisms to identify and manage the potential risks. If the main regulatory issues are extracted from this expansive discussion, they can be broadly classified into two main areas: (i) the adequacy of existing risk assessment approaches for predicting and understanding the potential impacts, and (ii) the appropriateness of national regulatory decision-making and governance processes. Broader cross-cutting issues that influence and shape regulatory policy developments include public health benefits, as well as broader social, economic, and ethical dimensions [20, 38, 52].

These regulatory and policy issues arise due to the primary distinction between LMOs containing engineered gene drives and more familiar LMOs, such as LM crops that have been released into the environment since the 1990s – the latter are not developed with the intention of dispersal throughout wild populations. Conversely, an LM crop is grown (and harvested) in a highly managed agricultural environment, with its ability to spread into unmanaged environments limited by control measures. The potential for LMOs

[2] CBD (adopted on June 5, 1992, entered into force on December 29, 1993) 1760 UNTS 79 (CBD).

[3] Analogous to the more commonly used term GMO; for consistency, the term LMO or LM is used throughout this chapter.

[4] See also: Document CBD/CP/RA/AHTEG/1/5, April 15, 2020, Annex I para 31; Document CBD/CP/RA/AHTEG/2020/1/INF/1.

containing engineered gene drives to spread throughout a population, potentially over regional and/or global scales (depending on the gene drive system), raises new considerations regarding the necessary information for environmental risk assessment (ERA), and where responsibility lies for authorizing releases of organisms that will not recognize political boundaries.

The aim of this chapter is to examine the applicable regulatory paradigm for the deployment of LM mosquitoes containing engineered gene drives for the public health purpose of controlling vector-borne diseases (VBD).[5] It begins with a review of the applicable international biotech/LMO regulatory framework, and then it considers developments that could result in new specific rules and procedures for LMOs containing engineered gene drives, regulatory and policy developments in other relevant governmental and inter-governmental fora, and parallel activities of the gene drive scientific community. The existing regulatory framework for LMOs is a pathway for the deployment of LM mosquitoes containing engineered gene drives, but international activities in this area have uncovered areas that need to be addressed in order to ensure comprehensive risk assessment and appropriate decision-making processes.

10.2 International Regulatory Framework and Current Developments

10.2.1 CBD

The CBD is an international environmental treaty with three primary objectives: the conservation of biological diversity, the sustainable use of its components, and the fair and equitable sharing of benefits arising from the use of genetic resources (Art 1). It is one of the most ratified of the UN treaties, with 196 member states (parties), which include all countries of the world except for the United States of America (USA) and the Holy See (see:

[5] This chapter focuses on the environmental release of these organisms and does not examine requirements related to research and development activities in containment.

https://www.cbd.int/information/parties.shtml). The CBD became legally binding on its parties on December 29, 1993, when it entered into force (see: https://www.cbd.int/intro/).

10.2.1.1 Biotechnology provisions

At the time the CBD was negotiated, the potential for biotechnology to contribute to the attainment of its objectives was recognized (see Art 16). However, as observed in the synthetic biology discussions of the present day, concerns were voiced about human and animal safety, potential adverse environmental impacts, ethical use, and the social and economic impacts of the resulting products [22]. To promote the potential benefits of biotechnology, the CBD includes specific provisions on participation in biotechnological research (Art 19.1), access to and distribution of the benefits of biotechnology (Art 19.2), and access to and transfer of technology including biotechnology (Art 16). In recognition of the potential risks, it also contains regulatory obligations (Art 8(g)).

The CBD defines biotechnology as "...*any technological application that uses biological systems, living organisms, or derivatives thereof, to make or modify products or processes for specific use*" (Art 2). Its parties are required to "*Establish or maintain means to regulate, manage, or control the risks associated with the use and release of LMOs resulting from biotechnology which is likely to have adverse environmental impacts that could affect the conservation and sustainable use of biological diversity, taking into account the risks to human health*" (Art 8(g)). The modalities for doing so are set out in a subsidiary treaty, the Cartagena Protocol on Biosafety to the CBD (Cartagena Protocol[6]) that entered into force in 2003 and currently has 173 parties (see: http://bch.cbd.int/protocol/parties/). The Cartagena Protocol also has a subsidiary agreement, the Nagoya-Kuala Lumpur Supplementary Protocol on Liability and Redress

[6]Cartagena Protocol on Biosafety to the CBD (adopted on January 29, 2000, entered into force on September 11, 2003) 2226 UNTS 208 (Cartagena Protocol). CBD Article 19(3) provides the legal basis for a protocol setting out these modalities.

to the Cartagena Protocol that entered into force in 2018.[7] The latter agreement is concerned with liability for damage caused by transboundary movements of LMOs, which has relevance to this topic but it will not be examined in this chapter.

10.2.1.2 Developments under the CBD related to gene drives

Discussions on the general topic of gene drives as a whole – i.e., types of drives, organisms, and potential uses – arose under the CBD agenda item of "synthetic biology", which has evolved over the past decade into a forum where the regulation of any new and emerging technology and resulting organisms in the field of biotechnology are debated. This is a contentious discussion largely driven by claims that the existing international biotech/LMO regulatory framework does not remain fit for purpose as technology evolves, with "gaps" that need to be addressed with new international rules [33]. At its most dramatic, this debate has featured campaigns by NGO coalitions for moratoria on releases of all organisms containing engineered gene drives as a measure necessitated by the precautionary principle until regulatory "gaps" are addressed [32]. Such a measure has been rejected by CBD parties as disproportionate; however, they have urged caution while research continues toward better understanding the technology, given its potential benefits [7, 8, 46].

The calls for moratoria climaxed during synthetic biology discussions at the most recent Conferences of the Parties ("COP"; COP13 in 2016 [7]; COP14 in 2018 [8–9]). The COP is the governing body of the CBD that meets biennially, and it has made decisions under its agenda item synthetic biology calling for increasingly prescriptive requirements for LMOs containing gene drives. The 2016 COP13 decision urged parties to take a "precautionary approach" with synthetic biology, including LMOs containing gene drives, and to establish or have in place effective risk assessment and management procedures and/or regulatory

[7]Nagoya-Kuala Lumpur Supplementary Protocol on Liability and Redress to the Cartagena Protocol on Biosafety (adopted on October 15, 2010, entered into force on March 5, 2018) UN Doc UNEP/CBD/BS/COP-MOP/5/17.

systems to regulate environmental releases.[8] That decision also noted that the general principles and methodologies for risk assessment under the Cartagena Protocol and existing biosafety frameworks provide a good basis for organisms developed through current applications of synthetic biology, but these may require updating.[9]

The 2018 COP14 decision again called for a precautionary approach, but specifically in regard to the "uncertainties regarding engineered gene drives".[10] It went further in specifically addressing gene drives, recognizing the "potential adverse effects", and that specific guidance may be needed to support risk assessment.[11] The decision also called for parties to only consider introducing organisms containing engineered gene drives into the environment when scientifically sound case-by-case risk assessments have been carried out; risk management measures are in place to avoid or minimize potential adverse effects, as appropriate; and where appropriate, the "prior and informed consent", the "free, prior, and informed consent" or "approval and involvement" of potentially affected IPLCs is sought or obtained.[12]

The governing body of the Cartagena Protocol, the Conference of the Parties serving as the Meeting of the Parties (COP-MOP), also meets biennially and concurrently with the CBD COP in an event called the United Nations Biodiversity Conference (UNCB). In parallel to COP14, the decision of COP-MOP9 reiterated that specific guidance may be useful to support risk assessment, and set out a program of work that included a detailed examination of this question (see 10.2.2.3).[13] Thus, currently there are parallel interconnected discussions and developments on this topic under the two treaties within the context of LMO regulation. The next UNBC, where COP15 and COP-MOP10 will be held, was scheduled for October 2020, but had to be delayed due to the COVID-19 pandemic and is currently expected to occur in late 2022. A

[8]Decision CBD/COP/DEC/XIII/17 paras 1 and 2; CBD/COP/DEC/XII/24 para 3.
[9]Decision CBD/COP/DEC/XIII/17 para 6.
[10]Decision CBD/COP/DEC/14/19 para 11.
[11]Decision CBD/COP/DEC/14/19 para 9.
[12]Decision CBD/COP/DEC/14/19 para 11.
[13]Decision CBD/CP/MOP/Dec/9/13.

decision is expected at the next UNBC regarding the need for development of additional risk assessment materials for LMOs containing engineered gene drives, and how ongoing work in that area will proceed.

10.2.2 Cartagena Protocol

10.2.2.1 Definitions & scope

The scope of the Cartagena Protocol is the transboundary movement, transit, handling, and use of all LMOs that may have adverse effects on the conservation and sustainable use of biological diversity, taking also into account human health (Art 4). The definition of LMO is "*any organism that possesses a novel combination of genetic material obtained through the use of modern biotechnology*" (Art 3(g)), with modern biotechnology broadly including "in vitro nucleic acid techniques" that overcome natural physiological reproductive or recombination barriers (Art 3(i)). In the CBD discussions, it is generally accepted that organisms containing engineered gene drives are LMOs within the scope of the Cartagena Protocol.[14]

It should be emphasized that while the Cartagena Protocol was drafted well before the emergence of engineered gene drives, it applies to almost all categories of LMOs,[15] not only those that were in existence at the time it was developed and for which it has predominantly applied to thus far, e.g., LM crops. Claims that international rules do not exist for other or "new" types of LMOs, including LMOs containing engineered gene drives, are misleading; there are rules, they are just not specific to any particular type of LMO. These rules include a general methodology for decision making based on scientifically sound risk assessment (Articles 10, 15, and Annex III), and the "advanced informed agreement procedure" that applies prior to the first intentional transboundary

[14]Document CBD/CP/RA/AHTEG/1/5, April 15, 2020, Annex I para 27(b); Document CBD/CP/RA/AHTEG/2020/1/4 para 20; see also EFSA GMO Panel 2020.

[15]The Cartagena protocol does not apply to transboundary movements of LMOs, which are pharmaceuticals for humans that are addressed by other international agreements or organizations (Art 5).

movement of an LMO into the environment of a party (Art 7). A fundamental principle of risk assessment enshrined in the Cartagena Protocol is the case-by-case approach, which provides flexibility and adaptability for different types of LMOs and uses. In practice, this means that the information required to support risk assessment will vary case-by-case depending on the LMO concerned, its intended use, and the likely potential receiving environment (Annex III, general principle 6).

10.2.2.2 National implementation

Parties to the Cartagena Protocol are required to implement a regulatory framework at the national level that is administered by a competent national regulatory authority. Such authorities conduct the necessary risk assessments and make decisions according to processes consistent with that set out in the Cartagena Protocol [39]. While the CBD is considered to have almost "universal" membership, the Cartagena Protocol does not, and there are several CBD parties that have experience with the regulation of LMOs who are not parties (e.g., Argentina, Australia, Canada, and Russian Federation; and the USA is not a party to either treaty). Many of these countries are major agricultural producers and exporters who established biotech regulatory frameworks prior to the Cartagena Protocol entering into force to facilitate development and commercial activities with LM crops, as well as other applications of biotechnology. Despite these countries not being parties, their regulatory processes, in particular their approaches to risk assessment, are largely consistent with the Cartagena Protocol, and these countries are able to trade LMO agricultural commodities with Cartagena Protocol parties.

It should also be noted that not all parties to the Cartagena Protocol have fully implemented its provisions at the national level, or they may not have experience in implementation. For example, the fourth national reports on the implementation of the Cartagena Protocol show that 15% of the 112 reporting countries have not established a mechanism to conduct risk assessments prior to taking decisions on LMOs. Notably, 21% of the reporting countries had no persons trained in risk assessment, risk management, and monitoring of LMOs, and 47% had not conducted any kind of LMO risk assessment in the reporting period.[16] The need for capacity building is frequently recognized in COP-MOP decisions, and such activities have been focussed on assisting developing parties to build the necessary administrative and technological base for conducting LMO risk assessments. Over time there has been some progress, but the most recent national report figures demonstrate that implementation and capacity issues remain important and require continued attention. In their 2016 report, NASEM [42] noted that the countries where gene drive research is predominantly occurring (such as the USA) have extensive regulatory expertise to offer, and there is a need for these countries to collaborate with the countries where releases of LM mosquitoes containing engineered gene drives are intended.

Decisions to authorize field testing of LM mosquitoes containing engineered gene drives will be the responsibility of the countries affected [10]. However, given the potential scale of the spread of LMOs containing engineered gene drives (depending on the type of drive), the CBD discussions have identified that a supranational, regional, or even an international approach to risk assessment and decision-making may be necessary.[17] This has been cited as justification for the development of new global rules; however, the Cartagena Protocol already provides a solution for such situations: *Parties may enter into bilateral, regional, and multilateral agreements and arrangements regarding intentional transboundary movements* of LMOs, provided the agreements that are consistent with the objectives of the Cartagena Protocol (Art 14(1)). The use of such agreements in anticipation of "unintentional" transboundary movements among a group of countries that wish to adopt gene drive technology has been suggested by members of the scientific community [10].

For some countries that have experience in LMO regulation, reports have been published navigating regulatory scope in regard to LMOs containing engineered gene drives, e.g., the USA [10, 40, 42,

---

[16]See: Report: CPB - Fourth National Report (replies to Articles 15 & 16 - Risk Assessment and Risk Management, Questions 61-84), Biosafety Clearing-House, https://beta.bch.cbd.int/reports/analyzer accessed on April 19, 2021.

[17]Document CBD/CP/RA/AHTEG/1/5, April 15, 2020, Annex I para 31; Document CBD/CP/RA/AHTEG/2020/1/INF/1 para 10(i).

46], Australia [4, 34], India, and Europe [34], and examining existing approaches to risk assessment [17, 42]. These reports point to the substantial prior relevant experience with releasing other types of LMOs, including LM insects, and with releases of insects for genetic and biological disease vector/pest control, and recommend drawing on this experience to inform ERA and post-release monitoring for LM insects containing engineered gene drives [17, 43, 47]. These reports also indicate that multiple national regulatory authorities will likely have overlapping responsibilities regarding releases of LM mosquitoes containing engineered gene drives. This arises because regulation applies beyond that which is LMO-specific; the use of the mosquitoes as a biocontrol agent or for VBD control will also be regulated. Coordination of these authorities and clarification of the applicable regulatory pathway will be needed for the deployment of the technology [10].

10.2.2.3 Risk assessment

Typically, in the development of an LMO, a structured, iterative, stepwise testing approach is applied, whereby technical and scientific data from each step informs the assessment and decision to progress to the next. The decision to release an LMO into the environment will primarily be based on the outcomes of an ERA conducted by a competent regulatory authority in the relevant jurisdiction. The established process of ERA is framed by problem formulation [12, 17, 53, 54]; this is a structured process involving the identification of the protection goals (e.g., the environmental elements to be protected from harm as set out in national regulations), and plausible pathways by which the proposed use (e.g., the environmental release of an LM mosquito containing an engineered gene drive) could potentially cause harm to those elements. This allows for testing of specific risk hypotheses and determination of the probability of harm occurring, and its severity if it does occur. Decisions can then be made based on the level of risk that is acceptable, which may or may not involve risk management measures and post-release monitoring [15, 47].

At COP-MOP9 in 2018, under the agenda item of "risk assessment", Cartagena Protocol parties decided on a program of work that included an assessment of the need for additional guidance materials to support ERA of LMOs containing engineered gene drives.[18] This included inviting submissions of information on the topic from parties, other governments, IPLCs, and relevant organizations,[19] as well as an online discussion series,[20] a commissioned study,[21] and deliberation of the information collected by an AHTEG.[22] The outcomes of this program of work have clarified several aspects of the CBD discussions on gene drives:

- Research on LMOs with engineered gene drives remains limited to laboratories, contained environments, and population modeling, with some applications potentially nearing release in trials.[23]
- There has not yet been an actual risk assessment conducted for an environmental release of organisms containing engineered gene drives.[24]
- The established risk assessment paradigm for LMOs remains applicable, but particular "technical and methodological challenges" require attention for organisms containing engineered gene drives.[25]
- Challenges include the sufficiency of modeling tools, and adequately trained personnel,[26] given that releases of organisms containing engineered gene drives may occur in countries that have less experience in conducting ERAs for LMOs.[27]
- There will be a need for post-release monitoring plans.[28]

---

[18]Decision CBD/CP/MOP/DEC/9/13.
[19]Notification SCBD/CPU/DC/MA/MW/87798; see: https://bch.cbd.int/ onlineconferences/submissions.shtml.
[20]See: http://bch.cbd.int/onlineconferences/forum_ra/discussion.shtml.
[21]Document CBD/CP/RA/AHTEG/2020/1/4, February 21, 2020; see: https://bch.cbd.int/onlineconferences/studies.shtml.
[22]Document CBD/CP/RA/AHTEG/1/5, April 15, 2020 (AHTEG Report).
[23]Document CBD/CP/RA/AHTEG/2020/1/4 para 5.
[24]Document CBD/CP/RA/AHTEG/2020/1/5, April 15, 2020, Annex I para 32.
[25]Document CBD/CP/RA/AHTEG/2020/1/5 Annex I para 29.
[26]Document CBD/CP/RA/AHTEG/2020/1/INF/1.
[27]Document CBD/CP/RA/AHTEG/2020/1/4 para 19; Document CBD/CP/RA/AHTEG/2020/1/5 Annex I para 26.
[28]Document CBD/CP/RA/AHTEG/2020/1/INF/1.

- Decision-making processes need to incorporate more extensive public consultations,[29] and benefits analyses.[30]
- There is much global activity in progress to prepare for environmental releases, with international expert meetings, problem formulation workshops, and development of risk assessment guidance.[31]
- Many of the concerns raised about "uncertainties" are not specific to LMOs containing engineered gene drives.[32]

The AHTEG recommendation to develop additional risk assessment guidance materials is endorsed in the draft recommendation of the 24th meeting of the Subsidiary Body on Scientific, Technical and Technological Advice (SBSTTA-24), which further recommends a process that includes the establishment of an AHTEG to support this.[33] The draft recommendation was prepared by SBSTTA-24 in a series of virtual meetings held in 2021 (due to the COVID-19 pandemic), and this will be the basis for deliberations and decision-making by Cartagena Protocol parties at COP-MOP10 later in 2022. The draft SBSTTA-24 recommendation indicates issues to resolve including diverging views on the exact process and the necessary scientific expertise to develop additional guidance, and if this needs to specifically focus on LM mosquitoes containing engineered gene drives[34] or address gene drives more broadly. Of note, statements made by NGO coalitions engaged in SBSTTA-24 meetings suggest a likely third gene drive moratorium campaign at the next UNBC [48, 49].

Many statements made in the SBSTTA-24 meetings called for a non-duplicative approach to this topic under the CBD and the Cartagena Protocol, and also for a non-duplicative and cooperative approach between these treaties and other relevant international organizations. Several parties pointed out that relevant risk assessment resources already exist (e.g., see 10.3 below), and recommended that the most appropriate path forward would be

---

[29]Document CBD/CP/RA/AHTEG/2020/1/INF/2.
[30]Document CBD/CP/RA/AHTEG/2020/1/5 Annex I para 25.
[31]Document CBD/CP/RA/AHTEG/2020/1/4 para 23.
[32]Document CBD/CP/RA/AHTEG/2020/1/4 para 23.
[33]Document CBD/SBSTTA/24/L6 paras 5 and 7, Annex.
[34]Document CBD/SBSTTA/24/L6 para 6, Annex paras 1(a) and (d).

adding any identified "needs" to those resources. This is consistent with recommendations from members of the scientific community, who have also urged that if additional guidance is developed for LM mosquitoes containing engineered gene drives, it would be more effective if led by the World Health Organization (WHO) [10], which has already developed a guidance framework for testing LM mosquitoes (see 10.3.1.1 and 10.3.1.3 below). The African Group (a UN regional group of more than 50 countries) expressed concern about an AHTEG being established for this purpose, due to existing processes not allowing for it to have a clear mandate.[35] The development of "guidance" by AHTEGs under the Cartagena Protocol has proven problematic in the past, with prior programs of work aimed at developing a "Guidance on Risk Assessment" for LMOs failing to deliver materials that were acceptable to the parties as an authoritative source, and while this document is publicly available it was not adopted or endorsed by the parties [29][36]. It has also not been widely employed used by parties in support of LMO risk assessments, with the fourth national report on the implementation of the Cartagena Protocol showing that 38% of reporting countries have used it while 62% have used other sources of guidance. Therefore, careful consideration is urged in regard to any calls for a moratorium or any other form of restriction on environmental releases pending the development of additional guidance materials under the CBD/Cartagena Protocol.

10.3 Regulatory and Policy Developments

10.3.1 International Developments

10.3.1.1 Scientific community

The proactive engagement of the gene drive scientific community to promote open discussion and rapid publication of developments in scientific journals is unprecedented. This is motivated by

[35]See statement 4 of May 24, 2021 (formal virtual session), accessible at: https://www.cbd.int/conferences/sbstta24-sbi3/sbstta-24/documents.

[36]See also: Decision CBD/CP/MOP/VII/12.

this community recognizing the need for the responsible and transparent development and management of this technology, with their work aimed at informing robust risk assessment, and improved coordination between different stakeholders including technology developers, regulators, and the public [23, 30]. Their "early" publications included commitments to the safe and responsible development of gene drive technology [3], the development of guiding principles gene drive research [18], recommendations for containment strategies for gene drive research [3–5], and endorsement of the development of a robust governance framework designed by all interested parties [16].

As technological developments have progressed, publications have included core commitments for field trials [38], safety and efficacy criteria for field trials [31], and examination of existing relevant guidance and needs [1, 10]. The latter work identified two primary areas where additional guidance is needed: updated *technical* guidance regarding best practices for field trials in confined cages and small-scale open field tests, and guidance on *community engagement* before, and during field testing [1]. In regard to technical guidance, as reflected in the CBD discussions, engagement of the WHO and updating of the 2014 *Guidance Framework for testing genetically modified mosquitoes* (*Guidance Framework* [54]) (see 10.3.1.3 below) has been strongly urged [1]. In response to these calls, the WHO released an updated second edition of the *Guidance Framework* in May 2021 [53], coinciding with virtual meetings of SBSTTA-24 where the draft recommendation was discussed.

A recurring theme in these publications is the need to engage the community in each stage of release of LM mosquitoes containing engineered gene drives, with societal acceptance recognized as an essential component of the pathway to deployment, beyond the need for regulatory authorization. Guidance in community engagement was considered to be the more critical need than technical guidance, with recommendations to build on previous initiatives [6, 37], and relevant experiences, such as that reported for the release of *Wolbachia*-infected mosquitoes in Australia (e.g., [35]; see 10.3.1.2 below) [1, 10]. This has been examined in detail by NASEM, who provided recommendations on considerations for engagement strategies at each step of the research and development pathway [42], and elaborated further by the WHO in their updated *Guidance framework* [53]. This has also been examined in the CBD context of "free and prior informed consent", and in regard to the engagement of potentially affected IPLCs [21].

More recent publications (e.g., [23])[37] are reporting on developments in mathematical modeling for predicting how gene drives will interact with ecosystems. These also detail the challenges with incorporating new qualitative analytical methods into regulatory infrastructures, with the need for regulators to adopt, validate and apply them, and then interpret the data to predict and evaluate risks. This contrasts with the approach typically used for current LMOs where empirical data is collected from several standardized experimental protocols and compared to other (usually non-modified) organisms with known risks. Therefore, for assessing LMOs containing engineered gene drives, there are capacity needs in regard to establishing how models will be used in regulatory processes, and also in the expertise of the regulators themselves [23].

10.3.1.2 LM (and non-LM) mosquitoes

A common misunderstanding in the general public is that LM mosquitoes that have already been released into the environment contain engineered gene drives, but this is not the case. Such mosquitoes have been assessed and authorized for release into the environment for field trials, e.g., Brazil [11], Burkina Faso [50],[38] Cayman Islands [25], Malaysia [36], Panama [24], and the USA;[39] or for unrestricted releases, via application of existing LMO

[37]See also: Gene Convene webinar series: https://www.geneconvenevi.org/gene-drive-and-genetic-biocontrol-webinars/#toggle-id-3.

[38]Decree No. 2018-453/MESRSI/SG/ANB authorizing the controlled release of genetically modified sterile male mosquitoes in the village of Bana or Souroukoudingan]. Ouagadougou: Ministry of Higher Education, Scientific Research and Innovation, National Biosecurity Agency; 2018.

[39]Environmental Protection Agency. Issuance of an experimental use permit. Federal Register 2020, 85(11): 35307. Available at: https://www.govinfo.gov/content/pkg/FR-2020-06-09/html/2020-12372.htm. See also https://www.epa.gov/pesticides/epa-approves-experimental-use-permit-test-innovative-biopesticide-tool-better-protect; https://www.oxitec.com/en/news/unanimous-decisions-by-

regulatory approaches. For example, Brazil (a Cartagena Protocol party) authorized the unconstrained release of LM *Aedes aegypti* (OX513A) in 2014 [13][40] and a second release (OX5034) in 2019.[41]

The decisions of these regulators reflect the application of the established structured problem formulation approach to ERA, and they concluded that the LM mosquito presented no greater risk than the non-modified mosquito. This approach is consistent with the recommendations of the WHO (see 10.3.1.3) that existing regulatory mechanisms should be adapted to the purpose rather than replaced for these organisms, and that internationally recognized risk assessment approaches should be applied, also taking into account the potential health benefits [56]. The process followed by Brazil for unconstrained releases also reflects the phased development and evaluation pathway developed by the WHO (see 10.3.1.3 below).

To add to the public confusion, another approach aimed at controlling VBD that does not involve LMOs but is sometimes described as a "gene drive-like" system, is the release of *Wolbachia*-infected mosquitoes into wild populations. *Wolbachia* is a maternally inherited intracellular bacterium that is able to spread through natural populations, and infected mosquitoes can interfere with disease transmission. *Wolbachia*-infected mosquitoes have been released into the environment in several countries, including Australia and Vietnam, for the control of dengue [28, 44]. The proposed release of these mosquitoes in 2010 in northern Australia presented a regulatory problem: there was no directly applicable regulatory scheme for this type of application since it did not involve the introduction of a new species into the country, nor did it involve an LMO. To address this a risk analysis was performed by independent experts, then again under the existing legislation for veterinary chemicals that was ultimately used for the formal regulatory authorization [14]. The releases of infected mosquitoes

us-epa-state-of-florida-approve-environmentally-sustainable-oxitec-friendly-mosquitoes-for-pilot-project.

[40]See: Technical opinion no. 3964/2014. Brasília: National Technical Biosafety Commission (2014). Available at: http://bch.cbd.int/database/attachment/?id=14514.

[41]See: CTNBio decision: https://www.in.gov.br/web/dou/-/extrato-de-parecer-tecnico-n-6.946/2020-258262552.

were preceded by an extensive period of community engagement, and had strong community support [28].

Notably, the first risk analysis conducted for the *Wolbachia*-infected mosquitoes in Australia followed the Risk Analysis Framework of the Office of the Gene Technology Regulator (for LMOs), as it was considered to be best practice for a novel system [41]. Australia is a CBD party but not a Cartagena Protocol party, and it has a comprehensive biotech regulatory framework that was in place prior to the Cartagena Protocol entering into force.[42] The Australian Academy of Science (AAS) has determined that releases of organisms containing engineered gene drives into the environment (field trial to unrestricted) are within the scope of this biotech regulatory framework [4]. They also made several recommendations, including that any decision for release into the environment being based on a risk assessment that includes environmental and evolutionary modeling, and involves public consultation, particularly with the affected communities [4].

10.3.1.3 World Health Organization (WHO)

The WHO is a specialized UN agency that serves as an authority to its 194 member states regarding interventions that have demonstrated public health value and should be considered for adoption by governments [42]. The use of biotechnology for the purpose of controlling VBD is supported by the WHO, in recognition of the increasing need and urgency for innovative supplemental or alternative tools to combat VBD. In their 2020 Position Statement, the WHO refers to engineered gene drives in mosquitoes as a tool that raises hopes for durable, affordable protection against disease transmission [56].

The WHO has developed resources relevant to developing and testing LM mosquitoes to be used for control of VBD, with the foundational document being the 2014 *Guidance Framework* [54]. This set out best practices for efficacy and safety testing, as well as fundamental considerations for addressing public engagement and transparency, all of which drew on experience with other

[42]See: http://www.ogtr.gov.au/internet/ogtr/publishing.nsf/Content/home-1.

new public health tools and relevant experience in biocontrol and agriculture. The recently released second edition incorporates developments in "self-sustaining" transgenic approaches that involve "gene drive-modified mosquitoes". These are differentiated from other "self-limiting" transgenic approaches by the drive mechanism and associated spread and persistence characteristics [53]. The *Guidance Framework* also recommends that decision-makers take into consideration both the safety (risk) of a tool and its efficacy (public health benefits) [30, 53, 54], and evaluation of economic benefits in comparison to alternative courses of action [53].

The WHO *Guidance Framework* sets out a phased development and evaluation pathway for LM mosquitoes. This involves stepwise progression from small-scale contained (physically confined) testing in a laboratory, insectary, or indoor cage facility, to physically (cages) or ecologically confined field testing, to staged field releases of increasing scale that enable observation of interactions with native mosquito populations and other elements of the local ecosystem [53, 54]. A phased testing pathway was supported as an appropriately precautionary approach for LMOs containing engineered gene drives in the analysis of NASEM [42]. The 2021 update of the *Guidance Framework* describes a modified testing pathway with additional precautions for LM mosquitoes containing engineered (low-threshold) gene drives, noting that this pathway may not be entirely linear. The additional considerations draw from the precedent of the testing pathway for exotic biocontrol agents that are expected to spread and persist after release into the environment [53].

The critical decision to progress from physically contained indoor testing for candidates that meet efficacy criteria to any level of field testing must be supported by risk assessment. According to the WHO, the appropriate safety criterion for the LMO is that it will do no more harm to human health than wild-type mosquitoes of the same genetic background, and no more harm to the ecosystem than other conventional vector control interventions [56]. This criterion was reiterated by the recent assessment by European Food Safety Authority (EFSA) for LM insects containing engineered gene drives [17]. Also, consistent with the scientific community and expert bodies [4, 42], the WHO emphasized the need for community acceptance of any intervention as being key to the successful deployment of LM mosquitoes [56]. They elaborate on this in their recently published WHO Guidance on ethics and vector-borne diseases [57], and emphasize the need for community engagement throughout the phased testing pathway in the 2021 update of the *Guidance Framework* [53]. The latter further emphasizes the importance of involving scientists in the countries of intended deployment in the research process to promote leadership and co-ownership of the technology [53].

A mechanism that facilitates access to vector control products is the WHO Vector Control Product Prequalification process (see: https://extranet.who.int/pqweb/vector-control-products). This is a voluntary process, the aim of which is to provide support to WHO member states in their decision making related to public health, and it is of greatest benefit to countries lacking capacity. This first requires a WHO policy recommendation, which the WHO does not currently have for LM mosquitoes (including those containing engineered gene drives) [56]. To develop one, the intervention would need to be assessed via a "new intervention pathway" procedure with the support of the Vector Control Advisory Group (VCAG) that is jointly managed by the Global Malaria Program, the Department of Control of Neglected Tropical Diseases, and the WHO Prequalification Team for vector control products. The procedure requires epidemiological data so that the public health value can be assessed [55]. Once a WHO policy recommendation is in place, the intervention will be classified into a "prequalification pathway", overseen by the WHO Prequalification Team for Vector Control Products (PQT-VCP). In this pathway, the safety, quality, and entomological efficacy of the intervention are assessed, and if demonstrated, the intervention will be added to the list of prequalified products by the PQT-VCP [56].

10.3.1.4 Organization for Economic Co-operation and Development (OECD)

The OECD Working Party on Harmonization of Regulatory Oversight in Biotechnology developed a consensus document on the biology of the mosquito species *Aedes aegypti* that was published in

2018 [45], with another in development for the mosquito species *Anopheles gambiae*. Many such documents have been developed by this Working Party for plant species,[43] whose membership includes national authorities who are responsible for conducting ERAs for LMOs. The documents are routinely used by regulators to inform certain requirements of ERAs for LMOs, as they provide relevant scientific information regarding the characteristics of the host organisms, such as its taxonomy, morphology, life cycle, reproductive biology, genetics, ecology, and interactions with other species and the environment.

10.3.2 Regional Developments

10.3.2.1 African Union (AU)

With the first deployment of LM mosquitoes containing engineered gene drives intended for malaria control in Africa, several activities of the gene drive scientific community have focussed on this region. Some of this work involves collaboration with the African Union Development Agency (AUDA) New Partnership for Africa's Development (NEPAD), which supports the comprehensive examination of the technology toward its development and deployment in Africa to control malaria [2]. In their High Level African Union Panel on Emerging Technologies (APET) Report of 2018 [2], AUDA-NEPAD emphasized the need for:

- A co-development approach, whereby African scientists are included in research and development of gene drive technology that is intended for deployment in Africa;
- The development of regulatory frameworks (where necessary) in African Union member countries;
- Strategies to address regulatory capacity needs in African Union member countries;
- Coordination of different regulatory agencies within countries where there are overlapping responsibilities;

[43]See: http://www.oecd.org/env/ehs/biotrack/consensusdocumentsfortheworkonharmonisationofregulatoryoversightinbiotechnologybiologyofcrops.htm.

- Regional harmonization of regulatory policies and approaches; and
- Strategies for public engagement, given its integral role in technology acceptance.

Most African countries are parties to the Cartagena Protocol, with LMOs regulated by a National Biosafety Authority or Committee. Several African countries also have experience in regulating LMOs, but this is predominantly for LM crops, and experience with LM insects is limited [30]. At the SBSTTA-24 meetings, the African Group highlighted the concern that some countries are yet to develop functional national biosafety frameworks and/or lack the capacity to conduct LMO risk assessments.[44] The need for regulatory capacity building is well recognized, with the intergovernmental agency NEPAD working with African countries to develop and implement functional regulatory systems (see: http://nepad-abne.net/). The APET report highlighted that three African countries – Burkina Faso, Mali, and Uganda – are in the process of building technical, regulatory, and institutional capacities in preparation for work with gene drives, and that these countries may lead the way for others on the African continent [2].

The APET report also highlighted the applicability of the phased testing pathway of the WHO *Guidance Framework*, and the crucial role of the WHO in providing guidance to African countries with its implementation. Their view is that the deployment of LM mosquitoes containing engineered gene drives would be in the remit of national malaria control programs [2], with coordination required between LMO/environment regulators and health regulators. This coordination is also required between regulators at national and regional levels, given the potential for transboundary movement [2] (see: https://www.nepad.org/content/scope-and-approach). There are various precedents for regional cooperation in Africa, including programs for mosquito control, pest eradication, and African Medicines Regulatory Harmonization (which includes malaria) [2, 10, 53]. This precedent is also noted in the 2021 update of the WHO

[44]See statement 4 of May 24, 2021 (formal virtual session), accessible at: https://www.cbd.int/conferences/sbstta24-sbi3/sbstta-24/documents

Guidance Framework that recommends a regional notification and agreement process for environmental introductions [53].

In relation to ERA in the African context, at the SBSTTA-24 meetings, the African Group acknowledged the existence of resources relevant to ERA of LMOs containing engineered gene drives, and the need for these to be adapted for local relevance, consistent with the Cartagena Protocol, and a scientifically sound case-by-case approach.[45] Relevant experience would include that with LM mosquitoes (not containing engineered gene drives), and detailed case studies have been published for releases of LM *Anopheles gambiae*, which only occurs on the African continent, e.g., a risk assessment for the hypothetical scenario of an escape of LM mosquitoes from an insectary in Western Africa [26], and a risk assessment for a proposed field release of LM mosquitoes in Burkina Faso based on data from insectaries [27]. Toward informing ERA of LM mosquitoes containing engineered gene drives, regional stakeholder consultations have been conducted to foster dialog on identifying relevant protection goals and potential pathways to harm for the environment [51], and a problem formulation analysis for releases of LM *Anopheles gambiae* containing an engineered gene drive in West Africa was recently published [12].

In anticipation of future deployment of LM mosquitoes containing engineered gene drives, a large multi-disciplinary working group examined the pathway for a low-threshold drive in LM *Anopheles gambiae* in sub-Saharan Africa, and they developed recommendations for safe and ethical testing [30]. The considerations raised and approach taken by the working group reflect and consolidate many of the approaches and recommendations described throughout this chapter: the phased development pathway of the WHO *Guidance Framework*, and the need to draw on prior experience with other vector control tools, LM organisms, and biocontrol agents. They also emphasized the crucial role of public engagement, with this requiring collaboration and partnering throughout the development process with government authorities, scientists, and

---

[45]See: statement 45 of February 19, 2021 (informal virtual session), accessible at: https://www.cbd.int/conferences/sbstta24-sbi3/sbstta-24-prep-03/documents.

social scientists, and ethicists in the countries of environmental releases [30].

10.4 Conclusion

This chapter presents a review of the applicable regulatory paradigm for LM mosquitoes containing engineered gene drives, and current developments in the international fora of the CBD and WHO that impact - and should be aimed at facilitating - a comprehensive examination of the potential of this technology. The international regulatory framework provided by the CBD/Cartagena Protocol addresses the fact that the mosquitoes are LMOs, while the guidance provided by the WHO addresses the use of the mosquitoes with engineered gene drives for VBD control.

The mix of issues and challenges presented by the development and deployment of LMOs containing engineered gene drives set them apart from current/previous LMOs that have been released in the environment on a large scale. While certain information, technical, and capacity needs have been identified in relation to conducting robust ERA for LM mosquitoes containing engineered gene drives, these can be addressed in parallel to technology development (in containment). However, the outcomes of the LMO ERA will not be the predominant consideration in decision-making; in this context, other important considerations include finding the appropriate balance between environmental protection and public health goals, and considering and comparing the impacts of available alternative solutions to the public health problem. As emphasized by the scientific community, this necessitates the involvement of international organizations with relevant expertise, and the review presented in this chapter points to the need for strong collaboration between the WHO and the CBD/Cartagena Protocol on this topic.

Beyond the regulatory pathway, an essential element for the successful the deployment of LM mosquitoes containing engineered gene drives is public support. This has been demonstrated with the deployment of LM mosquitoes and *Wolbachia*-infected mosquitoes. This need is well recognized by the gene drive scientific community, and their engagement in public outreach and partnering with scientists and governments in countries where the technology may be deployed in the future is another feature that sets gene drive technology apart from current/previous LMOs. This is likely to set a precedent for other applications of emerging biotechnologies, and may drive policy and regulatory developments in the areas of technical and scientific capacity building, and public engagement and participatory decision making.

Acknowledgments

The authors thank Camilla Beech of Cambea Consulting Ltd. for her critical reading of the manuscript and for providing helpful feedback, and Lysiane Snoeck for her assistance with formatting.

References

1. Adelman, Z., Akbari ,O., Bauer, J., Bier, E., Bloss, C., Carter, S.R., Callender, C., Costero-Saint Denis, A., Cowhey, P., Dass, B., Delborne, J., Devereaux, M., Ellsworth, P., Friedman, R.M., Gantz, V., Gibson, C., Hay, B.A., Hoddle, M., James, A.A., James, S., Jorgenson, L., Kalichman, M., Marshall, J., McGinnis, W., Newman, J., Pearson, A., Quemada, H., Rudenko, L., Shelton, A., Vinetz, J.M., Weisman, J., Wong, B. and Wozniak, C. (2017). Rules of the road for insect gene drive research and testing, *Nat. Biotechnol.*, **35**, pp. 716–718. doi: 10.1038/nbt.3926
2. African Union (AU) and New Partnership for Africa's Development (NEPAD) (2020). Gene drives for malaria control and elimination in Africa. High level APET report.
3. Akbari, O.S., Bellen, H.J., Bier, E., Bullock, S.L., Burt, A., Church, G.M., Cook, K.R., Duchek, P., Edwards, O.R., Esvelt, K.M., Gantz, V.M., Golic, K.G., Gratz, S.J., Harrison, M.M., Hayes, K.R., James, A.A., Kaufman, T.C., Knoblich, J., Malik, H.S., Matthews, K.A., O'Connor-Giles, K.M., Parks, A.L., Perrimon, N., Port, F., Russell, S., Ueda, R. and Wildonger, J. (2015). Safeguarding gene drive experiments in the laboratory, *Science*, **349**, pp. 927–929. doi: 10.1126/science.aac7932
4. Australian Academy of Science (AAS) (2017). Synthetic gene drives in Australia: Implications of emerging technologies. Discussion Paper, May 2017.

5. Benedict, M.Q., Burt, A., Capurro, M.L., De Barro, P., Handler, A.M., Hayes, K.R., Marshall, J.M., Tabachnick, W.J. and Adelman, Z.N. (2018). Recommendations for laboratory containment and management of gene drive systems in Arthropods, *Vector Borne Zoonotic Dis.*, **18**, pp. 2–13. doi: 10.1089/vbz.2017.2121
6. Brown, D.M., Alphey, L.S., McKemey, A., Beech, C. and James, A.A. (2014). Criteria for identifying and evaluating candidate sites for open-field trials of genetically engineered mosquitoes, *Vector Borne Zoonotic Dis.*, **14**, pp. 291–299. doi: 10.1089/vbz.2013.1364
7. Callaway, E. (2016). "Gene drive" moratorium shot down at UN biodiversity meeting, *Nature*. doi: 10.1038/nature.2016.21216
8. Callaway, E. (2018). UN treaty agrees to limit gene drives but rejects a moratorium, *Nature*. doi: 10.1038/d41586-018-07600-*w*
9. *Ca*llaway, E. (2018). Ban on "gene drives" is back on the UN's agenda - worrying scientists, *Nature*, **563**, pp. 454–455. doi: 10.1038/d41586-018-07436-4
10. Carter, S.R. and Friedman, R.M. (2016). Policy and regulatory issues for gene drives in insects, Report of a workshop held by UC San Diego and the J. Craig Venter Institute, 20–21 January 2016.
11. Carvalho, D.O., McKemey, A.R., Garziera, L., Lacroix, R., Donnelly, C.A., Alphey, L., Malavasi, A. and Capurro, M.L. (2015). Suppression of a field population of *Aedes aegypti* in Brazil by sustained release of transgenic male mosquitoes, *PLoS Negl. Trop. Dis.*, **9**, e0003864. doi: 10.1371/journal.pntd.0003864
12. Connolly, J.B., Mumford, J.D., Fuchs, S., Turner, G., Beech, C., North, A.R. and Burt, A. (2021). Systematic identification of plausible pathways to potential harm via problem formulation for investigational releases of a population suppression gene drive to control the human malaria vector *Anopheles gambiae* in West Africa, *Malar. J.*, **20**, p. 170. doi: 10.1186/s12936-021-03674-6
13. de Andrade, P.P., Aragão, F.J.L., Colli, W., Dellagostin, O.A., Finardi-Filho, F., Hirata, M.H., de Castro Lira-Neto, A., de Melo, M.A., Nepomuceno, A.L., da Nóbrega, F.G., de Sousa, G.D., Valicente, F.H. and Zanettini, M.H.B. (2016). Use of transgenic *Aedes aegypti* in Brazil: risk perception and assessment, *Bull. World Health Organ.*, **94**, pp. 766–771. doi: 10.2471/BLT.16.173377
14. De Barro, P.J., Murphy, B., Jansen, C.C. and Murray, J. (2011). The proposed release of the yellow fever mosquito, *Aedes aegypti* containing a naturally occurring strain of Wolbachia pipientis, a question of

regulatory responsibility, *Conference Proceedings "Decision Making and Science—the Balancing of Risk Based Decisions that In?uence Sustainability of Agricultural Production"*, 7–8 October 2010, Berlin, Germany, *J. Consum. Prot. Food S.* doi: 10.1007/s00003-011-0671-x

15. Devos, Y., Bonsall, M.B., Firbank, L.G., Mumford, J., Nogué, F. and Wimmer, E.A. (2020). Gene-drive modified organisms: Developing practical risk assessment guidance, *Trends Biotechnol.*, S0167–7799(20)30310–3. doi: 10.1016/j.tibtech.2020.11.015
16. Eckhoff, P.A., Wenger, E.A., Godfray, H.C.J. and Burt, A. (2016). Impact of mosquito gene drive on malaria elimination in a computational model with explicit spatial and temporal dynamics, *Proc. Natl. Acad. Sci.*, E255–E264. doi: 10.1073/pnas.1611064114
17. EFSA GMO Panel (EFSA Panel on Genetically Modified Organisms), Naegeli, H., Bresson, J-L, Dalmay, T., Dewhurst, I.C., Epstein, M.M., Guerche, P., Hejatko, J., Moreno, F.J., Mullins, E., Nogué, F., Rostoks, N., Sánchez Serrano, J.J., Savoini, G., Veromann, E., Veronesi, F., Bonsall, M.B., Mumford, J., Wimmer, E.A., Devos, Y., Paraskevopoulos, K. and Firbank, L.G. (2020). Scientific Opinion on the adequacy and sufficiency evaluation of existing EFSA guidelines for the molecular characterisation, environmental risk assessment and post-market environmental monitoring of genetically modified insects containing engineered gene drives, *EFSA J.*, **18**: 6297. doi: 10.2903/j.efsa.2020.6297
18. Emerson, C., James, S., Littler, K. and Randazzo, F. (2017). Principles for gene drive research, *Science*, **358**, pp. 1135–1136. doi: 10.1126/science.aap9026
19. Esvelt, K.M., Smidler, A.L., Catteruccia, F. and Church, G.M. (2014). Emerging technology: Concerning RNA-guided gene drives for the alteration of wild populations, *eLife*, **3**, e03401. doi: 10.7554/eLife.03401
20. European Group on Ethics in Science and New Technologies (2021). Ethics of genome editing, Opinion no. 32, European Commission, 19 March 2021. Available at: https://ec.europa.eu/info/sites/info/files/research_and_innovation/ege/ege_ethics_of_genome_editing-opinion_publication.pdf.
21. George, D.R., Kuiken, T. and Delborne, J.A. (2019). Articulating 'free, prior and informed consent' (FPIC) for engineered gene drives, *Proc. R. Soc. B*, **286**: 20191484. doi: 10.1098/rspb.2019.1484
22. Glowka, L., Burhenne-Guilmin, F., Synge, H., McNeely, J.A. and Gündling, L. (1994). A guide to the convention on biological diversity, IUCN Environmental Policy and Law Paper No. 30, p. 176.

23. Golnar, A.J., Ruell, E., Lloyd, A.L. and Pepin, K.M. (2021). Embracing dynamic models for gene drive management, *Trends Biotechnol.*, **39**, pp. 211–214. doi: 10.1016/j.tibtech.2020.08.011
24. Gorman, K., Young, J., Pineda, L., Márquez, R., Sosa, N., Bernal, D., Torres, R., Soto, Y., Lacroix, R., Naish, N., Kaiser, P., Tepedino, K., Philips, G., Kosmanna, C. and Cáceres, L. (2015). Short-term suppression of *Aedes aegypti* using genetic control does not facilitate *Aedes albopictus, Pest Manage. Sci.*, **72**, pp. 618–628. doi: 10.1002/ps.4151
25. Harris, A.F., Nimmo, D., McKeney, A.R., Kelly, N., Scaife, S., Donnelly, C.A., Beech, C., Petrie, W.D. and Alphey, L. (2011). Field performance of engineered male mosquitoes, *Nat. Biotechnol.*, **29**, pp. 1034–1037. doi:10.1038/nbt.2019.
26. Hayes, K., Barry, S., Beebe, N., Dambacher, J., Ford, J., Hosack, G., Peel, D., Ferson, S., Goncalves da Silva, A. and Thresher, R. (2016). Risk assessment for controlling mosquito vectors with engineered nucleases, Part I: Sterile Male Construct Final report. Commonwealth Science and Industrial Research Organization, Biosecurity Flagship.
27. Hayes, K.R., Hosack, G.R., Ickowicz, A., Foster, S., Peel, D., Ford, J. and Thresher, R. (2018). Risk assessment for controlling mosquito vectors with engineered nucleases: Controlled field release for sterile male construct. Risk assessment and final report. Commonwealth Science and Industrial Research Organization, Health and Biosecurity.
28. Hoffmann, A.A., Montgomery, B.L., Popovici, J., Iturbe-Ormaetxe, I., Johnson, P.H., Muzzi, F., Greenfield, M., Durkan, M., Leong, Y.S., Dong, Y., Cook, H., Axford, J., Callahan, A.G., Kenny, N., Omodei, C., McGraw, E.A., Ryan, P.A., Ritchie, S.A., Turelli, M. and O'Neill, S.L. (2011). Successful establishment of *Wolbachia* in *Aedes* populations to suppress dengue transmission, *Nature*, **476**, pp. 454–457. doi:10.1038/nature10356
29. Hokanson, K. (2019). When policy meets practice: The dilemma for guidance on risk assessment under the Cartagena Protocol on Biosafety, *Front. Bioeng. Biotechnol.*, **7**: 82. doi: 10.3389/fbioe.2019.00082
30. James, S., Collins, F.H., Welkhoff, P.A., Emerson, C., Godfray, H.C.J., Gottlieb, M., Greenwood, B., Lindsay, S.W., Mbogo, C.M., Okuma, F.O., Quemada, H., Savadogo, M., Singh, J.A., Tountas, K.H. and Toure, Y.T. (2018). Pathway to deployment of gene drive mosquitoes as a potential biocontrol tool for elimination of malaria in sub-Saharan Africa: Recommendations of a scientific working group, *Am. J. Trop. Med. Hyg.*, **98**, pp.1–49. doi: 10.4269/ajtmh.18-0083
31. James, S.L., Marshall, J.M., Christophides, G.K., Okumu, F.O. and Nolan, T. (2020). Toward the definition of efficacy and safety criteria for

advancing gene drive-modified mosquitoes to field testing, *Vector Borne Zoonotic Dis.*, **20**, pp. 237–251. doi: 10.1089/vbz.2019.2606

32. Kaebnick, G.E., Heitman, E., Collins, J.P., Delborne, J.A., Landis, W.G., Sawyer, K., Taneyhill, L.A. and Winickoff, D.E. (2016). Precaution and governance of emerging technologies, *Science*, **354**, pp. 710–711. doi: 10.1126/science.aah5125
33. Keiper, F. and Atanassova, A. (2020). Regulation of synthetic biology: Developments under the Convention on Biological Diversity and its Protocols, *Front. Bioeng. Biotechnol.*, **8**: 310. doi: 10.3389/fbioe.2020.00310
34. Kelsey, A., Stillinger, D., Pham, T.B., Murphy, J., Firth, S. and Carballar-Lejarazu, R. (2020). Global governing bodies: A pathway for gene drive governance for vector mosquito control, *Am. J. Trop. Med. Hyg.*, **103**, pp. 976–985. doi:10.4269/ajtmh.19-0941
35. Kolopack, P.A., Parsons, J.A. and Lavery, J.V. (2015). What makes community engagement effective? Lessons from the *Eliminate Dengue* Program in Queensland Australia, *PLoS Negl. Trop. Dis.*, **9**, e0003713. doi: 10.1371/journal.pntd.0003713
36. Lacroix, R., McKemey, A.R., Norzahira, R., Lim, K.W., Wong, H.M., Teoh, G.N., Siti Rahidah, A.A., Salman, S., Subramaniam, S., Nordin, O., Norhaida Hanum, A.T., Angamuthu, C., Mansor, S.M., Lees, R.S., Naish, N., Scaife, S., Gray, P., Labbé, G., Beech, C., Nimmo, D., Alphey, L., Vasan, S.S., Lim, L.H., Nazni Wasi, A. and Murad, S. (2012). Open field release of genetically engineered sterile male *Aedes aegypti* in Malaysia, *PLoS One*, **7**: e42771. doi: 10.1371/journal.pone.0042771
37. Lavery, J.V., Tinadana, P.O., Scott, T.W., Harrington, L.C., Ramsey, J.M., Ytuarte-Nuñez, C. and James, A.A. (2010). Towards a framework for community engagement in global health research, *Trends Parasitol.*, **26**, pp. 279–283. doi: 10.1016/j.pt.2010.02.009
38. Long, K.C., Alphey, L., Annas, G.J., Bloss, C.S., Campbell, K.J., Champer, J., Chen, C-H, Choudhary, A., Church, G.M., Collins, J.P., Cooper, K.L., Delborne, J.A., Edwards, O.R., Emerson, C.I., Esvelt, K., Evans, S.W., Friedman, R.M., Gantz, V.M., Gould, F., Hartley, S., Heitman, E., Hemingway, J., Kanuka, H., Kuzma, J., Lavery, J.V., Lee, Y., Lorenzen, M., Lunshof, J.E., Marshall, J.M., Messer, P.W., Montell, C., Oye, K.A., Palmer, M.J., Papathanos, P.A., Paradkar, P.N., Piaggio, A.J., Rasgon, J.L., Rašić, G., Rudenko, L., Saah, J.R., Scott, M.J., Sutton, J.T., Vorsino, A.E. and Akbari, O.S. (2020). Core commitments for field trials of gene drive organisms, *Science*, **370**, pp. 1417–1419. doi: 10.1126/science.abd1908
39. Mackenzie, R., Burhenne-Guilmin, F., La Viña, A.G.M., Werksman, J.D., Ascencio, A., Kinderlerer, J., Kummer, K. and Tapper, R. (2003) Explana-

tory guide to the Cartagena Protocol on Biosafety. IUCN Environmental Policy and Law Paper No. 46, p. 278.

40. Meghani, Z. and Kuzma, J. (2018). Regulating animals with gene drive systems, *J. Responsible Innov.*, **5**, pp. S203–S222. doi: 10.1080/23299460.2017.1407912
41. Murray, J.V., Jansen, C.C. and De Barro, P. (2016). Risk associated with the release of Wolbachia-infected Aedes aegypti mosquitoes into the environment in an effort to control Dengue, *Front. Public Health*, **4**: 43. doi: 10.3389/fpubh.2016.00043
42. National Academies of Science, Engineering and Medicine (NASEM) (2016). Gene drives on the horizon: Advancing science, navigating uncertainty, and aligning research with public values. Washington DC: National Academies of Sciences, Engineering and Medicine.
43. Netherlands Commission on Genetic Modification (COGEM), Rüdelsheim, P.L.J. and Smets, G. (2018). Experience with gene drive systems that may inform an environmental risk assessment, COGEM Report CGM 2018–03, June 2018.
44. Nguyen, T.H., Nguyen, H.L., Nguyen, T.Y., Vu, S.N., Tran, N.D., Le, T.N., Vien, Q.M., Bui, T.C., Le, H.T., Kutcher, S., Hurst, T.P., Duong, T.T.H., Jeffrey, J.A.L., Darbro, J.M., Kay, B.H., Iturbe-Ormaetxe, I., Popovici, J., Montgomery, B.L., Turley, A.P., Zigterman, F., Cook, H., Cook, P.E., Johnson, P.H., Ryan, P.A., Paton, C.J., Ritchie, S.A., Simmons, C.P., O'Neill, S.L. and Hoffmann, A.A. (2015). Field evaluation of the establishment potential of wmelpop *Wolbachia* in Australia and Vietnam for dengue control, *Parasites Vectors*, **8**: 563. doi:10.1186/s13071-015-1174-x
45. Organization for Economic Co-operation and Development (2018). Consensus document on the biology of mosquito *Aedes aegypti*. Series on Harmonization of Regulatory Oversight in Biotechnology No. 65. OECD Environment, Health and Safety Publications. Document ENV/JM/MONO(2018)23.
46. Oye, K.A., Esvelt, K., Appleton, E., Catteruccia, F., Church, G., Kuiken, T., Lightfood, S.B.-Y., McNamara, J., Smidler, A. and Collins, J.P. (2014). Regulating gene drives, *Science*, **345**, pp. 626–628. doi: 10.1126/science.1254287
47. Romeis, J., Collatz, J., Glandorf, D.C.M. and Bonsall, M.B. (2020). The value of existing regulatory frameworks for the environmental risk assessment of agricultural pest control using gene drives, *Environ. Sci. Policy*, **108**, pp. 19–36. doi: 10.1016/j.envsci.2020.02.016
48. Schabus, N. and Schröder, M. (2021). Summary of the 24th Meeting of the Subsidiary Body on Scientific, Technical and Technological Advice

of the Convention on Biological Diversity: May–June 2021. IISD Earth Negotiations Bulletin Vol. 9 No. 756.

49. Schabus, N. and Soubry, B. (2021). Summary of the informal meeting of the subsidiary body on scientific, technical and technological advice of the convention on biological diversity: 17–19 and 24–26 February 2021. IISD Earth Negotiations Bulletin Vol 9 No. 754.
50. Target Malaria (2021). Results of the small-scale release of non-gene drive genetically modified sterile male mosquitoes in Burkina Faso. Fact Sheet March 2021. Available at: https://targetmalaria.org/wp-content/uploads/2021/03/Development-pathway_FS_EN_Results-of-the-small-scale-release-of-non-gene-drive-genetically-modified-Burkina-Faso_March21.pdf
51. Teem, J.L., Ambali, A., Glover, B., Ouedraogo, J., Makinde, D. and Roberts, A. (2019). Problem formulation for gene drive mosquitoes designed to reduce malaria transmission in Africa: Results from four regional consultations 2016–2018, *Malar. J.*, **18**: 347. doi: 10.1186/s12936-019-2978-5
52. Thompson, P.B. (2018). The roles of ethics in gene drive research and governance, *J. Responsible Innov.*, **5**, pp. S159–S179. doi: 10.1080/23299460.2017.1415587
53. World Health Organization Special Program for Research and Training in Tropical Diseases, GeneConvene Global Collaborative (2021). Guidance framework for testing genetically modified mosquitoes, Second edition. Available at: http://apps.who.int/iris/bitstream/handle/10665/341370/9789240025233-eng.pdf?sequence=1&isAllowed=y
54. World Health Organization Special Program for Research and Training in Tropical Diseases and Foundation for the National Institutes of Health (FNIH) (2014). Guidance framework for testing genetically modified mosquitoes. Available at: http://www.who.int/tdr/publications/year/2014/guide-fmrk-gm-mosquit/en/
55. World Health Organization (WHO) (2017). Malaria vector control policy recommendations and their applicability to product evaluation. Geneva. Available at: http://www.who.int/malaria/publications/atoz/vector-control-recommendations/en/
56. World Health Organization (WHO) (2020). Evaluation of genetically modified mosquitoes for the control of vector borne diseases. Position Statement 13 October 2020. Available at: https://www.who.int/ publications/i/item/9789240013155
57. World Health Organization (WHO) (2020). Ethics and vector borne diseases: WHO Guidance.

Chapter 11

Gene Drive Mosquitoes: Ethical and Political Considerations

Daniel Edward Callies[a] and Athmeya Jayaram[b]
[a]*University of California, San Diego, Institute for Practical Ethics, California, USA*
[b]*Johns Hopkins University, Berman Institute of Bioethics, Maryland, USA*
danielcallies@gmail.com

11.1 Introduction

Ethics is concerned not so much with what people actually believe about values and how people actually act, but with what we have reason to believe and how we have reason to act. More directly, ethics is concerned with what we ought to believe and what we ought to do in certain circumstances—even if some of us (or even most of us) do not actually do so. What this means is that an ethical analysis of gene drive mosquitoes will not outline what people actually think about the project, but rather what they ought to think about it. There are various ethical traditions that one could use as a lens to evaluate the project. Here, we set those different traditions aside, and attempt to offer a brief overview of some of the ethical concerns such a project could raise. Rather than offering definitive answers to what one ought to think about, for instance, whether the

Mosquito Gene Drives and the Malaria Eradication Agenda
Edited by Rebeca Carballar-Lejarazú

ISBN 978-981-4968-33-1 (Hardcover), 978-1-003-30877-5 (eBook)
www.jennystanford.com

gene drive project exemplifies a hubristic attitude, we will simply outline the ethical concerns and then offer some reasons to think these concerns are weighty or not.

In what follows, we will examine eight ethical concerns. Does mere research into gene drive mosquitoes place us on a slippery slope toward their use? Given the uncertainties involved in releasing gene drive constructs into the wild, might we be better off simply playing it safe and abandoning the gene drive project? In pursuing a gene drive approach to malaria eradication, are we ignoring the success we have achieved with conventional vector control methods in favor of a techno-fix? Does releasing gene drives into the wild fail to show nature the proper respect it deserves? Are we acting hubristically by attempting to gain ever more control over the natural environment? How should the public be involved in decisions surrounding research and development? How can we fairly distribute the benefits and burdens from gene drive mosquitoes? And is there a dual-use concern that gene drives could be used as weapons of destruction? These are the questions this chapter explores.

11.2 Slippery Slope of Research

Gene drives are controversial. But it is not just the actual use or release of gene drive constructs in the wild that causes concern. Some would argue that even research into the technology is something that should be abandoned. One of the most oft-cited worries about research is the slippery slope worry. The thought is that once research starts it will continue inexorably until gene drives are released outside of laboratories. That is, taking the first step of the research will lead us inevitably down a slippery slope, at the bottom of which is the use of dangerous or morally fraught technology.

Slippery slope arguments all take the same form. If we allow *x*, then *y* will follow (the *empirical premise*). We do not want *y* to follow (the *normative premise*). Therefore, we should not allow *x* (the *conclusion*). Being instances of *modus tolens*, these arguments are valid; their premises, if true, logically support their conclusions.

The real question is whether these arguments are sound—whether, in addition to being valid, they, in fact, have true premises. And whether or not any particular slippery slope argument is sound will depend upon how the two premises are filled out and what relation they have to one another.

Consider an initial slippery slope worry about gene drive research:

P1: If we research gene drive mosquitoes, this will lead to their release in the wild.
P2: We do not want gene drive mosquitoes released in the wild.
C3: Therefore, we should not research gene drive mosquitoes.

As I mentioned above, this is a valid argument. But the soundness of the argument is in question. The first premise, the empirical premise, seems true. It seems very likely that research into gene drive mosquitoes will lead to their release in the wild. Maybe not tomorrow, maybe not this year (2021). But it seems likely that they will one day be released, and so the premise is likely true (though, we will have to wait to see about that). However, the truth-value of the second premise, the normative premise, is far less clear. That is, it is not clear that we do not want to release gene drive mosquitos in the wild. It may very well be the case they turn out to be a highly effective tool with which to help eradicate malaria. Our assessment is that it is currently too early to tell whether the release of gene drive mosquitoes to eradicate malaria is something we should welcome or avoid. This is because, if successful, gene drive mosquitoes could confer immense public health benefits to some of the most disadvantaged populations of the globe. So, it is too early to say that we should avoid their release in the wild. For this first slippery slope argument, the first (empirical) premise is likely true, but the second (normative) premise is suspect, which makes the soundness of the argument as a whole suspect.

But consider a more dystopian slippery slope worry about gene drive research.

P1: If we research gene drive mosquitoes, this will lead to a world full of gene-drive-modified crops, gene-drive-modified animals, and even gene-drive-modified humans.

P2: We do not want a world full of gene-drive-modified crops, animals, and humans.
C3: Therefore, we should not research gene drive mosquitoes.

This more dystopian slippery slope argument is, like the first, a valid argument. If the premises are true, then the conclusion necessarily follows. There are some who would not be deterred by the second, normative premise in this argument. But equally true is that many would be. Let us grant for the moment the second premise is true—let us grant that a world in which gene drives are nearly omnipresent is something to be avoided. Notice that the first, empirical premise in this more dystopian argument is far more questionable. It is not at all clear that researching gene drive mosquitoes for malaria eradication will lead to a completely engineered world, with gene drive constructs being nearly omnipresent.

One of the hallmarks of a troubling slippery slope is when there is no clear point on the slope at which to draw a line. That is, a slippery slope worry is appropriate when the slope is, in fact, slippery, and there are no points on the slope to draw a principled distinction. But notice that there are multiple such points on the slope in question in this argument. There are clear lines between researching gene drive mosquitoes and using gene drives in humans. As it stands, this more dystopian slippery slope argument is also unsound; this time because the first, empirical premise is likely false.

What the slippery slope worry should push us toward is a legitimate regulation of gene drives. With good regulation, we stand a better chance of safely and responsibly pursuing the benefits potentially associated with gene mosquitoes without worrying that this will lead to a world in which gene drives are omnipresent.

11.3 Precautionary Principle

Given that there is a non-zero chance that gene drives could cause significant environmental destruction, one might think the risk is simply not worthwhile. That is, one might think it better to err on the side of caution, and forgo the benefits that gene drives could deliver, in order to be certain that the destruction they could cause does not come about.

This kind of precautionary thinking is captured well by what has come to be known as the precautionary principle. The precautionary principle is a norm that aims to offer policy guidance in situations of uncertainty. The principle has been elaborated and endorsed by different governing bodies since at least the 1980s. The 1982 United Nations World Charter for Nature stated "Activities which are likely to pose a significant risk to nature shall be preceded by an exhaustive examination; their proponents shall demonstrate that expected benefits outweigh the potential damage to nature, and where potential adverse effects are not fully understood, the activities should not proceed" (United Nations, 1982). Put slightly differently the 1992 United Nations Rio Declaration states "Where there are threats of serious or irreversible damage, lack of full scientific certainty shall not be used as a reason for postponing cost-effective measures to prevent environmental degradation" (United Nations, 1992b). And the world's leading treaty on biological diversity—the United Nations Convention on Biological Diversity—reads: "Where there is a threat of significant reduction or loss of biological diversity, lack of full scientific certainty should not be used as a reason for postponing measures to avoid or minimize such a threat" (United Nations, 1992a).

As should be clear from the preceding paragraph, there is no one canonical precautionary principle. Rather, the precautionary principle gets at the idea contained in the adage of "better safe than sorry." Despite there being no singular version of the precautionary principle, Neil Manson has pointed out that each variant follows the same basic formula. According to Manson, there is always first a damage condition, second a knowledge condition, and finally a remedy. For example, take the formulation found in the Convention on Biological Diversity. "Where there is a threat of significant reduction or loss of biological diversity (the damage condition), lack of full scientific certainty (the knowledge condition) should not be used as a reason for postponing measures to avoid or minimize such a threat (the remedy)" (United Nations, 1992a).

What does the precautionary principle, as elaborated in the Convention on Biological Diversity, imply for the gene drive mosquito malaria eradication project? Given that we cannot be certain of the effects the gene drive mosquito project would have on ecosystems, we have to admit that there is a *threat* of a loss of biodiversity, even if the threat is a very low probability one. And the Convention on Biological Diversity notes that "lack of full scientific certainty should not be used as a reason for postponing measures to avoid or minimize such a threat." It would seem that the precautionary reasoning embedded in the Convention would preclude us from releasing gene drive mosquitos in the wild.

But many scholars, including Cass Sunstein (Sunstein, 2005), have noted how crippling—and, at times, incoherent—the precautionary principle is. The principle places an incredibly high burden of proof on new technologies. If we held all new technologies to the test that they must prove, beyond a reasonable doubt, that there is no threat of serious harm, then many things we currently rely upon— chlorine, airplanes, antibiotics, vaccines, X-rays, etc.— all would have been abandoned.

Ultimately, the advice of taking precaution, or attempting to avoid catastrophe, offers little to no guidance. The situation we are in is one in which the status quo is unacceptable. Millions of people are infected with malaria each year, and hundreds of thousands die from the disease. Simply recognizing that one ought to be cautious fails to deliver any sound policy recommendations. This being the case, it is unlikely that a crude version of the precautionary principle, like the wording found in the Convention on Biological Diversity, can help us reach a conclusion about the best course of action, all things considered.

11.4 Conventional Alternatives

Another ethical concern one might have about gene drives could come from suspicion about technocratic approaches to social problems. Some worry that a gene drive approach to malaria might be a quintessential technological fix or "techno-fix." As Alvin Weinberg has pointed out, a techno-fix occurs when a technological or engineering approach is taken to solve a problem instead of pursuing a social or behavioral change (Weinberg, 1967). For instance, one could view gastric band—or lap band—surgery as a technological fix. Rather than an individual pursuing a demanding behavioral change, they could instead opt to address their weight and heart issues with a surgery that constricts the amount of food their stomach can take in, thus reducing their food intake. We might be worried about society pursuing these kinds of technological approaches to problems, rather than changing institutions or our collective behavior. With respect to the gene drive mosquito project, one could worry that we are opting for a technological solution rather than the more demanding avenue of investing fully in conventional malaria control methods and health care infrastructure in poorer parts of the world.

There are two responses to the techno-fix concern that are worth pointing out. First, as Christopher Preston has argued (albeit in a different context), the moral status of techno-fixes is not entirely clear (Preston, 2013). That is, while the word has a negative connotation (Rosner, 2004), it is not clear that any particular "techno-fix" will be prima facie morally suspect. Whether or not we should have ethical reservations about a particular technological approach to a problem will depend upon the particulars of the technology, the magnitude of the problem it is attempting to address, and the specifics of the social and behavioral avenues that could also be pursued.

This leads directly into the second response to the techno-fix concern. It is not entirely clear that there is a socio-behavioral approach to the problem of malaria that is being overlooked in favor of a gene drive approach. There are, of course, a host of conventional malaria control methods that have been rather successful at eliminating much of the disease burden. Bednets, indoor residual spraying, the elimination of standing water, and more widely dispersed access to health care have certainly reduced the burden of malaria in many parts of the world (Callies, 2020). For example, malaria has been eradicated from countries like Algeria to Singapore with conventional vector control methods and investment in health care infrastructure. (World Health Organization 2017).

But it is also important to note that countries like Algeria and Singapore did not face anywhere near the disease burden faced in sub-Saharan Africa. The WHO (2019: 2) writes, "...even with our most optimistic scenarios and projections, we face an unavoidable fact: using current tools, we will still have 11 million cases of malaria in Africa in 2050." So, it is not entirely clear that there is an easy social change that could bring about the same potential reduction in malarial burden that could be achieved through gene drives.

One should also recognize that there are important differences between eliminating malaria with gene drives and doing so with increased investment in conventional methods. On the one hand, investing in health care infrastructure throughout sub-Saharan Africa would certainly help eliminate the burden of malaria. And this would have significant corollary benefits. But this would also be massively costly, and funding for malaria control and general development in this part of the world is already in short supply. The WHO estimated that 4.4 billion dollars was needed for conventional malaria control efforts in 2017, and yet only 3.1 billion dollars was provided. And, of course, this WHO estimate does not include massive investments in healthcare infrastructure.

Furthermore, some conventional control methods—like the draining of wetlands and the wide use of insecticides—have significantly deleterious ecological effects. Draining wetlands to prevent mosquitoes from breeding destroys entire ecosystems, and insecticides are similarly indiscriminate with the kinds of invertebrates they kill. So, more investment in conventional malaria control methods and much greater investment in health care infrastructure would certainly be desirable. But it may not be a sufficient response to the problem. And if there is not a clear socio-behavioral avenue to eliminating malaria, it is not clear how much weight the techno-fix worry should be given.

11.5 Environmental Ethics

Environmental ethics often asks us to think about things from outside of our own perspective. Imagine we were to go ahead with the gene drive mosquito project and everything was to go exactly as planned. Regardless of whether we used a suppression drive or a replacement drive, imagine we were able to reduce malaria to zero and there were minimal adverse side effects. Some environmental philosophers would still claim that we could be doing something wrong or failing to show nature the proper respect it deserves. There are different ethical outlooks in environmental ethics—with three being the most common.

Perhaps the most common is the anthropocentric outlook. Anthropocentrism, within environmental ethics, is the view that value is human-centered. Humans are the kinds of things that have intrinsic or final value, and anything else with value has its value as a means to some human end. But there are other outlooks on value as well. The biocentric outlook would place intrinsic or final value not just in humans, but in all individual forms of life. Any individual organism that counts as living is something that has intrinsic or final value. Someone like Holmes Rolston III would note that each species is a unique embodiment of the evolutionary process, and that this uniqueness confers intrinsic value. The ecocentric outlook would place intrinsic value not just in humans, and not just in individual forms of life, but in collectives of life or ecosystems. Ultimately, the kinds of things that have intrinsic or final value are natural communities or ecosystems. Ecocentrists, like Aldo Leopold, might note that protecting the value found in individuals could lead to significant harm to groups or ecosystems. For instance, individual members of invasive species can often damage ecosystem integrity. A biocentrist, someone who places value in individual organisms, may object to sacrificing one individual organism even for the sake of the whole. Whereas an ecocentrist, like Leopold, would focus on the group and say eliminating the one for the sake of the group is permissible, and sometimes even obligatory.

Each of these environmental outlooks has something to say for them. And it is not our intention to rehearse the merits and demerits of each here. Rather we want to note that the gene drive mosquito project could be seen as troubling from any of the three outlooks. It is easy to see how biocentrists and ecocentrists could take issue with the malaria eradication project were to it travel down the population suppression route. The biocentrist would object to the eradication of the particular individual specimens and the ecocentrist would worry about the removal of the species from the ecosystem. But even the most restrictive of these ethics, anthropocentrism, could find the gene drive malaria project troubling. Anthropocentrism identifies great value in nature, and that is in the form of nature's instrumental value. That is, an anthropocentric environmental ethic will still see nature as valuable because it is valuable to us.

There are at least two different aspects of nature that we value. On the one hand, we value nature's *autonomy*. The 18th-century philosopher John Stuart Mill noted that "nature" often refers to that which "takes place without the agency, or without the voluntary and intentional agency, of man" (Mill, 2008). The autonomy or independence of nature is something we value. In addition to its autonomy, there is a second sense of nature and naturalness that matters to us. For example, according to a founding document of the US National Park System, the primary goal of management should be that the "biotic associations within each park be maintained.... A national park should represent a vignette of primitive America" (Leopold et al. 1963). The key takeaway from the Leopold Report is that a sense of nature or naturalness is had in the *composition* of an ecosystem, of it bearing historical resemblance to the past. Following Gregory Cole and David Aplet, we can refer to this as nature's *ecological condition* (Cole & Aplet, 2010).

Importantly, these are not binary aspects of nature. Rather, they are continua. So, we can imagine the continuum of autonomy ranging from a completely human-controlled environment to a completely autonomous or self-willed environment. And we can imagine the continuum of ecological condition ranging from a completely novel ecosystem that bears little to no fidelity to the past, to a completely pristine ecosystem that is exactly as it would have been at some point in the past. We tend to value nature the more autonomous or wild it is, and we similarly tend to value nature the more its ecological condition maintains fidelity to some point in our history.

If we were to use a suppression drive to drive down or eliminate a particular mosquito population, we would be changing nature's ecological condition; we would be changing the composition of certain ecosystems, and doing so intentionally, thus diminishing nature's naturalness. Similarly, if we were to use a replacement drive, we would be affecting nature's autonomy; we would be intervening and controlling the organisms within an ecosystem, even if they maintained a kind of historical fidelity to previous ecosystems. Thus, even from an anthropocentric outlook, either malaria eradication approach will result in a loss of natural value of some kind.

Whether one subscribes to an anthropocentric, biocentric, or ecocentric outlook, either gene drive approach would carry with it a loss of some kind of value. But, importantly, it is doubtful that an anthropocentrist, biocentrist, or ecocentrist would think that this loss of value would necessarily make the agenda impermissible. Anthropocentrists find value in humans, and insofar as the gene drive malaria eradication project would produce significant anthropogenic benefits, the loss in natural value could be easily justified. And even biocentrists and ecocentrists note that our duties to the natural world are not absolute. Perhaps the most famous biocentrist of all time, Holmes Rolston III wrote: "The duty to species can be overridden, for example with pests or disease organisms." And when asked about mosquitoes, in particular, the famous ecologist E. O. Wilson said: "I would gladly throw the switch and be the executioner myself."

Basically, even if individual organisms or ecosystems have intrinsic value, so, too, do humans. Furthermore, organisms and ecosystems may have non-deontic intrinsic value (by which I mean value that does not create duties). And given the death and suffering engendered by malaria, the loss of natural value may be insufficient to make the agenda impermissible.

11.6 Hubris

The control that we can exhibit over the natural environment seems to be accelerating exponentially. And the ability to push specific alleles through a population and then subsequent generations is awesome, in the literal sense of the word. But one might wonder whether we are not overestimating our ability to control nature. Many ethical theories tell us to think about whether our actions are good or bad, permissible or impermissible. But a tradition dating back to Hellenic Greece and Aristotle pushes us to focus instead on the kinds of character traits our actions exhibit. Known as *virtue ethics* (Annas, 2005; Hursthouse, 2001; Nussbaum, 2013), this ethical theory places focus not so much on what we should do, but on how we ought to be (Driver, 2017).

In focusing on how we ought to be, virtue ethics pushes us to develop positive character traits—virtues—and avoid negative character traits—vices. There are myriad virtues and vices worthy of consideration, but one has been of particular focus when it comes to environmental matters: hubris. Hubris describes dangerous overconfidence in one's abilities. One need not look too far to find examples in which a hubristic attitude has led to environmental disasters.

Many households throughout the U.S. used Dichloro-diphenyl-trichloroethane (DDT) as a pesticide by the middle of the 20th century, and the U.S. military relied upon the chemical compound as a defoliant during the Vietnam War. It was not until the 1960's that the public became aware of DDT's harmful effects. The publication of Rachel Carson's *Silent Spring* brought to the fore the negative side effect DDT had on the environment, specifically the eggshell-thinning effect it had on many North American birds, including the iconic bald eagle (Carson, 2002). One might be tempted to refer to the unforeseen negative side effects of DDT and other environmental interventions not as manifestations of hubris, but rather as instances of insufficient risk assessment prior to the use of novel technologies. But Hoffmann-Riem and Wynne note that "unanticipated effects of novel technologies are not just possible but probable" (Hoffmann-Riem & Wynne, 2002). What this means with respect to the gene drive malaria eradication project is that we should count on there being unanticipated side effects. Though, it seems clear that we cannot have a good grasp of their magnitude *ex ante*.

But while we can count on there being unanticipated adverse side effects from gene drives, it is not clear whether those adverse side effects will be as detrimental as malaria they try to prevent or the conventional control methods already in use, like the use of undifferentiating insecticides. While we will want to adopt a humble attitude and avoid hubris when it comes to our attempts to control nature, there are other character traits equally important in the virtue ethics tradition. Adopting and acting in accordance with the virtues of compassion and beneficence, for example, should push us to refuse to accept the status quo in which hundreds of thousands of undeserving individuals fall victim to malaria each year. It is clear that we will want to avoid hubris and exemplify compassion and beneficence. But, and this is a common objection to virtue theory, it is not entirely clear what virtue theory recommends as the right course of action, recognition of these virtues and vices notwithstanding.

11.7 Public Participation

Amid all the questions over how to weigh and distribute the risks and benefits of gene drive mosquitoes (GDMs), perhaps the most important question is: who gets to make these decisions? Who gets to decide whether we develop or release GDMs, where we release them, or how they should be designed?

Some would argue that the decisions should be made by experts, based on the expected value of the outcomes or the demands of justice. However, there are two reasons to think that public participation should carry significant, or even decisive, weight in the final choice. First, experts may not have much of an advantage in making these decisions. The decision to release GDMs involves weighing unknown risks and potentially incommensurable values, so the public may be a better, and more legitimate, source of answers to such questions.

And second, the nature of the risks and benefits may trigger an ethical requirement for public influence (Callies, 2019). That is the claim of the "all-affected principle," which states that the people who are relevantly affected by a political decision should have influence over that decision. For example, imagine a country's legislature decides that the safest place to locate a nuclear power plant is near its border. Perhaps the legislature even fairly considers the interests of people across the border, whatever that entails. Still, it seems wrong that people on only one side of the border get to influence the decision, when both sides are equally affected.

The all-affected principle has an intuitive appeal, but it leaves open two crucial questions: (1) What does it mean to be "relevantly affected"? And, (2) How much influence does it entitle you to?

To answer the first question, we can consider what kind of effect would entitle me to participation rights. Clearly, there are some kinds of effects that would not entitle me to any influence. For instance, if Germany decided to require a visitors' visa for Americans, it may annoy me, but the effect is too trivial for Germany to be required to get my input. On the other hand, if I were a refugee seeking asylum in Germany, its decision would profoundly affect my life, so the all-affected principle would grant refugees some influence over that decision.

One theory of what it means for a decision to "relevantly affect" you are, therefore, that your basic interests are at stake—interests like security, nutrition, health, and education (Song, 2012). This is not the only plausible conception of "relevant effect." One might argue that decisions that affect the environment (like releasing GDMs) do not affect my basic interests, but still trigger a right to participate. For example, the decision to preserve or eliminate an animal species does not affect my basic interests, but it may still concern my value for nature in a way that gives me a right to weigh in. Nevertheless, we will focus here on the "basic interests" account as perhaps the strongest case for generating a right to participate.

People's basic interests are certainly at stake in the decision to develop and release GDMs. A best-case release of GDMs will remove the threat of malaria and improve health outcomes. A worst-case release of GDMs may worsen health outcomes or harm the environment in ways that affect lifestyles and livelihoods.

However, whose interests are at stake, and how much, depends on further complications in the theory, such as: should we consider any possible outcome of the decision to release GDMs or should we weigh the outcomes according to probability (Arrhenius, 2018)? Considering any possible outcome means that people who live in any mosquito habitat may be affected by releasing GDMs. If so, we might have to include half the globe in the decision-making process.

Weighing the outcomes according to probability, on the other hand, results in tiers of influence. The stakes would remain roughly the same everywhere, since the same basic interests are affected. But, we would multiply the stakes by the probability that they are affected, which would be the greatest closest to the initial release site and diminish as you get farther away. At some distance away, the probability may fall so low that we can consider the effect to be trivial, which would not trigger any participation requirement.

There are three major caveats to this analysis, however. First, if there is a slippery slope from research to initial deployment to widespread imposition, then the decision to research GDMs would significantly affect the decision to release them widely. Second, even if different communities are given an independent chance to decide on GDMs for themselves, research decisions affect the kinds of GDMs that they can decide upon. So, the malaria-suffering world may be entitled to a say in how GDMs are designed. And third, developing GDMs will make everyone who lives in mosquito habitats more susceptible to the malicious use of GDMs, which we discuss in the "dual use" section. All these factors may entitle the mosquito-suffering world to an influence over research decisions, as well as deployment.

There are other ways of fleshing out the all-affected theory but, however you do it, participation rights are unlikely to be exclusive to either the residents near the initial release site or to the citizens of a single country. The theory is also likely to grant more influence to people in malaria-suffering or mosquito-ridden areas, perhaps even over research decisions. Finally, if we accept the "basic interests" account of who is affected, then we should grant extra influence to the poor, whose life outcomes will vary most based on the success or failure of GDMs.

If those are the groups who ought to influence the decision, the next question is: how much influence should they have? There are two main views. The first is that everyone who has a stake above a certain threshold should have an equal influence on the decision, which is presumably an equal vote (Goodin, 2007). It is difficult to justify this option if we measure stakes in relation to the probability of outcomes, since different groups would then have very different stakes but the same amount of influence.

Instead, we may opt for Brighouse and Fleurbaey's view that people should have influence in proportion to their stake in the decision (Brighouse & Fleurbaey, 2010). One way to institutionalize that is by giving everyone a vote that is weighted by their stake in the decision, but that is not the only way. Many defenders of the all-affected principle are careful to note that the right to influence does not mean the right to a vote. We can offer influence in many other forms—consultation rights, the right to participate in deliberation, or even the right to relocate—depending on what would best protect people's interests or motivate their participation. On this view, the all-affected principle supplies the ideal, which we realize through institutional design and community engagement.

11.8 Distributive Justice

The primary purpose of GDMs is to benefit the developing world, so it may seem odd to worry about whether such a project is fair for the intended beneficiaries. However, because GDMs come with significant risks, and those risks fall disproportionately on some groups, we must consider whether they are distributed fairly.

One way to justify assigning a disproportionate risk to some is by citing the beneficial consequences for the whole. For example, if we initially release GDMs on an island, then we could lower the total risk of mortality to humanity, so the disproportionate risk to those islanders might be justified. Or, for that matter, if we release them in India, we could reduce the risk of mortality to South Americans. However, non-consequentialists would argue that the overall benefits do not justify assigning some people more than their fair share of risk.

Instead, we might argue that the benefits to the islanders (or the Indians) are sufficient to justify the risks. After all, if GDMs work, they could eradicate malaria, saving many people from death and the high costs of illness. On one hand, this seems like an important contribution to fairness; it is certainly worse to ask people to take on disproportionate risk if they see no benefit from it. But, even on an optimistic view of the risks, the benefits do not seem sufficient to justify the risks here. Consider that, if GDMs work as hoped, then they could soon be used everywhere, which means that everyone would soon enjoy the same benefit. There may be some value to getting that benefit sooner, but it is unlikely to justify bearing the brunt of the risk.

If the benefits alone do not justify the risk, then we might seek the consent of those who bear the risks. For instance, if the islanders vote to host the GDMs, then we have reason to think that they see the benefits as outweighing the costs and are, therefore, not unfairly burdened. However, consent loses its justificatory power when the conditions under which it is given are unfair. To the extent that the health or socioeconomic condition of the islanders is a product of previous injustice, the consent is not freely given and does not justify the disproportionate risk. Consent may still be a necessary condition, but a concern about background inequality makes it all the more important that the risks from GDMs are distributed fairly.

So, let us consider two theories of how to fairly distribute the risks: egalitarianism and prioritarianism. An egalitarian theory would require that the risks be distributed equally in some sense. In the case of GDMs, however, this is unappealing because it would greatly increase the total risk to humanity. For example, equality may mean that we release GDMs everywhere at once. Or, it may mean that we choose the initial release site by a lottery where each citizen has an equal chance of selection, making the most populous areas the most likely release sites.

A prioritarian theory, on the other hand, allows for unequal distributions as long as they are to the benefit of the least advantaged group. In this case, either the country (e.g., India) or the community around the release site would probably count as least advantaged among its peers. What arrangement would be to the advantage of the country or community that bears the risk?

One answer is to release GDMs in a developed country or a wealthy community. It would certainly be to the benefit of the least advantaged to find out if GDMs worked, without bearing any of the risks. Testing on more advantaged groups also alleviates our worries about consent and exploitation; the consent of rich communities will not be a product of background injustice or desperation. The problem with this strategy is that developed countries may not be scientifically suitable as test sites, and rich islanders are likely to block any such test. Neither development would be to the benefit of the least advantaged.

Instead, we could pose a similar question hypothetically: what it would it take for the rich to bear the risks from GDMs? This is essentially the strategy behind John Rawls' thought experiment, "the original position." In the original position, we consider the distribution of benefits and burdens in a society as if we did not know which social group we belonged to. If a reasonable person would be comfortable with belonging to any social group in a distributive scheme, then that distributive scheme is justified to all social groups.

So, the question becomes: what would it take for us to be indifferent to whether or not we belonged to the risk-bearing group? One possible answer is that nothing could make us accept the risks of GDMs. If the risks are high enough, a reasonable person might reject any arrangement under which she could have to bear those risks.

On the other hand, there might be conditions under which we would be indifferent between the risky and non-risky positions. For one, we would probably insist that every reasonable precaution was taken to minimize the risks of GDMs. And we might also require compensation for bearing those risks. The compensation might be a form of insurance for any harmful consequences, such as guaranteed medical care or relocation expenses. Or, it might be compensation for bearing the risks, whether or not the harms come to pass.

Of course, compensation is a controversial solution to this kind of injustice. Those who favor it argue that it gives all decision-makers an incentive to minimize risks. Without a compensation requirement, many groups stand to benefit from releasing GDMs without bearing any risk, such as the malaria-suffering nations far from the release site, or the developers and patent-owners of gene drive technology. But, if the beneficiaries have to compensate the risk-bearers, then the former has a reason to minimize the risks.

However, those who oppose compensation raise a series of difficult moral and practical questions: Is it fair for the rich to pay others to risk their health? Does it further undermine community consent if the community consents in order to receive compensation? What is the right level of compensation for risks that are hard to estimate or even anticipate? How can compensation address risks to future generations? Still, if GDMs are worth the risk, and the risk must be distributed unequally, then compensation, consent, and precautions are our only tools to justify that inequality (Hayenhjelm, 2012).

11.9 The Dual Use Dilemma

The "dual use dilemma" is an ethical problem for scientists, administrators, grantmakers, and all others involved in supporting research that could be used for great good or serious harm. The typical historical example is the research into nuclear fission, which scientists knew could be used to both generate energy and to make bombs. The dilemma is generally discussed today in connection with research into novel pathogens, which help prepare us for future diseases but also make their accidental or malicious spread more likely. Research into GDMs presents a similar problem. GDMs can be designed to prevent disease by making mosquitoes more resistant to spreading malaria, or they can be designed to spread disease by making them better hosts for malaria (or something more deadly).

Of course, most human creations can be used for both good and ill, so it is important to define the dual use dilemma more precisely. Thomas Douglas offers the following definition, which I have slightly amended (Douglas, 2013). A dual use dilemma arises when: (1) an agent is deciding whether to contribute to a scientific output; (2) that output could be used to produce significant benefit or harm; (3) the intended and primary use of the output is for good; (4) increasing the likelihood of the good use also increases the likelihood of the bad use; (5) there is no option to create an output that is useful but harmless; and (6) the potential benefit and harm are of comparable moral worth.

Given that GDMs fit the conditions for a dual use dilemma, is it ethical to pursue them? A natural way to answer that question is to compare the expected benefits of GDMs with the expected harms; are GDMs likely to be a net benefit to the world? The problem is that it is very hard to estimate GDMs' expected benefits or harms before releasing them. We do not know whether and how successful they will be in reducing the spread of disease, what kinds of known and unknown harms might result, and how likely they are to be used for malicious purposes.

One important difference between GDMs and dual use projects like pathogens is that, with GDMs, we at least know that there is a problem to be solved; millions of people die from malaria and dengue every year. This stands in contrast to pathogen research, where it is hard to know whether there will even be a problem to solve. What are the chances that, for instance, a more transmissible airborne smallpox will ever develop, accidentally or maliciously?

Still, it is hard to estimate the overall expected value of GDMs, which requires an understanding of infectious diseases, local environments, decision theory, and security risk, in addition to transgenic mosquitoes. So, however we estimate the expected value, we can at least conclude that it will require a variety of experts working together, perhaps, as some have proposed, in a biosecurity review board.

Rather than attempt any rough calculations of expected value here, we will briefly explore the ethical consequences of either a positive or negative result. Assume, first, that we arrive at a positive expected value; that is, we expect GDMs to do more good than harm. That is a strong *pro tanto* reason to develop them. But, can developing them still be wrong?

One could argue that, because their developers can foresee GDMs being used for harm, it is wrong to facilitate that harm, even if GDMs could also prevent a greater number of deaths. The typical response to this is to draw a distinction between foreseeing and intending a result. According to this response, it is permissible to develop GDMs if one merely foresees that they will be used for harm, as long as one does not intend that harm. For example, it is intuitively permissible for the military to bomb a weapons factory, even if it foresees that some civilians will die as a result. It is impermissible, however, for a terrorist to target the same number of civilians because, in that case, the terrorist intends their deaths.

Even if one accepts this distinction, however, it is important to note a relevant difference between the bombing case and GDMs. In the bombing case, there is (we assume) no way to mitigate the foreseeable casualties; once the bomb is released, the destruction follows. However, there are a number of options to possibly limit the harmful potential of GDMs, such as by developing reversible or self-limiting drives. If gene drive developers were to fail to mitigate foreseeable harm when possible, then they may be morally responsible for that harm, even if they did not intend it (Uniacke, 2013). And this could be true even if GDMs, overall, do more good than harm.

Now, let us consider the alternative. If we expected GDMs to do more harm than good; can it still be permissible to create them? Here is one argument that might absolve its creators of responsibility. They might say: "If I did not do it, someone else would have." Since we can assume that someone would have created GDMs regardless, their actual creators are not morally responsible for their harmful effects.

However, this defense seems mistaken from either of the two major ethical perspectives. From a consequentialist perspective, we judge ethical actions by their consequences, and we can grant that GDMs would exist no matter what their creators did. Still, the creators' actions do have consequences. If the creators had worked slowly or poorly, it would have at least delayed the harmful consequences of GDMs.

And, from a non-consequentialist perspective, it does matter if you contribute to a harmful result, even if the result would have been the same either way. Jonathan Glover names this principle after Alexandr Solzhenitsyn, who wrote: "Let the lie come into the world, even dominate the world, but not through me" (Glover & Scott-Taggart, 1975).

11.10 Conclusion

As should be clear from this discussion, making ethical choices on emerging technologies with profound social and environmental effects are both challenging and essential. The aim of this chapter has been to introduce some of the ethical concerns raised by the gene drive mosquito project. We have indicated where we think the balance of reason lies with respect to some of these issues, but most of them will require a closer examination of the arguments, and a broader discussion of our values. We hope this chapter will encourage those discussions among ethicists, scientists, and the public.

References

Annas, J. (2005). Virtue Ethics. In D. Copp (Ed.), *The Oxford Handbook of Ethical Theory* (pp. 515–536). Oxford University Press. https://doi.org/10.1093/0195147790.003.0019

Arrhenius, G. (2018). The Democratic Boundary Problem Reconstructed. *Ethics, Politics & Society, 1*, 34. https://doi.org/10.21814/eps.1.1.52

Brighouse, H., & Fleurbaey, M. (2010). Democracy and Proportionality. *Journal of Political Philosophy, 18*(2), 137–155. https://doi.org/10.1111/j.1467-9760.2008.00316.x

Callies, D. E. (2019). The Ethical Landscape of Gene Drive Research. *Bioethics*. https://doi.org/10.1111/bioe.12640

Callies, D. E. (2020). Bednets or Biotechnology: To Rescue Current Persons or Research for the Future? *Fudan Journal of the Humanities and Social Sciences, 13*(4), 559–572. https://doi.org/10.1007/s40647-020-00290-7

Carson, R. (2002). *Silent spring* (40th anniversary ed., 1st Mariner Books ed). Houghton Mifflin.

Cole, D. N., & Aplet, G. H. (2010). The Trouble with Naturalness: Rethinking Park and Wilderness Goals. In D. N. Cole & L. Yung (Eds.), *Beyond Naturalness: Rethinking Park and Wilderness Stewardship in an Era of Rapid Change.* (pp. 12–29). Island Press. http://qut.eblib.com.au/patron/FullRecord.aspx?p=3317526

Douglas, T. (2013). An Expected-Value Approach to the Dual-Use Problem. In B. Rappert & M. J. Selgelid (Eds.), *On the Dual Uses of Science and Ethics: Principles, Practices, and Prospects*. ANU Press. https://doi.org/10.22459/DUSE.12.2013.09

Driver, J. (2017). Virtue Ethics. In S. M. Cahn (Ed.), *Exploring Ethics: An Introductory Anthology* (Fourth Edition, pp. 145–147). Oxford University Press.

Glover, J., & Scott-Taggart, M. (1975). It Makes No Difference Whether or Not I Do It. *Aristotelian Society Supplementary Volume, 49*(1), 171–210. https://doi.org/10.1093/aristoteliansupp/49.1.171

Goodin, R. E. (2007). Enfranchising All Affected Interests, and its Alternatives. *Philosophy & Public Affairs, 35*(1), 40–68.

Hayenhjelm, M. (2012). What Is a Fair Distribution of Risk? In S. Roeser, R. Hillerbrand, P. Sandin, & M. Peterson (Eds.), *Handbook of Risk Theory* (pp. 909–929). Springer Netherlands. https://doi.org/10.1007/978-94-007-1433-5_36

Hoffmann-Riem, H., & Wynne, B. (2002). In Risk Assessment, One Has to Admit Ignorance. *Nature, 416*(6877), 123.

Hursthouse, R. (2001). *On Virtue Ethics*. Oxford University Press.

Mill, J. S. (2008). *On liberty and Other Essays*. Oxford University Press.

Nussbaum, M. C. (2013). Non-Relative Virtues: An Aristotelian Approach. In R. Shafer-Landau (Ed.), *Ethical Theory: An Anthology* (2nd ed., pp. 630–644). Wiley-Blackwell.

Preston, C. J. (2013). Ethics and Geoengineering: Reviewing the Moral Issues Raised by Solar Radiation Management and Carbon Dioxide Removal: Ethics & Geoengineering. *Wiley Interdisciplinary Reviews: Climate Change, 4*(1), 23–37. https://doi.org/10.1002/wcc.198

Rawls, J. (1999). *A Theory of Justice*. Harvard University Press.

Rosner, L. (Ed.). (2004). *The Technological Fix: How People Use Technology to Create and Solve Problems*. Routledge.

Song, S. (2012). The Boundary Problem in Democratic Theory: Why the Demos Should be Bounded by the State. *International Theory, 4*(1), 39–68. https://doi.org/10.1017/S1752971911000248

Sunstein, C. R. (2005). *Laws of Fear: Beyond the Precautionary Principle*. Cambridge University Press.

Uniacke, S. (2013). The Doctrine of Double Effect and the Ethics of Dual Use. In B. Rappert & M. J. Selgelid (Eds.), *On the Dual Uses of Science and Ethics: Principles, Practices, and Prospects*. ANU Press. https://doi.org/10.22459/DUSE.12.2013.10

United Nations. (1982). *World Charter for Nature*. http://www.un.org/documents/ga/res/37/a37r007.htm

United Nations. (1992a). *Convention on Biological Diversity* (p. Preamble). http://unfccc.int/essential_background/convention/items/6036.php

United Nations. (1992b). *Rio Declaration on the Environment and Development* (p. Principle 15). http://www.un.org/documents/ga/conf151/aconf15126-1annex1.htm

Weinberg, A. (1967). *Reflections on Big Science*. The MIT Press.

Index

Ae. aegypti 22, 24, 56, 96, 124, 131, 133, 167, 222–223
Africa 5, 128, 144, 147, 160, 164, 178, 244, 256, 318–319, 336
African countries 153, 319
African Union 318
African Union Development Agency (AUDA) 318
alleles 40, 46–47, 49–51, 54, 69–70, 117, 201–203, 339
  drive 44–45, 47–49, 54, 199, 201–202, 212, 223
  drive resistance 163
  drive-resistant 77, 82, 87, 95, 202, 224
  gene drive 201
  gRNA 203–204
  Medea 54
  resistant 93–94, 201, 218
  suppression gene drive 206
  wild-type 42–44, 48–49
  wildtype 200–202
An. albimanus 126
An. arabiensis 89, 154, 156–158
An. ardensis 155
An. brohieri 154–155
An. brunnipes 154–157
An. coustani 157–158
An. funestus 156–157
An. gambiae 10, 24, 26, 28–29, 47–49, 53, 75, 85–89, 94, 117, 123, 128, 131–132, 143, 151, 153–158, 161–163, 165, 168, 170–171, 175, 181, 204, 207, 210–211, 214, 222, 281, 318
  genomes 86, 174
  populations 168, 207
  zpg gene 88
An. stephensi 47, 81
Anopheles mosquitoes 4, 39–40, 42, 44, 46, 48, 50, 52, 54, 56
Anopheles species 40, 85, 161, 167
Anopheles species diversity 153, 162
Anopheline species 176
anthropocentrism 337
antibiotic resistance 21–22
antimalaria vaccine development 4
ATSBs, *see* attractive toxic sugar bait
attractive toxic sugar bait (ATSBs) 197, 214, 222
AUDA, *see* African Union Development Agency
Autonomous systems 73, 80, 92

biotechnology 300, 302–303, 305–306, 315, 317
bisazir sterilization 84
Burkina Faso 123, 168, 170, 172, 313, 319–320

Canary Islands 152, 173, 175
Cape Verde 151–152, 173, 175

carboxypeptidase genes 23
carboxypeptidase promoter 25–26
cargo gene 40, 42, 47
  antimalaria 39, 49
Cartagena Protocol on Biosafety 302–308, 310–311, 315, 319–320
Ceratitis 22
chromosome, wild-type 46, 80
control methods
  area-wide vector 277
  conventional genetic 118
  conventional malaria 335–336
  conventional vector 330, 335
  gene drive vector 270
  traditional genetic 115
control strategies
  genetic 65
  malaria 285
CRISPR-based gene drive technology 8
CRISPR-based homing gene drive systems 199
Culex quinquefasciatus 121
cytoplasmic incompatibility 55–56, 120

DDT, *see* dichlorodiphenyl-trichloroethane
dichlorodiphenyltrichloroethane (DDT) 5, 340
dispersal, anthropogenic 165, 167
DNA 21, 68, 70–71, 76, 79
  exogenous 68–69
DNA-break induced repair 71
DNA cleavage, nuclease-mediated 71
DNA sequencing 79
drive confinement 45–46
drive conversion efficiency 51
drive gene expression 24
drive mosquitoes, *An. gambiae* suppression 50
drive system 43, 46–47, 69–70, 73–74, 77, 81–82, 86, 88, 90, 92–93, 95, 207, 212, 214
  split 43–44, 203
Drosophila 22, 56, 97
  transgenic 23
Drosophila melanogaster 49–50, 66, 203
dual anti-malarial scFvs genes 95
dual use dilemma 347

effector genes 25–26, 31, 90, 92, 96, 142, 219
  antimalarial 91, 213
EIP, *see* extrinsic incubation period
EIR, *see* entomological inoculation rate
EMOD malaria model 207
endangered species 179, 181
endonuclease 71, 199
engineered gene 8, 305, 309, 313, 315, 321
Enolase–Plasminogen Interaction Peptide (EPIP) 26
entomological endpoints 146, 153, 160, 178
entomological inoculation rate (EIR) 146, 218
environmental risk assessment (ERA) 215, 220–221, 225, 243, 247–248, 301, 308, 314, 318, 320
enzymes 25–26, 86
epidemiological endpoints 146, 160
epidemiological impacts 91, 96, 160, 212, 217
  long-lasting 7
EPIP, *see* Enolase–Plasminogen Interaction Peptide

ERA, *see* environmental risk assessment
extrinsic incubation period (EIP) 213–214

Fault Trees 245, 251–252
female fertility genes 43, 49, 88
female-specific fertility genes 42
female-specific genes 42
field trials 83, 89, 119, 125, 142–143, 146, 179–180, 183, 199, 212, 216–217, 222, 225, 244, 274–275, 289, 312–313, 315
flightless phenotype 85
Florida 121–122
FOI, *see* force of infection
force of infection (FOI) 212–214

gametes 68, 126, 199, 202
GDMMs, *see* gene drive modified mosquitoes
GDMOs, *see* gene drive modified organisms
GDMs, *see* gene drive mosquitoes
GE, *see* genetically engineered
GEMs, *see* genetically engineered mosquitoes
gene
  disease-refractory 199, 223
  doublesex 50, 221
  endogenous 89, 95
  haplosufficient 40, 45
gene drive 39–56, 65–67, 69–70, 87–89, 92–94, 115–118, 141–142, 144, 146, 148, 150, 152, 154, 156, 158, 160, 162, 164, 166, 168, 170, 172, 174–176, 178, 180, 182, 206–208, 212, 242–243, 277
  applications 44, 65–66, 68, 70, 72, 74, 76, 78–98
  conventional 56
  efficient 92, 142
  engineered 56, 308, 320
  experimental 68
gene-drive, functional 93
gene drive
  homing-based 200–202
  invasive 143
  low-threshold 142, 167
  male-biased sex distorter 10
  male-biased sex-distorter 88
  modification type 56
  nuclease-based 68
  population suppression 220
  transposon-based 47
gene drive-based technologies 3–4, 6, 8, 10, 12
gene drive cage trials 216
gene drive chromosomes 67
gene drive inheritance 216, 221
gene drive insects 44
gene-drive males 94
gene drive modified mosquitoes (GDMMs) 242–247, 255, 257, 316
gene drive modified organisms (GDMOs) 242, 257
gene drive mosquito projects 197–198, 200, 202, 204, 206, 208, 210, 212, 214–216, 218, 220, 222, 224–225, 334–337, 349
  non-localized 219
gene drive mosquitoes (GDMs) 82, 93, 116, 118, 120, 122, 124, 126, 128, 130, 132, 134, 198, 215, 223–224, 329–332, 334, 336, 338, 340–350
  large cage trials of 116, 118, 120, 122, 124, 126, 128, 130, 132, 134
  male 92
gene drive system 65, 74, 77, 79, 82–83, 86, 88, 93, 96

coupled effector 91
CRISPR-based homing 199
homing-based 199, 207
low-threshold 216–217, 219, 225
non-localized 212
nonlocalized 222
population suppression 75, 220, 223
prototype 92
self-sustaining 221
split 214
suppression 210
synthetic 86, 242, 245, 256
user-defined 207
gene drive technology 7–9, 78, 99, 211, 270, 272, 274, 286, 300, 307, 312, 318, 322, 346
gene drive transgene 116
gene editing 66, 68–69, 90, 96
gene editing research 67
gene flow 142, 147, 170, 174, 176
inter-population 147
gene frequencies 200
gene promoters 87, 91–92
gene transfer, horizontal 252
Generalized Linear Models (GLMs) 251–252
genes, foreign 29
genetic control method 131
genetic isolation 147, 165, 168, 170, 172–174
genetic load 42, 51, 68, 75–76, 81, 83, 93–94
genetic population suppression system, non-drive 85
genetic variation 127, 218
genetically engineered (GE) 10, 21, 141, 153, 254, 270, 277
genetically engineered microorganisms 89
genetically engineered mosquitoes (GEMs) 10, 141–144, 146–147, 153, 160, 163, 165, 173–175, 178–180, 254, 270, 277, 279, 281, 285, 290, 292
genetically modified organisms (GMOs) 9, 30, 180, 241–244, 248, 299
regulated 257
genomes 47, 69, 71, 74, 76, 78, 117, 171
genotypes 87, 116–117, 127, 170, 203
germline 49, 68, 74, 80–81, 199–200
germline transformation 22
GFP, *see* green fluorescent protein
Glacial Maximum Mainland Connection (GMMC) 148, 150
GLMs, *see* Generalized Linear Models
GMMC, *see* Glacial Maximum Mainland Connection
GMOs, *see* genetically modified organisms
gonotrophic cycle 205
green fluorescent protein (GFP) 22, 72, 79, 86
gRNA 48, 50–51, 54–55, 71, 74, 80, 86, 88, 92–93, 98, 203
multiplexed 48, 50, 55
gRNA locus 203–204

HDR, *see* homology-directed repair
HEGs, *see* homing endonuclease genes
hemocoel 23–24
heterozygotes 42, 45, 47, 49, 54, 73, 199–201
heterozygous gene-drive parents 87
Hidden Markov Models 252
homing endonuclease genes (HEGs) 71, 86, 88, 98
homology-directed repair (HDR) 47–50, 80, 92, 199, 202, 218

human malaria 96, 144, 213
humans 4, 91, 162, 198, 212–214, 218, 246, 275, 305, 332, 337, 339
  gene-drive-modified 331
  susceptible 213–214

IGDs, *see* integral gene drive
IIT, *see* incompatible insect technique
incompatible insect technique (IIT) 115
insect dsx genes 87
insecticide resistance 6, 178, 181
integral gene drive (IGDs) 95
International Union for Conservation of Nature (IUCN) 179
irradiation 122, 124
IUCN, *see* International Union for Conservation of Nature

Lake Victoria Islands 148, 151, 161, 175–176
larvae 72, 82, 153, 204, 208
  anopheline 209
larval competition 209
larval development time 209
larval habitat 211–212
larval stages 10, 204
Last Glacial Maximum (LGM) 148
LGM, *see* Last Glacial Maximum
livestock 214–215, 243
living modified organisms (LMOs) 300–303, 305–311, 313–316, 318–321
  regulation 304, 306–307
LLINs, *see* long-lasting insecticide-treated nets
LM *Anopheles gambiae* 320
LM crops 300, 305–306, 319
LM insects 308, 316, 319
LM mosquitoes 301, 307–308, 310–313, 316–321
LMOs, *see* living modified organisms
long-lasting insecticide-treated nets (LLINs) 197, 214, 223

malaria 3–5, 19, 21–24, 26, 28, 30–31, 39–40, 48, 90, 99, 115, 141, 178, 197, 204, 217, 220, 256, 281, 285, 319, 331, 334–336, 339–342, 344, 347
  treatment of 4
malaria control 3–7, 9–11, 44, 65–66, 68, 70, 72, 74–76, 78, 80, 82, 84, 86, 88, 90, 92, 94, 96, 98, 142, 153, 220, 277, 281, 285, 289, 292, 318, 336
  conventional 336
  large-scale 142
  mosquito-targeted 5
  self-sustainable 10
  wide-scale 200
malaria control approaches, novel 7
malaria control programs 11, 180, 319
malaria control strategies 8, 83, 142
malaria elimination 6, 90, 198
malaria epidemiology 96, 205, 224
malaria eradication 147, 199, 225, 277, 330, 332
malaria parasites 8–9, 27–28, 40, 89, 219
  development 26, 30
  human 25
  mosquito vectors of 66, 69
  target model system 91
  transmission 65
malaria prevalence 6, 11

malaria prevention by novel control methods 6
malaria transmission 40, 50, 56, 91, 118, 146, 160–161, 179, 198, 213–214, 216, 218, 220, 222, 285, 287
malaria transmission models 212, 215, 218
malaria vector control 7
malaria vector control interventions 218
malaria vector gene drive projects 199
malaria vector mosquitoes 70, 78, 85
malaria vectors 10, 162, 199
mark-release-recapture (MRR) 205
Medea 54–55, 98
microhomology-mediated end-joining (MMEJ) 202
midgut epithelium 24–25
midgut peptide 25–26
MMEJ, *see* microhomology-mediated end-joining
mosquito control 178, 319
mosquito ecology 200, 216, 221–222
mosquito ecology models 221–222
mosquito gene-drive systems 71
mosquito gene drives 3, 21, 39, 65, 115, 141, 197, 241, 269–270, 272, 274, 276, 278, 280, 282, 284, 286, 288, 290, 292, 299, 329
mosquito habitats 342–343
mosquito life cycle 204
mosquito population model 204
mosquito population modification 94
mosquito population suppression 8, 77, 83, 220
mosquito populations 7–8, 11, 27–28, 30, 39, 65, 83, 165, 178, 198–199, 207, 211, 213
  age-structured 221
  conspecific 167
  indigenous 180
  malaria-transmitting 8
  native 316
  natural 205
  suppressed wild-type 86
  wild 205
mosquito resistance 27, 29
mosquito transgenesis 22, 24, 76
mosquito vector ecology 198, 200
mosquitoes
  adult 81, 120, 205, 210, 213
  gene drive-modified 198, 218–219, 221–222
  genetically-engineered 99
  infected 314
  wild-type 81–82, 92, 153, 254, 316
mosquitos, transgenic 25–26, 129
MRR, *see* mark-release-recapture

NEPAD, *see* New Partnership for Africa's Development
New Partnership for Africa's Development (NEPAD) 318
NGOs, *see* non-governmental organizations
NHEJ, *see* non-homologous end-joining
non-governmental organizations (NGOs) 180, 277, 287, 291, 299
non-homologous end-joining (NHEJ) 77, 202, 218
nucleotide diversities 165

ovengensis 154, 156
overexpression 9

ovipositions 120, 125, 128

P. falciparum 26–29
P. falciparum infection 26–27
paratransgenesis 21–22, 24, 26, 28–30
pathogens 10, 23, 25, 89, 212, 347
PCA, *see* principal components analysis
PCR, *see* polymerase chain reaction
Plasmodium 9, 23
polymerase chain reaction (PCR) 79, 161
population genetics models 200, 207
population modification 7–9, 11, 43, 51, 89, 93, 95, 118, 142, 146, 281
  gene drive-based 11
principal components analysis (PCA) 168, 170, 172

randomized controlled trials (RCTs) 223
RBM, *see* relationship-based model
RCTs, *see* randomized controlled trials
reaction-diffusion models 208
relationship-based model (RBM) 277–292
resistance alleles 48–49, 51, 202, 224
  functional 48–51
  nonfunctional 50–51
risk assessment 198, 239, 242–247, 249–250, 253, 255, 257, 271, 291, 304, 306–309, 315–316, 320
  ecological 119, 241
  independent 270, 272
  probabilistic 242, 248–251, 253, 257
  qualitative 242, 248–249, 257
  quantitative 248, 256–257
risk assessment models 167, 253
RNAi 41, 54
RNAi toxin 53–54

SD, *see* segregation distortion
SDGD, *see* sex-distorter gene drive'
segregation distortion (SD) 71
sex-distorter gene drive' (SDGD) 10, 88–89
sgRNA 71, 74
SGV, *see* standing genetic variation
single nucleotide polymorphism (SNPs) 70, 77, 168, 174, 211
SLMP, *see* surrounding land mass proportion
SNPs, *see* single nucleotide polymorphism
split drive 43–44, 80, 203, 222
sporozoite infection 161–162
sporozoites 23, 26
SSMs, *see* State Space Models
standing genetic variation (SGV) 163, 202, 218
State Space Models (SSMs) 251–252
sterile insect technique 10, 115
sub-Saharan Africa 5–6, 141–142, 148, 180, 210–211, 320, 335–336
surrounding land mass proportion (SLMP) 150, 152, 174
symbiotic bacteria, engineered 28

TALENs, *see* transcription activator-like effector nucleases
target product profile (TPP) 91, 94, 199, 212, 216–217, 245, 255, 257
tetracycline 10, 85, 127

TPP, *see* target product profile
transcription activator-like effector nucleases (TALENs) 67
transgene fixation 11
transgene inheritance 144
transgene insertion 117
transgene phenotype 120
transgenes 7, 24, 29, 68, 80, 115–120, 124, 127–129, 131, 133, 144, 153, 175, 179, 210, 216, 220–221
  gRNA/cargo 80
transgenesis 21–23, 25, 29–30, 66, 69, 83

Uganda 159, 162, 164, 168, 170–171, 319
UNEP, *see* United Nations Environment Program
United Nations Environment Program (UNEP) 150–151

VBD, *see* vector-borne diseases
vector-borne diseases (VBD) 39, 301, 314–315, 317
vitellogenin gene 24

West Africa 158–159, 162, 212, 320
WHO, *see* World Health Organization
Wolbachia-infected *Ae. aegypti* 222–223
Wolbachia-infected mosquitoes 312, 314–315, 321
World Health Organization (WHO) 90, 144, 221, 257, 281, 290, 311–317, 319, 321, 335–336

X-shredder 51, 53
X-shredder population suppression strategy 88

Zanzibar 158, 168, 176–177
ZFNs, *see* zinc-finger nucleases
zinc-finger nucleases (ZFNs) 67